Biological and Medical Physics, Biomedical Engineering

Aaron Lewis, Department of Applied Physics
Hebrew University
Jerusalem, Israel

Stuart M. Lindsay, Department of Physics and Astronomy
Arizona State University
Tempe, USA

Xiang Yang Liu, Department of Physics, Faculty of Sciences
National University of Singapore
Singapore, Singapore

David Mauzerall, Rockefeller University
New York, USA

Eugenie V. Mielczarek, Department of Physics and Astronomy
George Mason University
Fairfax, USA

Markolf Niemz, Medical Faculty Mannheim
University of Heidelberg
Mannheim, Germany

V. Adrian Parsegian, Physical Science Laboratory
National Institutes of Health
Bethesda, USA

Linda S. Powers, University of Arizona
Tucson, USA

Earl W. Prohofsky, Department of Physics
Purdue University
West Lafayette, USA

Tatiana K. Rostovtseva, NICHD
National Institutes of Health
Bethesda, USA

Andrew Rubin, Department of Biophysics
Moscow State University
Moscow, Russia

Michael Seibert, National Renewable Energy Laboratory
Golden, USA

Nongjian Tao, Biodesign Center for Bioelectronics
Arizona State University
Tempe, USA

David Thomas, Department of Biochemistry
University of Minnesota Medical School
Minneapolis, USA

This series is intended to be comprehensive, covering a broad range of topics important to the study of the physical, chemical and biological sciences. Its goal is to provide scientists and engineers with monographs- both, authored and contributed volumes to address the growing need for information. The fields of biological and medical physics and biomedical engineering are broad, multidisciplinary and dynamic. They lie at the crossroads of frontier research in physics, biology, chemistry, and medicine.

Books in the series emphasize established and emergent areas of science including molecular, membrane, and mathematical biophysics; photosynthetic energy harvesting and conversion; information processing; physical principles of genetics; sensory communications; automata networks, neural networks, and cellular automata. Equally important is coverage of applied aspects of biological and medical physics and biomedical engineering such as molecular electronic components and devices, biosensors, medicine, imaging, physical principles of renewable energy production, advanced prostheses, and environmental control and engineering.

Nesrin Hasirci • Vasif Hasirci
Editors

Hydrogels and Bioinks in Tissue Engineering

Springer

Editors
Nesrin Hasirci
BIOMATEN Center of Excellence in
Biomaterials and Tissue Engineering
Middle East Technical University
Ankara, Türkiye

Vasif Hasirci
Department of Biomedical Engineering
Acibadem University
Istanbul, Türkiye

ISSN 1618-7210 ISSN 2197-5647 (electronic)
Biological and Medical Physics, Biomedical Engineering
ISBN 978-3-031-90965-8 ISBN 978-3-031-90966-5 (eBook)
https://doi.org/10.1007/978-3-031-90966-5

Translation from the Turkish language edition: "Doku Mühendisliği Hidrojeller ve Biyomürekkepler 3 - Boyutlu Organ Yapımı" by Nesrin Hasirci and Vasif Hasirci, © METU Press 2023. Published by METU Press. All Rights Reserved.

English translation of the original Turkish edition published by METU Press 2023

This Springer imprint is published by the registered company Springer Nature Switzerland AG
The registered company address is: Gewerbestrasse 11, 6330 Cham, Switzerland

Preface

These are times during which science and technology progress faster than everything else. In parallel to this, today the everyday life is also changing faster. As always, the most important thing for us is to have a healthy body and a productive life. The field of biomaterials is a result of a collective attempt in achieving this goal by producing devices for the treatment of the damaged or unfunctional tissues or organs. The field of biomaterials was in our lives since the beginning of human existence on earth and has become an increasingly productive discipline after the 1970s. It has developed so much that we are now doing research even under a variety of extreme conditions such as microgravity for use in space and planning to produce spare body parts and organs.

We, ourselves, entered this field about 50 years ago and have been modest contributors of the groundbreaking developments in the field through research, publications, and training of younger scientists.

We decided to publish this edited book ***Hydrogels and Bioinks in Tissue Engineering*** to reflect the contribution of the highly exciting new interdisciplinary field of additive manufacturing or 3D printing, because it is one of the most critical techniques used in the production of tissues and organs through tissue engineering. Each chapter was written by scientists who stepped into the biomaterials field at BIOMATEN, the Center of Excellence in Biomaterials and Tissue Engineering at the Middle East Technical University (METU) in Ankara, Türkiye. The chapters together constitute a whole in covering the significant amount of information developed on the materials, products, and use of subject.

This book was earlier published by METU in Turkish. The present English version is published by Springer Nature as a result of an agreement and collaboration between these two institutions. We hope it will be a very informative introduction to this exciting field.

We would like to thank all the contributors, most of whom are former students and colleagues who added their knowledge and experience to the preparation of this book. We extend our gratitude to our family for their continuous support during the very busy times of the whole process.

We would like to thank Springer Nature for undertaking the printing process and the assistance provided in every possible way.

Ankara, Türkiye Nesrin Hasirci
Istanbul, Türkiye Vasif Hasirci

Contents

Chapter 1
Tissue Engineering

Fatma Nese Kok, Vasif Hasirci, and Nesrin Hasirci

Abstract The active lives of the urban people, increasing traffic accidents, and extended lifespan lead to various defects in the body, such as fractures and non-functioning organs. Use of different materials available in the environment to treat the damages is not new; they were applied even in the old ages. Today, with the developments in science and engineering, it became possible to manufacture devices in our laboratories that are fully biocompatible and able to substitute for damaged organs and tissues in performing their functions quite satisfactorily. In general, novel metal and ceramic products are used in the treatment of damaged hard tissues such as hip prostheses or dental implants. Polymers are mostly used in softer tissues and organs or in modern tissue engineering applications as scaffolds or in the delivery of bioactive agents. Hydrogels, which can hold large amounts of water and mimic the extracellular matrix of the tissues, can be successfully used in 3D bioprinting biomaterials carrying cells. Today, studies are intensely focused on the production of various tissues such as bone, cartilage, nerve, skin, etc., are tools of the

F. N. Kok
Department of Molecular Biology and Genetics, Faculty of Arts and Sciences, Istanbul Technical University, Istanbul, Türkiye

V. Hasirci
BIOMATEN Center of Excellence in Biomaterials and Tissue Engineering, Middle East Technical University, Ankara, Türkiye

Department of Medical Engineering, Acibadem Mehmet Ali Aydinlar University, Istanbul, Türkiye

Biomaterials Center, Acibadem Mehmet Ali Aydinlar University, Istanbul, Türkiye

N. Hasirci (✉)
BIOMATEN Center of Excellence in Biomaterials and Tissue Engineering, Middle East Technical University, Ankara, Türkiye

Department of Chemistry, Middle East Technical University, Ankara, Türkiye

Department of Bioengineering, Near East University, Nicosia, TRNC, Türkiye
e-mail: nhasirci@metu.edu.tr

N. Hasirci, V. Hasirci (eds.), *Hydrogels and Bioinks in Tissue Engineering*, Biological and Medical Physics, Biomedical Engineering,
https://doi.org/10.1007/978-3-031-90966-5_1

new field of research called "Tissue Engineering" and are used in what we now call "Regenerative Medicine."

Keywords Tissue engineering · Regenerative medicine · Tissue engineered medical products · TEMPs · Stem cells · Bioactive molecules · Growth factors · Scaffolds

1.1 Development of the Tissue Engineering Field

In prehistoric times, people used many natural things like leaves, fruit extracts, or roots for therapeutic purposes. Later on, different materials present on earth (e.g., ores, metals, ceramics, wood, etc.) were used to treat damaged or incapacitated organs. Although these materials helped improve the living conditions of the patients to some extent, they also had undesirable effects. Metals did rust, ceramics broke, and polymers disintegrated under load application or unsuitable environmental conditions [1]. Over the centuries humans learned a lot and developed technologies. In the 1970s it became possible to use stem cells, blood plasma (platelet-rich plasma, PRP), or biological agents such as growth factors to positively influence the regeneration of tissues [2]. Such treatments helped relieve the patients by repairing minor- or moderate-level damages, but they were not very successful in the treatment of large tissue damages or loss of organs. In time, the prosthetic devices started to be replaced by tissues obtained from patients themselves or from other individuals as autografts or as allografts. This was not always a perfect solution because the supply was not in sufficient amount or quality or led to undesirable consequences. Then new techniques were developed aimed at producing "living tissues" in the laboratory. Researchers added cells on porous materials, scaffolds, mostly made of polymers, and let the cells grow in the culture media supported by various factors. Scaffolds mimicked the extracellular matrix (ECM) in a tissue, serving as a platform for the cells to adhere, grow, and increase in number. As the cells proliferated, they created their own natural ECM as a fibrous network consisting of a variety of macromolecules specific for the cell types grown. The scaffolds are usually fibrous or spongy structures made from synthetic or biological polymers. Eventually the scaffolds, with mature and proliferated cells in the culture medium, were placed at the defect site of the patients [3, 4].

In the early years of tissue engineering research, neutral and inactive scaffolds having no interaction with the environmental living tissue and stay intact were preferred. These scaffolds were placed and initiated regeneration of the defect site with the help of growth factors secreted by the cells or added into the culture media. In time, the scaffolds were made of biodegradable materials or activated by addition of functional materials to enhance the compatibility with natural tissue. Biodegradable materials are broken down over time by chemicals and enzymes secreted by cells and are replaced by the newly developing tissue. In the 1990s, novel biodegradable materials were developed so that as the cells proliferated and tissues formed, the scaffold material decomposed and was excreted from the body. At the end, no foreign material would remain and its place would be taken by the newly regenerating tissue. Such studies eventually were known as 'Tissue Engineering', a sub-branch of the biomaterials field [5]. Tissue engineering made it possible to produce tissues and, lately, organs. In this way, the problems arising from lack of donor tissues or organs started to be solved.

As can be seen from the introduction, tissue engineering is an interdisciplinary field. Scientists from different branches of science (chemistry, biology, mathematics, etc.), engineering (electrical, mechanical, chemical, materials, etc.), medicine (histology, hematology, neurology, etc.), and new areas such as nanotechnology, artificial intelligence, etc. are collaborating, bringing their knowledge and know-how together to better mimic the natural tissues (Fig. 1.1).

Cellular therapies have also gained importance alongside the tissue engineering studies in the attempt to create tissues and organs in the laboratory as they both have a significant cross section, the cells. Use of blood obtained from healthy people and given to patients is known for many years, especially after the identification of the presence of blood groups and their differences. On the other hand, knowledge about stem cells and their use in medical treatments have gained importance later, in the last two to three decades. Scaffolds are not part of cell therapies unlike tissue engineering where scaffolds are used to create a frame to build a new 3D tissue. Thus, there is no physical structure to fill a certain volume and support or limit the movement of cells. With tissue engineering it is possible to fill large defects but with cell therapy the therapeutic tool, the cells, are free to move and reach sites which a cell attached to a scaffold cannot achieve. Over time, these fields, namely Biomaterials, Tissue Engineering, and Cellular Therapies, began to be recognized as "Regenerative

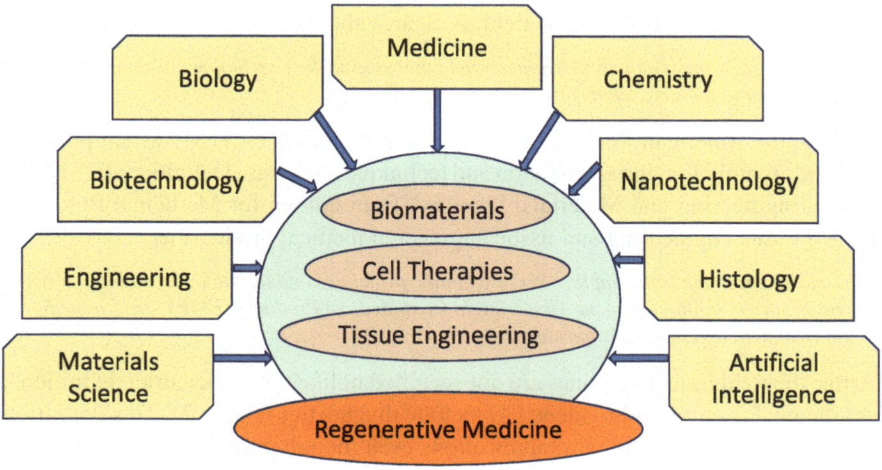

Fig. 1.1 Regenerative medicine and related fields

Medicine." In fact, Regenerative Medicine is recognized as a new interdisciplinary subject, which is based on the knowledge of biology, chemistry, materials science, engineering, and clinical medicine [6].

As stated above, tissue engineering is an interdisciplinary area that deals with the design and production of tissues and organs under in vitro conditions using scaffolds, cells, and some bioactive agents together to be implanted in the patient to take the place of the non-functioning or harmed tissue or organ of the organism [7].

Tissue engineered medical products have three basic components [8]:

1. Cells (they proliferate and form the tissue),
2. Scaffolds (behave as the ECM for the growing cells),
3. Bioactive agents (e.g., growth factors, drugs, etc., that guide and support the maturation of the cell-seeded scaffolds into tissues).

1.2 Definition of Tissue Engineering

The definition of tissue engineering changed and was modified over the years due to differences and developments in the fields of the international institutions and organizations.

A 1987 report from a workshop supported by the American National Science Foundation (USA-NSF) defined it as:

> *Tissue engineering is the application of principles and methods of engineering and life sciences to understand the basic functioning of biological tissue elements and to improve, maintain and enhance human tissue functions.*

In this definition, NSF mentions the related fields involved in the studies but does not refer to the repair or regeneration of tissues and organs.

The later definition in 1993 made by Robert Langer and Charles Vacanti, who are considered as the pioneers of the field, is clearer about the goal of the studies:

> *Tissue engineering is an interdisciplinary field that aims to develop biological elements that can be used to replace, repair or heal dysfunctional tissue.*

In 2020, the American Society for Testing and Materials (ASTM) which prepares standards mainly for characterization and techniques for tests (Definition: F2312-11, Tissue Engineering and Materials. Standard Terminology for Medicinal Products) defined tissue engineering and tissue engineered medical products as:

> *Tissue engineering is the application of scientific principles and technologies under in vivo and in vitro conditions to make Tissue Engineered Medical Product (TEMP) used for medical treatment and diagnosis purposes.*

In this, the fields and disciplines are not specified unlike the earlier strict definitions, and the goals were also broadened to include diagnostics (Fig. 1.2). This definition also covers in vivo and in vitro applications even though ultimate goal is a medical product as stated in the definition. The target in the therapeutic definition is much

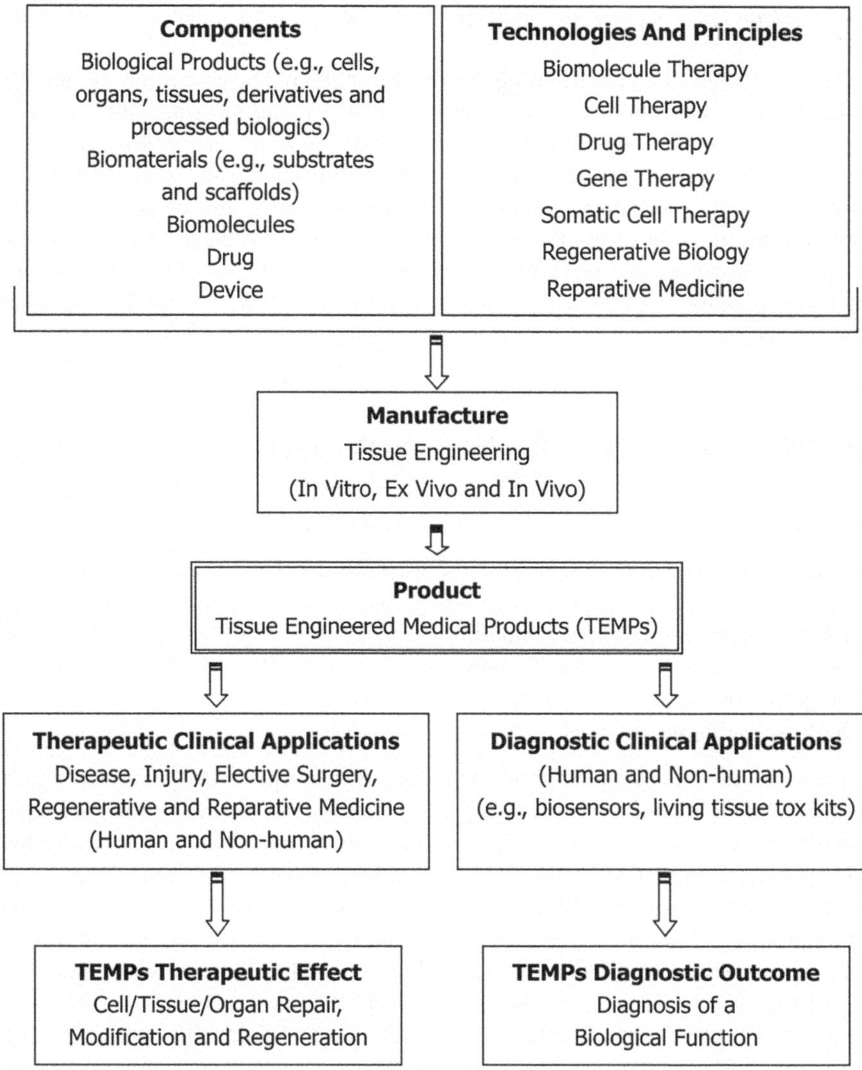

Fig. 1.2 Tissue engineering components, production, and applications [9]

clearer: "*TEMP*" *is a medical product that repairs, replaces or renews the structure, function, or both of a patient's cells, tissues and organs.*"

This definition also includes the renewal of functions of cells, tissues, and organs instead of replacing or treating them. In all, it actually defines regenerative medicine as we know today. However, the statement "*Regenerative medicine is a branch of medical science that applies the principles of regenerative biology to specifically repair or rebuild the structure and function of human cells, tissues, and organs that cannot adequately regenerate.*" narrows the field again to make it a "medical" activity.

As stated above the definitions change based on the field, occupation, or interest of the defining institution. The main purpose of tissue engineering products is to regenerate damaged tissues and organs. An important use of tissue engineering is to serve as model systems to study the mechanisms of diseases and the comparison of the contribution of the three components (materials, cells, and bioactive components) on the function of the medical product, especially in place of in vivo tests, and also as diagnostic devices for pharmaceutical trials. Of course, their main function is regeneration of the defective tissues and organs followed by gradual removal from the site of application [10, 11].

1.3 The Uses of Tissue Engineering Products

Tissue engineering products are generally used for three purposes:

1. Serving as tissue models (cell clusters or organoids) for purposes such as diagnosing diseases, determining drug doses, and substituting for the animals normally used in in vivo experimentation
2. Serving as "tissue engineered medical product" (TEMP) to be implanted in the body in order to heal and regenerate tissues and organs.
3. Serving as diagnostic tools

Tissue engineered systems for use with diagnostic purposes are generally placed on microfluidic devices or microchips, in addition to tissue culture set ups. Appropriate cells to represent the organ in question are introduced to these systems and kept under culture conditions (pH, T) to initiate growth of the tissue. Diagnostic tissue systems are often used for research purposes and mostly in place of in vivo testing [10, 11]. One can design a metastasis model using tissue engineered cancer tissue mimic and the recipient such as a healthy bone. Then the concentration or the type of the drug to be used can be applied with various frequencies, doses, drug combinations, and responses such as effective dose (ED_{50}), cell viability, gene expression, differentiation, and mobility are measured. Changes in the increase of expression from different genes upon these applications are measured [12].

If tissue engineering is performed for tissue and organ regeneration, the autologous cells (patient's own cells) or cells that will not cause an immune reaction need to be grown on the scaffold, and the rate of degradation should be parallel to the tissue formation at the defect site. Progress is being made in all tissues, because new techniques such as microfluidic systems, 3D printing, and electrospinning are making the tissue engineered products much more accurately than simple molding technique [6, 13].

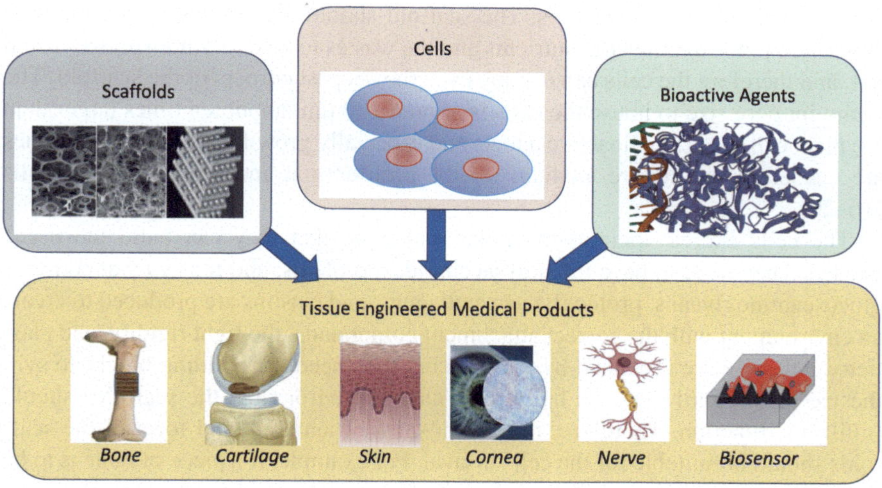

Fig. 1.3 Components of tissue engineering products

1.4 Tissue Engineering Components

Tissue engineered medical devices have three components: these are the cells that will form the living component, the tissue, then the scaffolds that substitute for the ECM for the cells to attach and multiply in, and the bioactive agents including growth factors that help the development of the cells and make the three-component structure one unified tissue piece (Fig. 1.3).

1.4.1 The Cell Types and the Environment

Cells constitute the living components in the tissue engineered product. They adhere to the scaffold surface, live and secrete their own extracellular matrix, and multiply and convert the artificial scaffold medium into a piece of living tissue. Since this product is already placed at the defect site, new tissue that formes fills the gap. The environment is very critical for the cells as they need to have the correct nutrients, ions, pH, and temperature to multiply. In addition, the surface properties of the scaffold need to entice cells to adhere. For example, extra smooth or very wet scaffolds are not suitable for the cells to attach to. The number of cells seeded is also critical because cells are in continuous contact with each other and if only a few are seeded or few cells were able to attach due to the surface properties, cells do not make contact and are not viable. The immune responses of the host tissue is important because when the immune response is triggered the newly introduced cells do not survive at their attachment sites. The scaffold should also be highly porous with connected pores to allow the nutrients and the wastes to be easily transported in and out, and therefore the cells can occupy even the deepest corners of the scaffold. The optimum pore size to house the cells is around 200 μm, about ten times larger than a typical cell [14, 15]. Bioactive agents and especially growth factors and hormones are needed in the culture medium for the proliferation and function of the cells [16–20].

The cells excrete their extracellular matrix so that they can build the tissue needed. They need to have the polysaccharides, proteins, and ions so that complex glucoseaminoglycans, proteoglycans, collagens, and elastins are produced to create an environment with the correct attachment compounds, the right rigidity, and elasticity. All these are produced by the cells but they need the building blocks to synthesize them. Until the cells build their natural environment the scaffolds should fulfill this function. Therefore, the composition, chemistry, and form of the scaffolds should be suitable for the cell survival. For example, if a piece of bone is to be tissue engineered, the scaffold needs to have the right rigidity, flexibility, surface properties, and an appropriate degradation rate. These requirements are quite different than those when a tissue engineered ear is to be prepared. In the first one, compressive properties are important while in the second, it is flexibility that is needed. As a result, the material selection in scaffold preparation necessitates the development of both strong and soft materials and the correct methods of production.

In the meantime, a number of natural polymers are frequently used as scaffolds (due to their appropriate biodegradation rates), their blends and composites with other natural or synthetic polymers, or ceramics and metal particles are also used to optimize their properties and biocompatibility [21–24].

Cells are the only living component of the tissue engineering process. The cells used in tissue engineering can be stem cells which have the ability to differentiate into a number of mature cell types depending on their source or they can be fully differentiated, mature primary cells whose tissue type is well defined. The cell types used are presented below.

1.4.1.1 Primary Cells

Primary cells are mature cells which have differentiated to become specific for certain tissues. They are obtained from the patient (autologous), from another human (allogeneic), or from an animal (xenogenic).

- **Autologous cells:** As stated above, these are cells isolated from the corresponding tissues of the patient. In order to isolate cells from tissue, a small piece of biopsy (patient's tissue) is chopped into smaller pieces and then immersed in a growth medium where the cells leave the tissue and move into the culture medium and multiply. These cells do not lead to immune responses because their source is the same host. The only disadvantage of this procedure is the new, self-inflicted scar created in the body of the patient who is already suffering from another.

- **Allogeneic cells:** These cells are obtained from another human and have the risk of pathogen transmission and also immune response. The main advantage is that the cells are not obtained from the patient who already is in need and may not tolerate extra pain.
- **Xenogenic cells:** These cells are obtained from non-human donors but their use carries a number of risks. Among these are transmission of human and non-human disease and immune response. As a result, these cells are often used in scientific research, in the testing of new designs, and in new experimental conditions such as new scaffolds and effects of bioactive agents. The risk of transmission of unknown pathogens is very serious, especially due to absence of treatments for some. After the initial optimization studies tests are repeated with human cells or cell lines.

1.4.1.2 Stem Cells

Stem cells need special emphasis as they are primary cells that can differentiate into cells of various tissues or organs. They are undifferentiated and have the ability to differentiate into a number of specialized cells as determined by their origin. They also can undergo multiple cell divisions and renew themselves infinitely.

Their unlimited ability to divide enables them to regenerate limitless times. The most important source of stem cells is bone marrow. There are three types of stem cells:

- Totipotent: These stem cells are the first four cells obtained from the embryo (after two divisions) and are able to differentiate into any cell in the body. As such, it is possible to create a whole human using these cells.
- Pluripotent: These stem cells are cells obtained from the undifferentiated cells in a human embryo. As such, they can differentiate into cells of the ectoderm, mesoderm, and endoderm of the embryo, but they cannot differentiate into all cells in the human body like the totipotent cells. In that sense they are limited by the embryonic source they come from.
- Multipotent: These cells are more limited in their differentiation than the pluripotent cells; they can only differentiate into a limited number of cell types. They are found in the bone marrow, pelvis, brain and liver. They can give rise to a number of differentiated cell types, all within a separate tissue, organ, or physiological system. Mesenchymal stem cells (MSCs) are of the multipotent family, and they can also differentiate into a limited number of cells such as those of bone, cartilage, and fat cells (Fig. 1.4). Osteocytes (bone), chondrocytes (cartilage), adipocytes (fat), hepatocytes (liver), myocytes (muscle), neurons (nerve), and cardiomyocytes (heart muscle) are all based on MSCs. Another cell type that is used in research is hematopoietic stem cells that can differentiate into all adult blood components.

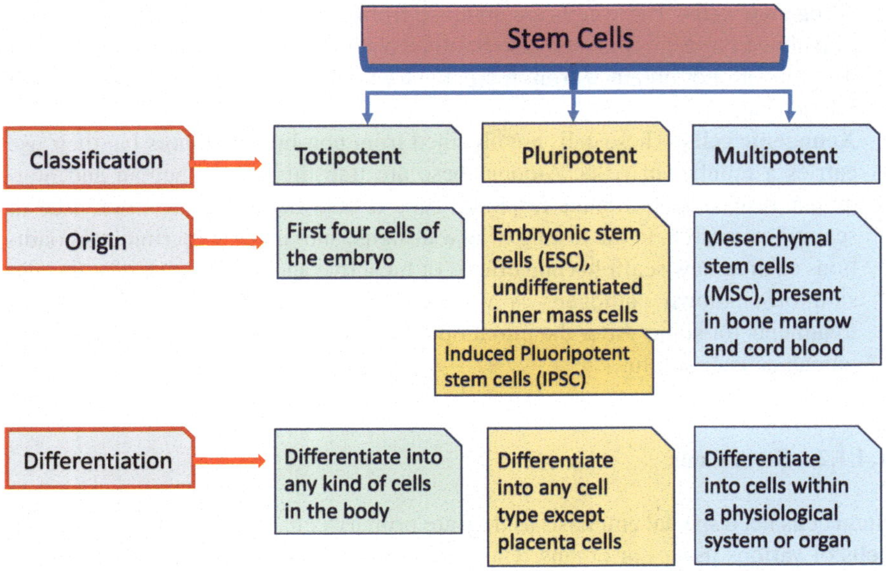

Fig. 1.4 Stem cell categories and properties

1.4.1.3 Cell Lines

Stem cells can divide indefinitely but when they mature, they can only divide a limited number of times due to cellular senescence (aging). However, under certain conditions cells may mutate and become immortal. These gain the ability of limitless division again and are called "cell lines". Some cell lines are relatively easy to obtain, especially from rodents. HeLa is an immortalized cell line from humans. This cell line was obtained from a cervical cancer patient (Henrietta Lacks) after whom it is named and is used in many studies as a source for cancer cells.

Researchers prefer to use freshly isolated cells in their experiments [25]. They do not accept that cell lines fully represent a human primary cell. On the other hand, cell lines are frequently used since they are able to divide many times, less sensitive than primary cells and therefore easier to culture, commercially available, and more importantly the results from different labs can be easily compared when obtained using the same cell line.

1.4.2 Bioactive Molecules

In order to activate the cells for growth and differentiation, certain bioactive molecules are added to the culture medium. These molecules, the growth factors, are generally proteins [14] which attach to the membranes of the cells and carry signals between cells. Growth factors bind to the receptors on the cell membranes activate certain pathways. In tissue engineering growth factors are added to the culture medium to regulate the activities including proliferation and differentiation. The most commonly used growth factors are presented in Table 1.1.

1.4.3 *Scaffolds*

The scaffolds are the main extracellular matrix mimic that simulates the tissue in which the cells proliferate, differentiate, and function to produce the tissue missing due to a defect. They are planned to be temporarily available at the site and to

Table 1.1 Some growth factors and their functions [26]

Name	Code	Function
Fibroblast growth factor	FGF	FGF regulates a wide range of biological functions, including cellular proliferation, survival, migration, and differentiation
Epidermal growth factor	EGF	EGF controls cell growth by stimulating cell division
Nerve growth factor	NGF	NGF supports the life of neurons which are communicating and the life and differentiation of sensory neurons
Vascular endothelial growth factor	VEGF	VEGF supports angiogenesis and endothelial cell proliferation, plays a role in the regulation of vascularization, controls vasodilatation, cell migration and apoptosis, and induces migration and proliferation of endothelial cells
Platelet derived growth factor	PDGF	PDGF is a powerful stimulator of mesenchymal cells, smooth muscle cells and glial cells, induces the proliferation of undifferentiated mesenchymal stem cells and some basic cells, and directs the differentiation of some specialized cells
Insulin and insulin-like growth factor	IGF	IGFs control cell growth and cell survival, affect the proliferation and differentiation of osteoblast cells, and function in the regional regulation of bone formation
Transforming growth factor	TGF	TGF induces osteoblast proliferation and differentiation and regulates bone formation
Bone morphogenic proteins (BMP)	BMP1	BMP1 affects osteoblast proliferation and differentiation, regulates bone formation, is involved in ECM formation and regulation of other growth factors
	BMP2	BMP2 has a very important role in bone repair and regeneration
	BMP3 BMP4	BMP3 and BMP4 stimulate ALP activity and collagen synthesis, play a role in sulfated proteoglycan synthesis and cartilage repair
	BMP5 BMP6	BMP5 and BMP6 are effective in chondrocyte differentiation and growth of long bones
	BMP7	BMP7 is involved in the differentiation and migration of bone-forming cells
	BMP8	BMP8 plays an important role in embryonic development and the formation of the central nervous system
	BMP9	BMP9 increases the expression of aggrecan and cartilage oligomeric matrix proteins, also important for the development of nerve cells
Tumor necrosis factor	TNF	TNF-α causes necrosis in normal tissues but contributes to liver repair and regeneration

disappear gradually as the fresh tissue is produced. It is expected that the scaffolds mimic the physical, mechanical, and chemical properties of the tissue in which they are introduced. If the scaffold is for bone tissue engineering, it is then expected to have the appropriate mechanical strength, especially against compression, and the cells introduced should be bone cells (osteoblasts or osteocytes) or stem cells that can differentiate into bone cells. The growth factors used for this purpose are mainly the bone morphogenic proteins (BMPs). The shapes of the scaffolds (fibrous, porous, 3D printed, and engineered surface) and their chemistry (hydrophobic, hydrophilic, and reactive) are very important for the optimum performance of the implant.

The scaffolds are made of biodegradable materials as we expect them to be replaced by the newly formed tissue [6, 21, 22]. The more critical aspect is the processing technique with which the scaffold is produced in the form that is optimum for that tissue or organ. For example, a scaffold for skin needs to be not rigid, but elastic, highly porous or spongy. How does one produce a scaffold in this shape and from what kind of a material? A ceramic or a metal is not suitable for this application, and therefore, natural or synthetic macromolecules, polymers, are more appropriate. Will they be of biological polymer such as collagen, gelatin, or a polysaccharide or use a biodegradable polymer such as poly(lactic acid) or its copolymers? How will they be produced in a spongy form? Pour some solution into a mold and evaporate the solvent, or 3D print using a model made by scanning the normal tissue of the patient, or by using electrospinning on a model? As can be seen from the complexity of the choices to be made tissue engineering is not as simple as using an off-the-shelf implant.

The form of the scaffolds is an important aspect to consider for a proper scaffold performance. Some insist that the scaffold must be a highly porous in order to allow the cells to penetrate and then occupy the whole volume, while others state that the scaffolds should have well-defined designs that guide the cells to proliferate in an anisotropic way, especially for use in the formation of nerves or muscles [27–29]. Researchers, as a result, use many different porous designs that allow cell migration and proliferation as their final implants require [30–32]. Pharmaceutical agents, such as drugs can be applied directly on the scaffolds or within micro or nanocarriers which are placed into scaffolds to enhance or control the performance of the medical device [33–36].

Another important point of the biomaterials sometimes considered is how their production influences the nature or in other words the eco-friendliness of the production cycle. Scientists prepared 3D scaffolds by Digital Light Processing technique from polyethylene glycol diacrylate having aramid nanofiber, and found that the fabrication process of the scaffolds might negatively influence global warming [37].

1.5 Tissue Engineered Medical Products (TEMPs)

Tissue engineered medical products (TEMPs) are used for medical treatments and as diagnostic devices. These devices are also manufactured for the treatment and/or diagnosis of non-human organisms, e.g., animals. In the last decades, especially cell-carrying scaffolds are used for the production of TEMPs. Therefore, scientists from different areas (almost all branches of science, engineering, and medicine) should collaborate and combine their knowledge and technology in the production of TEMPs.

The ASTM definition of TEMPs is given as:

TEMPs may be a cell-biomaterial construct or a drug-cell formulation that is topically applied, implanted, injected, or used as an ex vivo device.

ASTM committee (F04 TEMPs standard effort), grouped TEMPs in to ten organ and tissue systems as follows [38]:

1. ***Cardiovascular*** (heart, valves, arterial and venous blood vessels, and microvasculature and cardiac muscle)
2. ***Digestive*** (oral cavity, tongue, teeth, salivary glands, pharynx, tonsils, esophagus, stomach, small intestine, colon, pancreas, biliary tract, gall bladder, liver, appendix, and recto-anal canal)
3. ***Endocrine*** (Pancreas/islets, pituitary, parathyroid, thyroid, adrenal, and pineal body)
4. ***Hematopoietic*** (Blood and bone marrow, lymph nodes, spleen, thymus, and lymphatic vessels)
5. ***Integument*** (skin (epidermis and dermis), hair, nails, sweat glands, and sebaceous glands)
6. ***Musculoskeletal*** (Tendons, ligaments, bone structures, cartilage structures (elastic, hyaline, fibrous cartilage), bone (compact and spongy), skeletal, and smooth muscles)
7. ***Respiratory*** (nasal cavity and sinuses, trachea, larynx, and lungs)
8. ***Nervous*** (spinal cord, ganglion, brain (e.g., cerebellum, cerebrum), eyes, inner ear (sensory systems), nerve fibers, and nerve bodies of any type)
9. ***Reproductive*** (Male: ducts, sex glands (e.g., prostate), testes, epididymis, penis; Female: mammary gland and nipples, ovary, uterus, vagina, uterine tubes, and, in cases of pregnancy, the placenta)
10. ***Urinary*** (kidneys, bladder, ureters, and urethra)

Since there are quite a number of TEMPs, their classifications depend on the product components, site of action, therapeutic target, therapeutic effect, mode of action, duration of therapy, and lifetime as shown in Fig. 1.5 [38].

When a TEMP is manufactured and approved by authorities for its safety and effectiveness, still there are some additional considerations in clinical applications due to human ethics. Some patients may refuse to have a new therapy, and some are willing to try a new device. In clinic, TEMPs must be applied with the patient's knowledge and consent.

The following chapters will provide information about polymeric structures, hydrogels, bioinks, 3D printing techniques, and studies carried out to create certain TEMPs using hydrogels and bioinks.

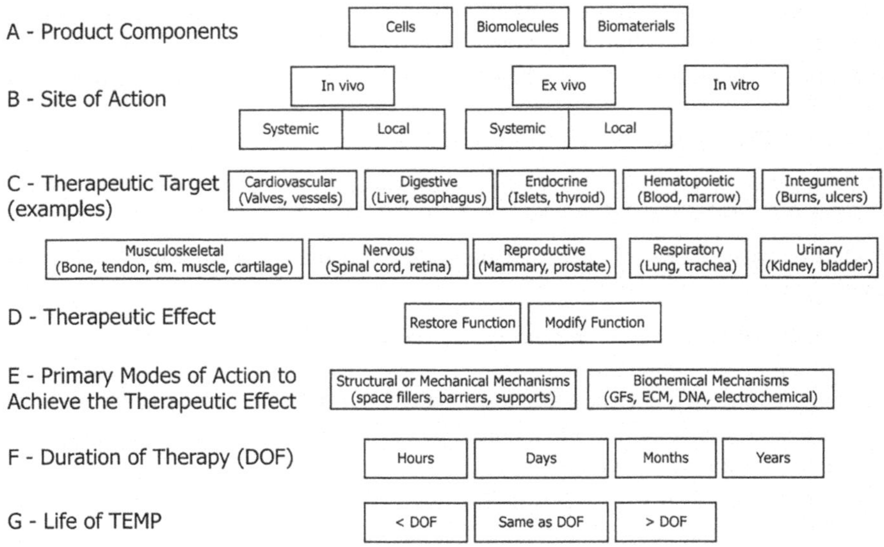

Fig. 1.5 Aspects of TEMPs [38]

1.6 Conclusion

Biomaterials, tissue engineering, and regenerative medicine applications are wonderful solutions to a majority of the health problems that the public is confronted with. Donor tissues or implants are partial solutions with their own limitations. Even though the materials, cell and bioactive agent requirements, and their medical approval for use make the whole tissue engineering process a very complex issue, the final solution is to make the patient whole again and that is incomparable with the other options.

References

1. Hasirci, V., Hasirci, N.: Fundamental of Biomaterials, 2nd edn, pp. 1–25. Springer. Ch 1. ISBN 978-1-4939-8856-3. https://doi.org/10.1007/978-3-031-54046-2_1
2. Vacanti, J., Morse, M., Saltzman, M., Domb, A.J., Atayde, A.P., Langer, R.: Selective cell transplantation using bioabsorbable artificial polymers as matrices. J. Pediatr. Surg. **23**, 3–9 (1988). https://doi.org/10.1016/S0022-3468(88)80529-3
3. Kilic, C., Dursun, T., Hasirci, N., Hasirci, V.: Engineered hydrogels. In: Nardin, C., Schlaad, H. (eds.) Biological Soft Matter: Fundamentals, Properties and Applications. Wiley (2021). Chapter 4, 288 Pages. https://doi.org/10.1002/9783527811014.ch4
4. Hasirci, N., Kilic, C., Kömez, A., Bahcecioglu, G., Hasirci, V.: Hydrogels in regenerative medicine. In: Demirci, U., Khademhosseini, A. (eds.) Gels Handbook: Fundamentals, Properties and Applications (3 volumes). Volume 2—Applications of Hydrogels in

Regenerative Medicine, pp. 1–52. World Scientific Publishing Company (2016). https://doi.org/10.1142/9789813140394_0001

5. Tian, A., Yi, X., Sun, N.: Application of mesenchymal stem cells combined with nano-polypeptide hydrogel in tissue engineering blood vessel. Regen. Ther. **21**, 277–281 (2022)
6. Tamay, D., Hasırcı, N.: Bioinks—materials used in printing cells in designed 3D forms. J. Biomater. Sci. Polym. Ed. **32**(8), 1072–1106 (2021). https://doi.org/10.1080/09205063.2021.1892470
7. Langer, R., Vacanti, J.: Advances in tissue engineering. J. Pediatr. Surg. **51**(1), 8–12 (2016)
8. Dursun, T., Kilic, C., Hasirci, N., Hasirci, V.: Engineered biopolymers. In: Nardin, C., Schlaad, H. (eds.) Biological Soft Matter: Fundamentals, Properties and Applications. Wiley (2021). Ch 3. ISBN: 978-3-527-34348-5. https://doi.org/10.1002/9783527811014.ch3
9. Standard Terminology Relating toTissue Engineered Medical Products, ASTM: Designation: F2312–11 (Reapproved 2020), TEMPs. https://www.astm.org/Standards/F2312.htm
10. Chahsetareh, H., Yazdian, F., Pezeshki-Modaress, M., Aleemardani, M., Hassanzadeh, S., Najafi, R., Simorgh, S., Nooshabadi, V.T., Bagher, Z., Davachi, S.M.: Alginate hydrogel-PCL/gelatin nanofibers composite scaffold containing mesenchymal stem cells-derived exosomes sustain release for regeneration of tympanic membrane perforation. Int. J. Biol. Macromol. **262**(Part 2), 130141 (2024). https://doi.org/10.1016/j.ijbiomac.2024.130141
11. Bostancı, N.S., Büyüksungur, S., Hasırcı, N., Tezcaner, A.: Bioprinted scaffolds assembled as synthetic skin grafts by natural hydrogels containing fibroblasts and bioactive agents. Int. J. Polym. Mater. Polym. Biomater. **73**(11), 927–945 (2024). https://doi.org/10.1080/00914037.2023.2230339
12. Komez, A., Buyuksungur, A., Antmen, E., Swieszkowski, W., Hasirci, N., Hasirci, V.: A 2-compartment bone tumor model to investigate interactions between healthy and tumor cells. ACS Biomater. Sci. Eng. **15**(3), 035007 (2020). https://doi.org/10.1088/1748-605X/ab6b31
13. Eke, G., Mangır, N., Hasirci, N., MacNeile, S., Hasirci, V.: Development of a UV cross-linked biodegradable hydrogel containing adipose derived stem cells to promote vascularization for skin wounds and tissue engineering. Biomaterials. **129**, 188–198 (2017). https://doi.org/10.1016/j.biomaterials.2017.03.021
14. Yilgor, P., Yilmaz, G., Onal, M.B., Solmaz, I., et al.: An in vivo study on effect of scaffold geometry and growth factor release on the healing of bone defects. J. Tissue Eng. Regen. Med. **7**, 687–696 (2013). https://doi.org/10.1002/term.1456
15. Schumann, P., Tavassol, F., Lindhorst, D., Stuehmer, C., et al.: Consequences of seeded cell type on vascularization of tissue engineering constructs in vivo. Microvasc. Res. **78**(2), 180–190 (2009). https://doi.org/10.1016/j.mvr.2009.06.003
16. Cesur, O., Tanir, T.E., Celepli, P., Ozarslan, F., Hucumenoglu, S., Karaibrahimoglu, A., Hasirci, N.: Enhancing esophageal repair with bioactive bilayer mesh containing FGF. Sci. Rep. **11**, 19203 (2021). https://doi.org/10.1038/s41598-021-98840-w
17. Ates, U., Gollu, G., Bingol-Kologlu, M., Billur, D., Kucuk, G., Oruklu, N., Bagrıacik, U., Hasırcı, N., Dindar, H.: The effects of local and systemic growth hormone treatment on germ cell population and fertility in an experimental unilateral testicular torsion and orchiectomy model. J. Urol. **194**, 1816–1822 (2015). https://doi.org/10.1016/j.juro.2015.07.087
18. Bilgic, H., Demiriz, M., Ozler, M., Ide, T., Dogan, N., Gumus, S., Kiziltay, A., Endogan, T., Hasirci, V., Hasirci, N.: Effects of EGF containing gelatin sponges as wound healing scaffolds. J. Biomater. Tissue Eng. **3**(2), 205–211 (2013)
19. Uguralp, S., Bay Karabulut, A., Mizrak, B., Kaymaz, F., Kiziltay, A., Hasirci, N.: The effect of sustained and local administration of epidermal growth factor on improving bilateral testicular tissue after torsion. Urol. Res. **32**, 323–331 (2004)
20. Saygun, I., Bal, V., Kızıltay, A., Yilgor Huri, P., Avcu, F., Üstün, K., Hasirci, V., Hasirci, N.: Proliferation and differentiation of mesenchymal stem cells in chitosan scaffolds loaded with nanocapsules containing bone morphogenetic proteins-4, platelet-derived growth factor and insulin-like growth factor 1. J. Biomater. Tissue Eng. **4**, 181–188 (2014). https://doi.org/10.1166/jbt.2014.1153

21. Spicer, C.D.: Hydrogel scaffolds for tissue engineering: the importance of polymer choice. Polym. Chem. **11**, 184–219 (2020). https://doi.org/10.1039/c9py010
22. Shimojo, A.A.M., Rodrigues, I.C.P., Perez, A.G.M., Souto, E.M.B., Gabriel, L.P., Webster, T.: Scaffolds for tissue engineering: a state-of-the-art review concerning types, properties, materials, processing, and characterization. In: Li, B., Moriarty, T., Webster, T., Xing, M. (eds.) Racing for the Surface. Springer, Cham (2020). https://doi.org/10.1007/978-3-030-34471-9_23
23. Bektas, C.K., Kimiz, I., Urkmez, A.S., Hasirci, V., Hasirci, N.: A bilayer scaffold prepared from collagen and carboxymethyl cellulose for skin tissue engineering applications. J. Biomater. Sci. Polym. Ed. **29**(14), 1764–1784 (2018). https://doi.org/10.1080/09205063.2018.1498718
24. Isikli, C., Hasirci, V., Hasirci, N.: Development of porous chitosan-gelatin/hydroxyapatite composite scaffolds for hard tissue engineering applications. J. Tissue Eng. Regen. Med. **6**(2), 135–143 (2012). https://doi.org/10.1002/term.406
25. Mandatori, D., D'Amico, E., Romasco, T., Gatto, M.L., Notarangelo, M.P., Mangano, C., Furlani, M., Penolazzi, L.: A 3D in vitro model of biphasic calcium phosphate (BCP) scaffold combined with human osteoblasts, osteoclasts, and endothelial cells as a platform to mimic the oral microenvironment for tissue regeneration. J. Dent. **151**, 105411 (2024). https://doi.org/10.1016/j.jdent.2024.105411
26. Ren, X., Zhao, M., Lash, B., Martino, M.M., Julier, Z.: Growth factor engineering strategies for regenerative medicine applications. Front. Bioeng. Biotechnol. Sec. Nanobiotechnol. **7**, 469 (2019). https://doi.org/10.3389/fbioe.2019.00469
27. Hasirci, V., Hasirci, N.: Fundamentals of Biomaterials, 2nd edn, pp. 313–334 (2024). Ch 19. ISBN 978-1-4939-8856-3. https://doi.org/10.1007/978-3-031-54046-2_19
28. Li, Y., Ma, Z., Ren, Y., Lu, D., Li, T., Li, W., Wang, J., Ma, H., J.: Zhao tissue engineering strategies for peripheral nerve regeneration. Front. Neurol. **12**, 768267 (2021). https://doi.org/10.3389/fneur.2021.768267
29. Namjoo, A.R., Abrbekoh, F.N., Saghati, S., et al.: Tissue engineering modalities in skeletal muscles: focus on angiogenesis and immunomodulation properties. Stem Cell Res Ther. **14**, 90 (2023). https://doi.org/10.1186/s13287-023-03310-x
30. Nochalabadi, A., Khazaei, M., Rezakhani, L.: Exosomes and tissue engineering: a novel therapeutic strategy for nerve regenerative. Tissue Cell. **93**, 102676 (2025). https://doi.org/10.1016/j.tice.2024.102676
31. Bahcecioglu, G., Hasirci, N., Bilgen, B., Hasirci, V.: Hydrogels of agarose, and methacrylated gelatin and hyaluronic acid are more supportive for in vitro meniscus regeneration than three dimensional printed polycaprolactone scaffolds. Int. J. Biol. Macromol. IJBIOMAC. **122**, 1152–1162 (2019). https://doi.org/10.1016/j.ijbiomac.2018.09.065
32. Wu, G., Wang, L., Cao, Y., Wang, M., Yang, C., Zhang, J.: 4D bioprinting of transformable living constructs with sustained local growth factor presentation for advanced tissue engineering applications. Colloids Surf. B Biointerfaces. **248**, 114484 (2025). https://doi.org/10.1016/j.colsurfb.2024.114484
33. Ucar, S., Yilgor, P., Hasirci, V., Hasirci, N.: Chitosan based wet spun scaffolds for bioactive agent delivery in bone tissue engineering. J. Appl. Polm. Sci. **130**(5), 3759–3769 (2013). https://doi.org/10.1002/APP.39629
34. Mangır, N., Eke, G., Hasirci, N., Chapple, C.R., Hasirci, V., MacNeil, S.: An estradiol releasing proangiogenic hydrogel candidate material for use in soft tissue interposition. Neurourol. Urodyn. **38**(5), 1195–1202 (2019). https://doi.org/10.1002/nau.23971
35. Malikmammadov, E., Hasirci, N.: Dual and multi-stimuli responsive polymers for biomedical applications. In: Aguilar, M.R., Román, J.S. (eds.) Smart Polymers and Their Applications, 2nd edn, pp. 255–278. Woodhead Elsevier Pub (2019). (Page Count: 650). eBook ISBN: 9780081024171; Paperback ISBN: 9780081024164. https://doi.org/10.1016/B978-0-08-102416-4.00008-9
36. Khodaverdi, K., Naghib, S.M., Mozafari, M.R., Rahmanian, M.: Chitosan/hydroxyapatite hydrogels for localized drug delivery and tissue engineering: a review. Carbohydr. Polym. Technol. Appl. **9**, 100640 (2025). https://doi.org/10.1016/j.carpta.2024.100640

37. Tajurahim, N.A.N., Mahmood, S., Ngadiman, N.H.A., Sing, S.L.: Biomaterials for tissue engineering scaffolds: balancing efficiency and eco-friendliness through life cycle assessment. Clean. Environ. Syst. **16**, 100253 (2025). https://doi.org/10.1016/j.cesys.2024.100253
38. Standard Classification forTissue-Engineered Medical Products (TEMPs), ASTM, F2211−13. https://compass.astm.org/document/?contentCode=ASTM%7CF2211-13R21%7Cen-US&proxycl=https%3A%2F%2Fsecure.astm.org&fromLogin=true (2021)

Chapter 2
Tissue Scaffold Preparation Techniques

Deniz Sezlev Bilecen

Abstract Tissue scaffolds are artificial structures that function as carriers for cells while providing the microenvironment necessary for the proliferation of these cells and supporting three-dimensional tissue formation. Tissue scaffolds that are developed using biocompatible and generally biodegradable biomaterials are placed in the body either alone or together with cells or bioactive molecules for the damaged tissue to regain its function. In this aspect, cell viability and cell-to-cell connection within the scaffold are crucial for the success of the tissue scaffold. This success is dependent on several factors, such as biomaterial chemistry, degradation rate, and compatibility of the mechanical properties of the biomaterial to the target tissue, surface properties of the scaffold, pore size, and pore distribution. Biomaterials used in the field of tissue engineering can be classified as ceramic, metal, synthetic and biological polymers and the combinations of these that are called composites.

Keywords Scaffolds · Electrospinning · Particle leaching · Solvent casting · Melt molding · Phase separation · Gas foaming · Freeze drying

2.1 Tissue Scaffold Preparation Techniques

In the scaffold preparation for medical applications, the type of biomaterials and the techniques to be used are chosen based on the properties of the target organ. Depending on the preparation technique, the scaffold may be two dimensional (2D) or three dimensional (3D). 2D scaffolds are advantageous due to the ease of controlling the physical properties, imaging, and characterization of the final product. 3D

D. Sezlev Bilecen (✉)
Department of Molecular Biology and Genetics, Faculty of Agriculture and Natural Sciences, Konya Food and Agriculture University, Konya, Türkiye
e-mail: deniz.bilecen@gidatarim.edu.tr

N. Hasirci, V. Hasirci (eds.), *Hydrogels and Bioinks in Tissue Engineering*, Biological and Medical Physics, Biomedical Engineering,
https://doi.org/10.1007/978-3-031-90966-5_2

scaffolds (such as sponges and 3D printed structures), on the other hand, are advantageous for better tissue integration and being more compatible with the mechanical properties of the target tissue. Scaffolds that are used in tissue engineering applications can be produced by using different techniques that lead to different scaffold structures (fibers, sponges, or films). The commonly used scaffold production techniques are electrospinning, wet spinning, particle leaching, phase separation, gas foaming, freeze drying, hydrogel preparation and 3D printing.

2.1.1 Electrospinning

Electrospinning is one of the scaffold preparation techniques with which tissue scaffolds consisting of either micron- or nano-sized fibers are obtained. The nanofibers, especially in the size range of 50–500 nm, can mimic the natural tissue due to their resemblance to collagen fibers found in the extracellular environment of almost every tissue. In addition, due to their nano-size, their porosity and surface area-to-volume ratio are quite high. This property facilitates cell attachment and enables diffusion of nutrients through the scaffold. These types of tissue scaffolds, therefore, bring together the important factors for the field of tissue engineering, such as maintenance of high cell viability and proliferation [1].

The components of electrospinning are a viscous solution of synthetic or natural polymer, a syringe pump, a power supply for the generation of high-voltage electrical field, and a collector (disk, drum or plate). The electrical field generated between the syringe and the collector leads to the formation of a droplet at the tip of the needle, called the "Taylor cone." When the applied voltage overcomes the surface tension of the polymer solution, the droplet is ejected to the collector. During ejection, the solvent in the polymer solution evaporates, and the solidified polymer is collected on the collector in the form of fibers (Fig. 2.1).

The morphological and mechanical properties of the electrospun fibrous scaffolds depend on several factors such as applied voltage, polymer flow rate, the distance between the needle and the collector, temperature, the viscosity of the polymer solution, and volatility of the solvent. For instance, with increasing polymer flow rate, the fiber diameter would increase due to ejection of more polymer solution from the Taylor cone. Similarly and in general, increasing the polymer concentration would increase the fiber diameter due to an elevation in solution viscosity.

Although electrospinning technique is considered advantageous due to control of porosity and availability of the different polymers applicable for the process, it is not a suitable method for obtaining thick 3D structures. To produce tissue scaffolds with electrospinning technique, both natural polymers such as chitosan, collagen, or gelatin and synthetic polymers such as polycaprolactone (PCL) or poly(lactic acid-co-glycolic acid) (PLGA) can be used. The scaffolds prepared with this technique are mostly formed as thin sheets. For example, collagen sheets were used as skin substitutes [2]. Wise and his colleagues used electrospun and oriented polycaprolactone (PCL) scaffolds for the regeneration of articular cartilage [3]. Recently, Heydari

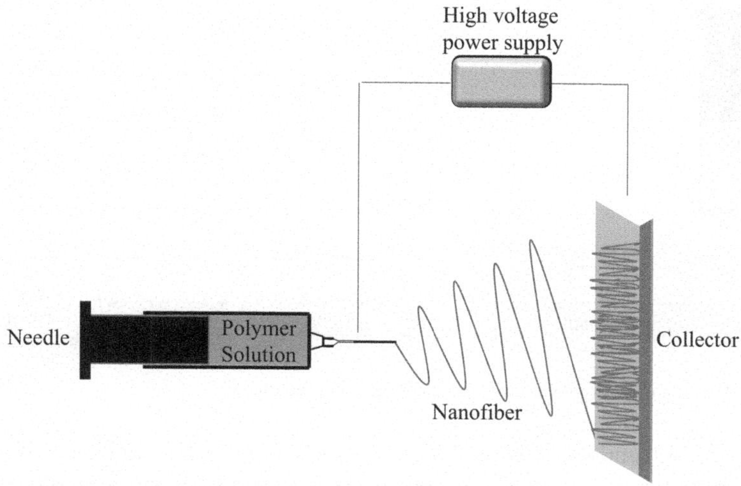

Fig. 2.1 Schematic presentation of nanofiber production with electrospinning process

et al. prepared wound dressing by using platelet-rich plasma-enriched polyglycerol sebacate/polylactic acid electrospun nanofibers [4]. In addition to these, thin sheets obtained by using polymers may be combined with other biomaterials for the formation of 3D structures. For instance, Xue and his colleagues used electrospun membranes composed of gelatin and polycaprolactone polymers in the form of fibers. The sheet was seeded with chondrocytes which were rounded onto an ear-shaped porous titanium alloy mold [5].

2.1.2 *Wet Spinning*

Unlike electrospinning, in wet spinning technique polymer solution is injected directly into the coagulation bath, which is a solution miscible with polymer solvent. When the polymer solution is extruded into the bath, the polymer begins to solidify in the form of fiber [6]. The fibers can be passed through washing or stabilizing baths to remove any residues and can be collected and dried on rotating cylinders (Fig. 2.2).

One of the advantages of this technique is that the entire process can be carried out at room temperature, making it possible to encapsulate heat-sensitive drugs or cells into the fiber structure. In addition, this technique produces fibers much faster than electrospinning, and both synthetic and natural polymers can be used for the process. The other important advantage of the technique is the application of tensile force to the formed fibers, which improves the molecular orientation and increases the mechanical properties of the fibers.

The parameters affecting the fiber morphology can be listed as the viscosity of the polymer solution, the molecular weight of the polymer, the properties of

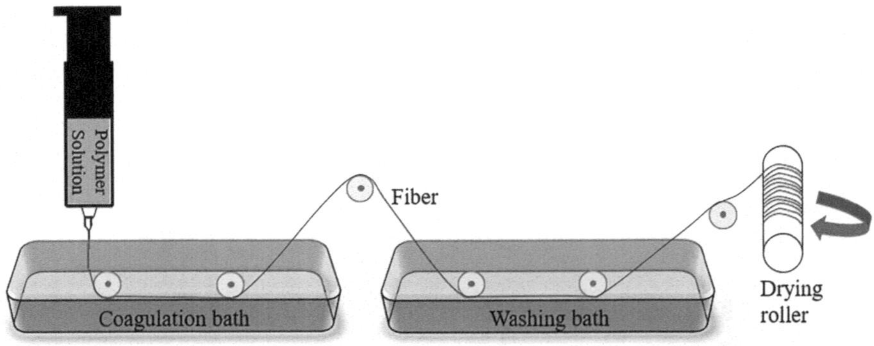

Fig. 2.2 Process of fiber production with wet spinning technique

solvent/non-solvent, the flow rate of polymer solution, and the diameter of syringe tip. Different combinations and optimizations of these parameters would result in fibers with different mechanical properties or porous fiber structures. For instance, the increase in the viscosity of polymer solution would result in denser, less porous and thicker fibers. The reason of this is that, as the polymer viscosity increases, it becomes more difficult for the solvent of the polymer solution to move away from the polymer in the coagulation bath, and it also becomes difficult for the coagulation liquid to diffuse into the polymer.

To produce tissue scaffolds by using wet spinning technique, polyesters such as poly(lactic acid-co-glycolic acid) (PLGA), poly(lactic acid) (PLA), poly(caprolactone) (PCL), polysaccharides such as cellulose, chitin, chitosan, hyaluronic acid, alginate, or proteins such as collagen, gelatin, and silk can be used [6]. Examples of such scaffolds may be given from bone tissue engineering applications. For example, the use of PCL with ceramic materials that are quite similar to bone structure (hydroxyapatite—HAp) [7] or the use of poly(3-hydroxybutyrate-co-3-hydroxyvalerate) (PHBV) coated with peptide sequence that induces vascularization [8]. Nerve guides modified with bone marrow stem cell derived ECM were developed with PLGA hallow fibers with wet spinning technique [9]. Other than these, chitosan fibers for the production of sutures [10], fibrin and alginate fibers for mimicking the muscle tissue [11] were obtained with the wet spinning technique.

2.1.3 Particle Leaching

Particle leaching method is based on the principle of dispersing a solid substance in the polymer solution, and then removing it by using its solvent which results in the formation of pores in the final structure. This technique is the most used technique in tissue scaffold preparation. Particle leaching can be performed either by solvent casting or melt molding (Fig. 2.3). Synthetic polymers such as poly(lactic acid) (PLA), poly(3-hydroxybutyrate) (PHB), and poly(lactic acid-co-glycolic acid)

(PLGA) are used with this technique for the preparation of different tissue engineering applications. For instance, PHB polymer is used in the production of scaffolds for bone or cartilage engineering [12]. Elastic copolymers (poly(glycolic acid-co-caprolactone) (PGCL) are used in the scaffolds designed for smooth muscle engineering [13], graphite/polyvinyl alcohol (PVA) scaffolds for artificial cornea skirt engineering [14], melt-molded poly (glycerol-sebacate) (PGS)-poly (caprolactone) (PCL)—gelatin for the fabrication of elastomeric biomimetic heart valve [15], or

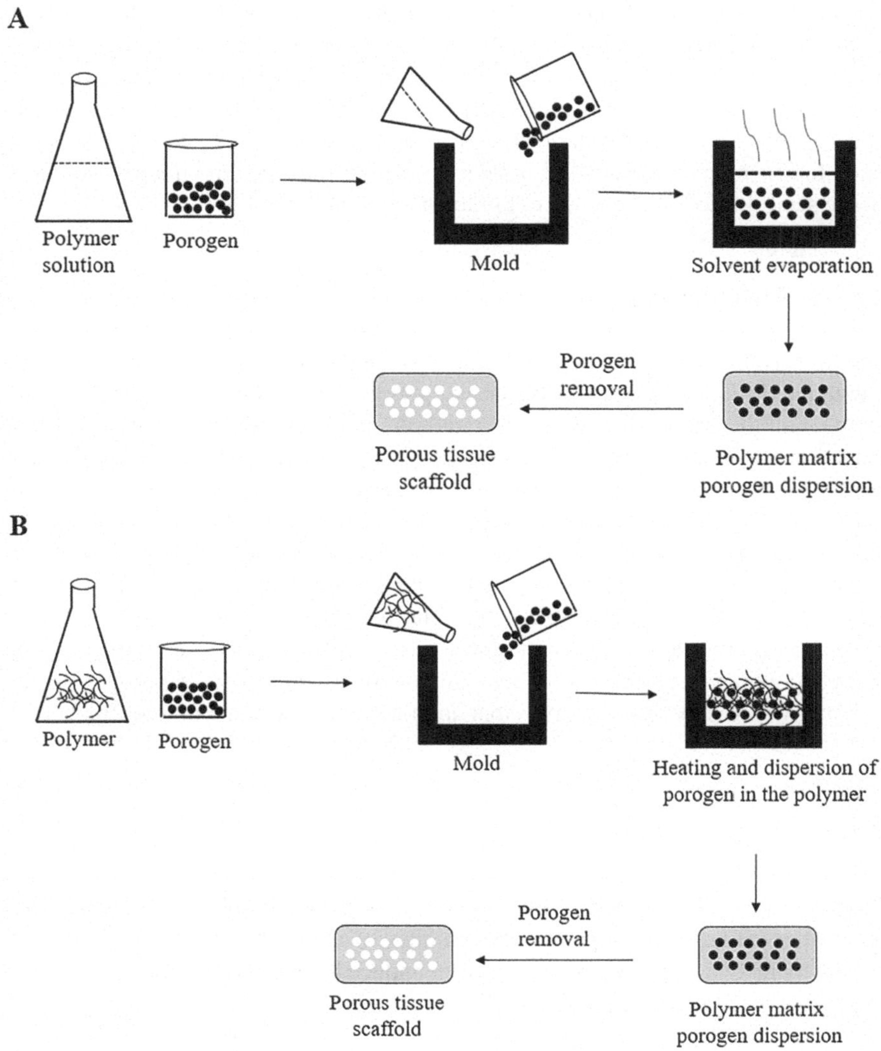

Fig. 2.3 Schematic presentation of tissue scaffold preparation by using particle leaching methods. (**a**) Solvent casting, (**b**) Melt molding

microsphere embedded melt extruded PCL-PEG sutures [16] are other examples of products fabricated with this technique.

2.1.3.1 Solvent Casting

A polymer solution prepared by dissolving the polymer in an organic solvent is mixed with materials that can be dissolved in water such as salt or sugar (act as porogen). The dispersion is poured into a mold and the organic solvent is evaporated. The remaining polymer matrix-porogen dispersion is washed with the solvent of the porogen (water), resulting in the removal of porogen and formation of pores at these sites [17] (Fig. 2.3a). Pore size can be adjusted based on the size of the porogen crystal and the ratio of polymer/porogen. Although the technique is easy to perform, there are some limitations, such as aggregation or precipitation of porogen particles, incomplete removal of organic solvent or porogen, and the production of tissue scaffolds as 2D films due to limited membrane thickness.

2.1.3.2 Melt Molding

Unlike solvent casting method, 3D scaffolds can be produced with melt molding technique. In this technique, polymer and porogen are poured into a mold in solid form. With the application of heat to mold, the polymer melts and porogen is dispersed in the melted polymer. With temperature reduction, polymer solidifies and porogen is removed by washing with appropriate solvent leading to porous structure (Fig. 2.3b). By using different types and adjusting the amounts of the porogen, the shape and size of the pores and the porosity can be controlled [18]. In addition, porous tissue scaffolds can be prepared in the desired shape using different molds. However, the use of high temperatures creates the risk of degradation of polymer or bioactive agent encapsulated into the structure. Apart from this, the viscosity of the melted polymer may cause non-homogenous distribution of the porogen in the polymer which may result in pores that are not interconnected with one another.

2.1.4 Phase Separation

Phase separation is achieved by first mixing two immiscible solvents and separating them afterwards. Phase separation techniques to obtain porous scaffolds can be classified in two approaches: non-solvent induced phase separation (NIPS) and thermal induced phase separation (TIPS).

2.1.4.1 Non-solvent Induced Phase Separation (NIPS)

In NIPS technique, polymer is dissolved in its solvent and poured into a mold which is then treated with coagulator (non-solvent) to precipitate the polymer. This results in the formation of two phases: the polymer-rich phase and the polymer-poor phase. Polymer-rich phase forms the matrix, while polymer-poor phase forms the pores of the tissue scaffold [19]. NIPS-generated scaffolds have limited use in tissue engineering applications because of the heterogenous porosity composed of large pores.

2.1.4.2 Thermal Induced Phase Separation (TIPS)

In TIPS technique, polymer solution is prepared in a latent solvent which acts as a solvent at high temperatures and as a non-solvent at low temperatures. The polymer is dissolved in high temperature and the solution is then cooled down which leads to phase separation (into polymer-rich and polymer-poor phases). The polymer-rich phase forms the main structure (matrix) and polymer-poor phase forms the pores in the structure. Phase separation occurs in two ways: solid-liquid phase separation and liquid-liquid phase separation. In solid-liquid (polymer-solvent) phase separation, first solid polymer is dissolved in its solvent to form a solution, and then the temperature of the polymer solution is lowered leading to crystallization of the solvent. The crystals are then washed and/or removed for the formation of porous structure. In liquid-liquid phase separation (polymer-solvent/non-solvent), the polymer solution is heated above the critical temperature and polymer-rich and polymer-poor phases are formed. With the evaporation of solvent, porous structures are formed [20]. The biggest advantage of this technique is that a strong polymer network can be obtained with an easily applicable method. In addition, depending on the experiment conditions, open or closed porous structures, or scaffolds with different forms such as fibrous or sheet-like morphologies can be obtained.

Both synthetic polymers such as (poly(lactic acid-co-glycolic acid) (PLGA), polyurethane (PU), poly(3-hydroxybutyrate-co-3-valerate) (PHBV), polycaprolactone (PCL) composites, or natural polymers (chitosan, collagen, fibrin, hyaluronic acid) may be used. Wound dressings [21], scaffolds for heart wall regeneration [22], heart patches [23], bridging materials for cuff tear regeneration [24], and conduits for sciatic neuron regeneration [25] are the examples of tissue scaffolds prepared by using this technique.

2.1.5 Gas Foaming

In this technique, gas is passed through the polymer solution. The gas bubbles lead to the formation of porous or tunnel-like structures in the scaffolds. The gas used in the process is called 'blowing agent" which can be classified as physical and chemical agents. Physical foaming agents are inert gases, such as argon or volatile liquids.

These physical agents cause the foam formation, gaining volume during evaporation. Chemical agents, on the other hand, react with polymer and cause the formation of CO_2 or N_2 gases. Since this technique requires the foaming of a hydrophobic gas in the polymer aqueous solution, the choice of polymer is restricted to hydrophilic polymers. The salt formation and difficulty of achieving the desired amount of gas formation are the other disadvantages of the technique. The use of physical agents, on the other hand, allows the control of the amount of applied gas which makes the optimization of pore size and porosity easier. It is also an important advantage that the 2D tissue scaffolds can be transformed into 3D structures by gas foaming method [26]. In addition to this, since the use of organic solvent is not obligatory, it provides an advantage for the encapsulation of cells or active agents without giving any harm to them. The porosity and pore size, however, would depend on the used biomaterial, which are important parameters for cell attachment and proliferation. For instance, tissue scaffolds of 93% porosity with an average pore size of 100 μm can be obtained with this technique [27].

With the use of this technique for scaffold fabrication, peripheral nerve conduits [28] and scaffolds for bone tissue regeneration [17] were prepared from pure chitosan or in combination with agarose. The scaffolds obtained by mixing the synthetic polymer poly(propylene carbonate) (PPC) with natural polymers and bioglass ceramics were used for joint implants [29]. 3D scaffolds containing pullulan, derived from 2D electrospun fiber membranes, were used as scaffold structure for skin tissue engineering [30] or the scaffolds obtained with gelatin polymer were fabricated to be used as wound dressing [31] or cartilage tissue engineering [32].

2.1.6 Freeze Drying

Freeze drying (lyophilization) technique is commonly used in scaffold fabrication, and it provides an advantage particularly for the use of heat-sensitive materials during scaffold construction. The polymer dissolved in a suitable solvent is frozen in a mold at either −20 °C, −80 °C or in liquid nitrogen (−196 °C). The ice crystals in the structure acts as porogen and are sublimized in low temperature and pressure, resulting in the formation of porous tissue scaffold [17] (Fig. 2.4). With this technique, tissue scaffolds with interconnected pores of size up to 200 μm and up to 90% porosity can be obtained [33].

The size of the pores mostly depends on the formed ice crystals. Therefore, freezing temperature, applied vacuum, concentration of polymer solution, type of polymer and solvent, and the presence of crosslinkers are important determinants of the pore size. The important advantage of freeze-drying is that, if the biomaterial used for tissue scaffold fabrication is not soluble in water or if the environment does not contain an enzyme that would degrade the polymer, the structure of the obtained tissue scaffold can be protected in that aqueous environment, which leads to long-term usage of the scaffold.

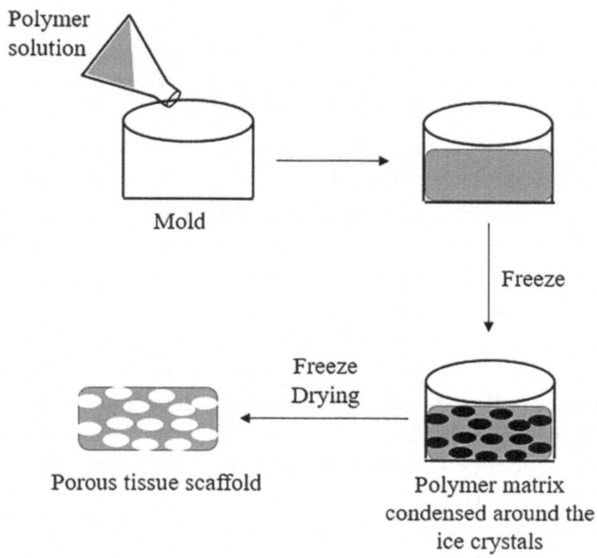

Fig. 2.4 Schematic representation of scaffold fabrication with lyophilization (freeze drying) method

This technique allows the usage of both natural and synthetic polymers. Tissue scaffolds formed from collagen has been used in bone or dental pulp tissue regeneration [34]. Gelatin scaffolds were evaluated for their use in vascular grafts [35], and polyurethane scaffolds prepared by using this technique were used in meniscus prosthesis [36]. Highly porous silk fiber-reinforced chitosan scaffolds were prepared by freeze-drying technique, and promising results were obtained for their angiogenesis and osteogenesis abilities [37]. In addition, freeze-drying technique is commonly used in tendon or bone allograft preparations in tissue engineering applications [38].

2.1.7 Hydrogels

Hydrogels are 3D networks of hydrophilic polymers that possess high water absorption capacity. Hydrogel preparation techniques are extensively used in tissue engineering because they allow the formation of highly porous scaffolds with interconnected structures, which is very important in cell migration and proliferation. The interaction between the polymer chains with one another determines the amount of water that the hydrogel can hold without disrupting the structure and thus the swelling of the hydrogel. The interaction of chains in this network structure can be obtained both physically and with chemical linkers. Chemical crosslinking can be achieved with the use of functional groups in the polymer chain (such as Schiff base reaction and click chemistry) in addition to the use of crosslinkers such as genipin or glutaraldehyde [39]. The disadvantage of crosslinking is that covalent bonds formed in the structure make the hydrogel susceptible to stresses such as bending and twisting. In addition, the remaining unused crosslinkers may leach out and cause cytotoxicity. Physical interactions, on the other hand, might be due to entanglement of polymer chains or through non-covalent bonding such as hydrogen bonding or electrostatic interactions. Therefore, the hydrogel structures fabricated with physical interactions may change depending on the pH, ionic strength or temperature of the environment. Hydrogels can be fabricated both from synthetic polymers such as polyvinyl alcohol (PVA), polyethylene glycol (PEG), and polylactic acid (PLA) and from natural polymers such as polypeptides and polysaccharides, collagen, gelatin, chitosan, and alginate. Hydrogels prepared with natural polymers are more advantageous to that of synthetic polymers because of their resemblance to extracellular matrix (ECM) and biocompatibility. There are commercialized network-structured hydrogels prepared by crosslinking natural and synthetic polymers for their use as wound dressings [40]. In addition, poly(2-hydroxymethylmethacrylate) (PHMEA) hydrogels have been developed as contact lenses [41], and hydrogels of natural polymers such as chitosan or hyaluronic acid have been used in corneal wound healing [42]. Recently, silver nanoparticle grafted methacrylated gelatin (GelMa) hydrogels were synthesized for enhancing the attachment of bone cells, resulting in regeneration of the defect site in rabbits [43].

Another network-structured scaffold, cryogels, resemble hydrogels; however, they have a different fabrication technique. Generally, water is used as a solvent for cryogel production. The aqueous solutions of polymer and crosslinker mixtures are frozen (−5 to −20 °C). Polymer chains are crosslinked around the formed ice crystals and gelation takes place. When the temperature of the cryogel is increased, the melting of ice leads to the formation of scaffolds with interconnected macropores [27]. Cryogels fabricated from synthetic PEG, PHEMA or natural alginate or gelatin polymers are being developed to be used as dermal fillers in cosmetics; injectable cryogels produced from modified chitosan and carbon nanotubes are developed to be used for hemorrhage control [44]. In addition, cryogels constructed from chitosan-calcium phosphate crystals are tested for their application in the treatment of bone defects [45].

2.1.8 Three-Dimensional Printing (3DP)

Three-dimensional printing technology allows the precise control of porosity and pore size along with the arrangement of distance between the pores. In addition, patient-specific tissue scaffolds can be produced due to the possibility of using digital methods such as computed tomography (CT), computer-aided design (CAD), ultrasound, or magnetic resonance imaging (MRI) [20]. Selective laser sintering (SLS) and fused deposition modeling (FDM) are some examples of the techniques for tissue scaffold fabrication used in 3D printing. Metals such as gold, steel, and titanium; ceramics such as calcium phosphate; and polymers such as poly (methyl methacrylate) (PMMA), polycaprolactone (PCL), and polyurethane (PU) can be used for the construction of scaffolds. The properties of scaffolds are dependent on the technology used for printing, type of polymer (synthetic or natural), temperature, number of printed layers, and printing speed [1]. These techniques are explained in detail in the following chapters.

2.2 Conclusion

Tissue scaffolds, being one of the three components of the field of tissue engineering, are structures fabricated from biomaterials such as natural or synthetic polymers, ceramics, metals, or composites. Upon implantation of tissue scaffolds combined with cells and growth factors have the potential to heal and restore the function of the diseased or damaged tissue. The success of the scaffold also depends on the physicochemical properties, such as compatibility of the mechanical properties of the biomaterial with that of the target tissue, porosity, pore size, and the interconnections in the porous structure. All these properties ultimately determine cell attachment, cell proliferation, differentiation, and migration, resulting in the integration of the scaffold with the tissue. Therefore, tissue scaffolds produced using the different techniques mentioned in this section may reduce the requirement of donors to restore the functionality of the patients' damaged tissue.

References

1. Kumar, A., Jacob, A.: Techniques in scaffold fabrication process for tissue engineering applications: a review. J. Appl. Biol. Biotechnol. **10**, 163–176 (2022)
2. Law, J.X., Liau, L.L., Saim, A., Yang, Y., Idrus, R.: Electrospun collagen nanofibers and their applications in skin tissue engineering. Tissue Eng. Reg. Med. **14**, 699–718 (2017)
3. Wise, J.K., Yarin, A.L., Megaridis, C.M., Cho, M.: Chondrogenic differentiation of human mesenchymal stem cells on oriented nanofibrous scaffolds: engineering the superficial zone of articular cartilage. Tissue Eng. A. **15**(4), 913–921 (2009). https://doi.org/10.1089/ten.tea.2008.0109
4. Heydari, P., Zargar, A.K., Shariati, L.: Enhanced wound regeneration by PGS/PLA fiber dressing containing platelet-rich plasma: an in vitro study. Sci. Rep. **14**(1), 12019 (2024)
5. Xue, J., Feng, B., Zheng, R., Lu, Y., Zhou, G., Liu, W., Cao, Y., Zhang, Y., Zhang, W.: Engineering ear-shaped cartilage using electrospun fibrous membranes of gelatin/polycaprolactone. Biomaterials. **34**(11), 2624–2631 (2013)
6. Rohani Shirvan, A., Nouri, A., Sutti, A.: A perspective on the wet spinning process and its advancements in biomedical sciences. Eur. Polym. J. **181**, 111681 (2022)
7. Dini, F., Barsotti, G., Puppi, D., Coli, A., Briganti, A., Gianness, E., Miragliotta, V., Mota, C., Pirosa, A., Stornelli, M.R., Gabellieri, P., Carlucci, F., Chiellini, F.: Tailored star poly (ε-caprolactone) wet-spun scaffolds for in vivo regeneration of long bone critical size defects. J. Bioact. Compat. Polym. **31**(1), 15–30 (2016)

8. Alagoz, A.S., Rodriguez-Cabello, J.C., Hasirci, V.: PHBV wet-spun scaffold coated with ELR-REDV improves vascularization for bone tissue engineering. Biomed. Mater. **13**(5), 055010 (2018)
9. Wu, M., Wang, H., Xu, K., Mei, J., Wang, Z.: Enhancing neurite growth and neural functions on polymeric nerve conduit with BMSC-derived ECM coating. Biomed. Mater. **19**(5), 055031 (2024)
10. da Silva, M.C., da Silva, H.N., Cruz, C.A.L., Amoah, S.K.S., Silva, S.M., Fook, M.V.L.: N-acetyl-D-glucosamine-loaded chitosan filaments biodegradable and biocompatible for use as absorbable surgical suture materials. Materials. **12**(11) (2019)
11. Ergene, E., Sezlev Bilecen, D., Kaya, B., Yilgor Huri, P., Hasirci, V.: 3D cellular alignment and biomimetic mechanical stimulation enhance human adipose-derived stem cell myogenesis. Biomed. Mater. **15**(5), 055017 (2020)
12. Bakhtiari, S.S.E., Karbasi, S., Toloue, E.B.: Modified poly(3-hydroxybutyrate)-based scaffolds in tissue engineering applications: a review. Int. J. Biol. Macromol. **166**, 986–998 (2021)
13. Lee, S.H., Kim, B.S., Kim, S.H., Choi, S.W., Jeong, S.I., Kwon, I.K., Kang, S.W., Nikolovski, J., Mooney, D.J., Han, Y., Kim, Y.H.: Elastic biodegradable poly(glycolide-co-caprolactone) scaffold for tissue engineering. J. Biomed. Mater. Res. A. **66A**, 29 (2003). https://doi.org/10.1002/jbm.a.10497
14. Liu, K., Li, Y., Xu, F., Zuo, Y., Zhang, L., Wang, H., Liao, J.: Graphite/poly (vinyl alcohol) hydrogel composite as porous ringy skirt for artificial cornea. Mater. Sci. Eng. C. **29**(1), 261–266 (2009)
15. Atari, M., Labbaf, S., Javanmard, S.H.: Fabrication and characterization of a 3D scaffold based on elastomeric poly-glycerol Sebacate polymer for heart valve applications. J. Manuf. Process. **29**(102), 350–364 (2023)
16. Deng, X., Gould, M.L., Katare, R.G., Ali, M.A.: Melt-extruded biocompatible surgical sutures loaded with microspheres designed for wound healing. Biomed. Mater. **19**(5), 055007 (2024)
17. Suamte, L., Tirkey, A., Barman, J., Jayasekhar Babu, P.: Various manufacturing methods and ideal properties of scaffolds for tissue engineering applications. Smart Mater. Manuf. **1**, 100011 (2023)
18. Chimerad, M., Barazesh, A., Zandi, M., Zarkesh, I., Moghaddam, A., Borjian, P., Chimehrad, R., Asghari, A., Akbarnejad, Z., Khonakdar, H.A., Bagher, Z.: Tissue engineered scaffold fabrication methods for medical applications. Int. J. Polym. Mater. Polym. Biomater. **72**, 1455 (2022). https://doi.org/10.1080/00914037.2022.2101112
19. Bhushan, S., Singh, S., Maiti, T.K., Sharma, C., Dutt, D., Sharma, S., Li, C., Eldin, E.M.T.: Scaffold fabrication techniques of biomaterials for bone tissue engineering: a critical Review. Bioengineering. **9**(12), 728 (2022)
20. Hasirci, V., Hasirci, N.: Fundamentals of Biomaterials. Springer., ISBN: 978-1-4939-8854-9 (2018)
21. McKenna, E., Klein, T.J., Doran, M.R., Futrega, K.: Integration of an ultra-strong poly(lactic-co-glycolic acid) (PLGA) knitted mesh into a thermally induced phase separation (TIPS) PLGA porous structure to yield a thin biphasic scaffold suitable for dermal tissue engineering. Biofabrication. **12**(1) (2020)
22. Fang, Y., Zhang, T., Zhang, L., Gong, W., Sun, W.: Biomimetic design and fabrication of scaffolds integrating oriented micro-pores with branched channel networks for myocardial tissue engineering. Biofabrication. **11**(3), 035004 (2019)
23. Xu, C., Okpokwasili, C., Huang, Y., Shi, X., Wu, J., Liao, J., Tang, L., Hong, Y.: Optimizing anisotropic polyurethane scaffolds to mechanically match with native myocardium. ACS Biomater Sci. Eng. **6**(5), 2757–2769 (2020)
24. Wang, L., Kang, Y., Chen, S., Mo, X., Jiang, J., Yan, X., Zhu, T., Zhao, J.: Macroporous 3D scaffold with self-fitting capability for effectively repairing massive rotator cuff tear. ACS Biomater Sci. Eng. **7**(3), 904–915 (2021)

25. Farzamfar, S., Salehi, M., Tavangar, S.M., Verdi, J., Mansouri, K., Ai, A., Malekshahi, Z.V., Ai, J.: A novel polycaprolactone/carbon nanofiber composite as a conductive neural guidance channel: an in vitro and in vivo study. Prog. Biomater. **8**(4), 239–248 (2019)
26. Xia, D., Chen, J., Zhang, Z., Dong, M.: Emerging polymeric biomaterials and manufacturing techniques in regenerative medicine. Aggregate. **3**(5) (2022)
27. Vesvoranan, O., Anup, A., Hixon, K.R.: Current concepts and methods in tissue interface scaffold fabrication. Biomimetics. **7**(4), 151 (2022). https://doi.org/10.3390/biomimetics
28. Rao, F., Yuan, Z., Li, M., Yu, F., Fang, X., Jiang, B., Wen, Y., Zhang, P.: Expanded 3D nanofibre sponge scaffolds by gas-foaming technique enhance peripheral nerve regeneration. Artif. Cells Nanomed. Biotechnol. **47**(1), 491–500 (2019)
29. Manavitehrani, I., Le, T.Y.L., Daly, S., Wang, Y., Maitz, P.K., Schindeler, A., Dehghani, F.: Formation of porous biodegradable scaffolds based on poly(propylene carbonate) using gas foaming technology. Mater. Sci. Eng. C. **1**(96), 824–830 (2019)
30. Bianchi, E., Ruggeri, M., Vigani, B., Totaro Fila, C., IcaroCornaglia, A., Boselli, C., Viseras, C., Rossi, S., Sandri, G.: Gas foamed scaffolds as smart 3D structures in skin tissue engineering. J. Drug Deliv. Sci. Technol. **1**, 95 (2024)
31. Poursamar, S.A., Hatami, J., Lehner, A.N., da Silva, C.L., Ferreira, F.C., Antunes, A.P.M.: Potential application of gelatin scaffolds prepared through in situ gas foaming in skin tissue engineering. Int. J. Polym. Mater. Polym. Biomater. **65**(6), 315–322 (2016)
32. Chen, Y., Xu, W., Shafiq, M., Tang, J., Hao, J., Xie, X., Yuan, Z., Xiao, X., Liu, Y., Mo, X.: Three-dimensional porous gas-foamed electrospun nanofiber scaffold for cartilage regeneration. J. Colloid Interface Sci. **603**, 94–109 (2021)
33. Collins, M.N., Ren, G., Young, K., Pina, S., Reis, R.L., Oliveira, J.M.: Scaffold fabrication technologies and structure/function properties in bone tissue engineering. Adv. Funct. Mater. **31** (2021). https://doi.org/10.1002/adfm.202010609
34. Dayı, B., Sezlev Bilecen, D., Eröksüz, H., Yalçın, M., Hasırcı, V.: Evaluation of a collagen-bioaggregate composite scaffold in the repair of sheep pulp tissue. Eur. Oral. Res. **55**(3), 152–161 (2021)
35. Norouzi, S.K., Shamloo, A.: Bilayered heparinized vascular graft fabricated by combining electrospinning and freeze drying methods. Mater. Sci. Eng. C. **94**, 1067–1076 (2019)
36. Heijkants, R.G.J.C., Tienen, T.G., Groot, J.H., Pennings, A.J., Buma, P., Veth, R.P.H., Schouten, A.J.: Preparation of a polyurethane scaffold for tissue engineering made by a combination of salt leaching and freeze-drying of dioxane. J. Mater. Sci. **41**(8), 2423–2428 (2006)
37. Xu, Z., He, F., Yu, J., Yang, Z., Zhu, Y., Liao, R., Lyu, R., Yang, M., Zhu, L., Yang, M.: From common biomass materials to high-performance tissue engineering scaffold: biomimetic preparation, properties characterization, in vitro and in vivo evaluations. J. Bioresour. Bioprod. **9**(2), 185–196 (2024)
38. Ariffin, A.A., Chan, H.H., Yusof, N., Mohd, S., Ramalingam, S., Ng, W.M., Mansor, A.: Establishing freeze drying process for cortical and cancellous bone allograft cubes. Original report. JUMMEC. (2019). https://doi.org/10.22452/jummec.vol22no1.10
39. Xue, X., Hu, Y., Wang, S., Chen, X., Jiang, Y., Su, J.: Fabrication of physical and chemical crosslinked hydrogels for bone tissue engineering. Bioact. Mater. **12**, 327 (2022). https://doi.org/10.1016/j.bioactmat.2021.10.029
40. Caló, E., Khutoryanskiy, V.: Biomedical applications of hydrogels: a review of patents and commercial product. Eur. Polym. J. **65**, 252 (2015). https://doi.org/10.1016/j.eurpolymj.2014.11.024
41. J. F. Kunzler, P. Nandu Mahendra. Polymer Compositions for Contact Lenses, 1998
42. Khosravimelal, S., Mobaraki, M., Eftekhari, S., Ahearne, M., Seifalian, A.M., Gholipourmalekabadi, M.: Hydrogels as emerging materials for cornea wound healing. Small. **17**, e2006335 (2021). https://doi.org/10.1002/smll.202006335
43. Abdelmoneim, D., Coates, D., Porter, G., Schmidlin, P., Li, K.C., Botter, S., et al.: In vitro and in vivo investigation of antibacterial silver nanoparticles functionalized bone grafting substitutes. J. Biomed. Mater. Res. A. **112**, 2042 (2024)

44. Zhao, X., Guo, B., Wu, H., Liang, Y., Ma, P.X.: Injectable antibacterial conductive nanocomposite cryogels with rapid shape recovery for noncompressible hemorrhage and wound healing. Nat. Commun. **9**(1), 2784 (2018)
45. Abueva, C.D.G., Padalhin, A.R., Min, Y.K., Lee, B.T.: Preformed chitosan cryogel-biphasic calcium phosphate: a potential injectable biocomposite for pathologic fracture. J. Biomater. Appl. **30**(2), 182–192 (2015)

Chapter 3
Three-Dimensional Printing Techniques

Onur Demirak, Seyda Gokyer, Emre Ergene, and Pinar Yilgor

Abstract Three-dimensional (3D) printing is an advanced manufacturing process, and it uses computer-aided design (CAD) for the production of any 3D object layer by layer. This technique has been used in the medical industry since the early 1990s, enabling the production of specially designed implants and prostheses that are adapted to perfectly fit the target area using patient medical imaging data. Furthermore, it has become possible to produce personalized functional tissues and organs in the laboratory through 3D bioprinting of cell-laden biomaterials, known as "bioink." 3D modeling and 3D printing techniques can also be used to create surgical planning models and surgical guides that provide ease of application during surgery, tailored to each individual patient. This chapter describes how 3D printing technology is applied in the medical field, including clinical practice, experimental research, and medical education. The anticipated developments for the near future are also covered.

Keywords 3D printing · Stereolithography · Digital light processing · Selective laser sintering · Selective laser melting · Droplet-based 3D printing · Bioinks · SLA · DLP · SLS · SLM

3.1 Steps for Three-Dimensional (3D) Printing Process

3D printing is an umbrella term that describes additive, or layered, manufacturing of solid 3D objects according to data driven from a digital file. Manufacturing of a 3D object through 3D printing principles includes the following steps: (i) creation/obtaining of 3D digital data, (ii) image processing, (iii) modeling, and (iv) 3D printing.

O. Demirak · S. Gokyer · E. Ergene · P. Yilgor (✉)
Department of Biomedical Engineering, Ankara University, Ankara, Türkiye
e-mail: phuri@ankara.edu.tr

N. Hasirci, V. Hasirci (eds.), *Hydrogels and Bioinks in Tissue Engineering*, Biological and Medical Physics, Biomedical Engineering,
https://doi.org/10.1007/978-3-031-90966-5_3

3D printing starts with a 3D model, and generating 3D digital data is the first step toward obtaining this. There are two main methods for creating 3D digital data: (i) CAD softwares and (ii) scanners, including surface scanners or medical imaging tools like computed tomography (CT), magnetic resonance imaging (MRI), or ultrasonography (USG).

In the second stage of the production process, volumetric 3D data is generated through the creation and segmentation of 3D images, which are then exported in a compatible file format, such as standard tessellation language (STL). Incorporating the pertinent structure into the model meant for 3D printing is known as segmentation. Technically, segmentation defines the criteria for the voxels (3D equivalents of pixels) to be included in the 3D model. The subsequent step in the 3D printing process involves slicing the 3D digital object into layers and then generating geometric codes (G-codes) for each layer, thereby establishing the workflow for 3D printing. Settings such as resolution, size, and internal structure can be adjusted in line with the intended specifications for the printed object. The accuracy and resolution in 3D printing depend on factors such as the characteristics of the technique used by the 3D printer (e.g., laser, digital light, and nozzle), the minimum achievable layer thickness, *XY* axis resolution, and the diameter of the printing component.

The final phase of the 3D production process involves additive operations carried out by equipment known as 3D printers. The fundamental idea behind additive manufacturing is the sequential addition of material layers until the product is fully built. Each layer can be seen as a thin cross-section of the object. Numerous methods for 3D printing have been developed to date, which will be briefly covered in the upcoming subsections.

3.2 3D Printing Methods

3D printing methods can be broadly categorized into four main groups: (i) resin-based, (ii) powder-based, (iii) extrusion-based, and (iv) droplet-based production techniques. Furthermore, a distinct category might be created for 3D bioprinting, a method for producing systems that mimic biological structures and contain living cells.

3.2.1 Resin-Based 3D Printing Methods

Vat photopolymerization is a general term for resin-based 3D printing, and this technique can be subdivided into (i) stereolithography (SLA) and (ii) digital light processing (DLP). A detailed examination of these techniques is provided below.

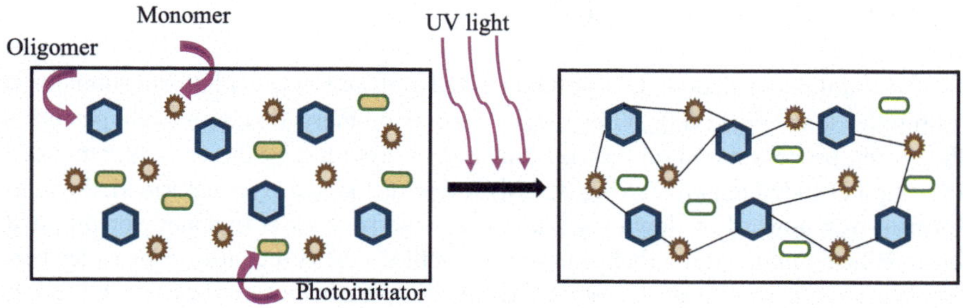

Fig. 3.1 Schematic illustration of the working principle of the vat photopolymerization method

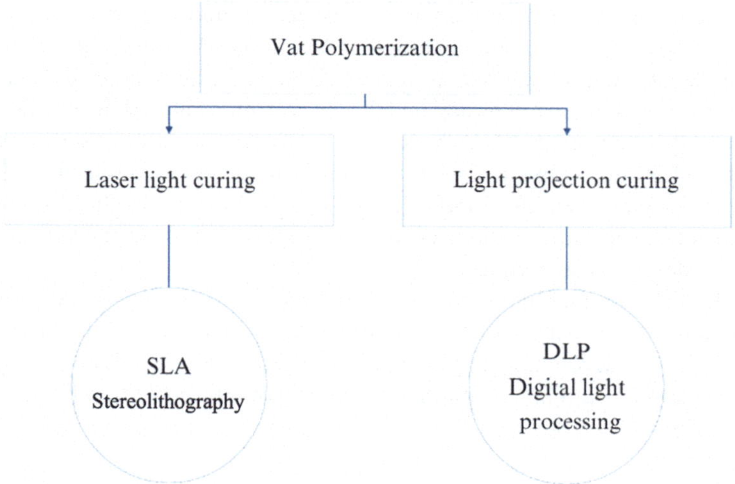

Fig. 3.2 Classification of vat photopolymerization-based 3D printing techniques

3.2.1.1 Vat Photopolymerization

Vat photopolymerization is a resin-based additive manufacturing method, and it involves polymerization of a liquid resin induced by light, in a layer-by-layer manner within a vat. This process uses a light source, typically visible or ultraviolet (UV) light. In vat photopolymerization, a light-sensitive polymer in liquid form undergoes cross-linking and solidification within a vat when exposed to curing light (Fig. 3.1). This irreversible chemical reaction results in a solid polymer that cannot revert to liquid. As the light source solidifies the resin, the build platform moves upward or downward, repeating the process for each layer [1].

3D printing methods based on vat photopolymerization are classified according to their curing techniques (Fig. 3.2). Among these, the most used method is SLA and it employs laser light for curing, whereas DLP uses light projections for the same purpose.

3.2.1.1.1 Stereolithography (SLA)

SLA technology utilizes a photopolymer-filled tank and a build platform submerged within this tank (Fig. 3.3a). While the platform is submerged, a laser beam performs the curing process based on the coordinate data provided by the G-code generated in computer-aided manufacturing (CAM) software. Once a layer is formed, the platform moves upward or downward, and the same process is repeated for the next layer. When compared to other 3D printing methods, SLA technology provides better accuracy and resolution. Common SLA device parameters are shown in Table 3.1.

SLA technique has revolutionized medical 3D printing applications and bioprinting by enabling the creation of precise and complex structures. SLA achieves exceptional resolution during the manufacturing due to the nature of the laser-induced photopolymerization, making it ideal for producing customized medical devices, prosthetics, and anatomical models for pre-surgical planning. This precision is particularly advantageous in regenerative medicine, where SLA has been used to develop vascularized tissue models and scaffolds tailored to specific patient anatomies. In bioprinting, SLA's precision facilitates the fabrication of intricate scaffolds and cellular constructs, closely mimicking the microarchitecture of native tissues [2]. By incorporating living cells and hydrogel materials, SLA supports the development of cell-laden constructs, such as cartilage and bone substitutes [3, 4], with high fidelity to native tissues.

Recent research highlights SLA's contributions to pediatric care, including growth-adaptive implants, which are particularly valuable for children requiring replacements that accommodate growth [5, 6]. Furthermore, its ability to integrate flexible electronics into biocompatible matrices shows promise for creating neural

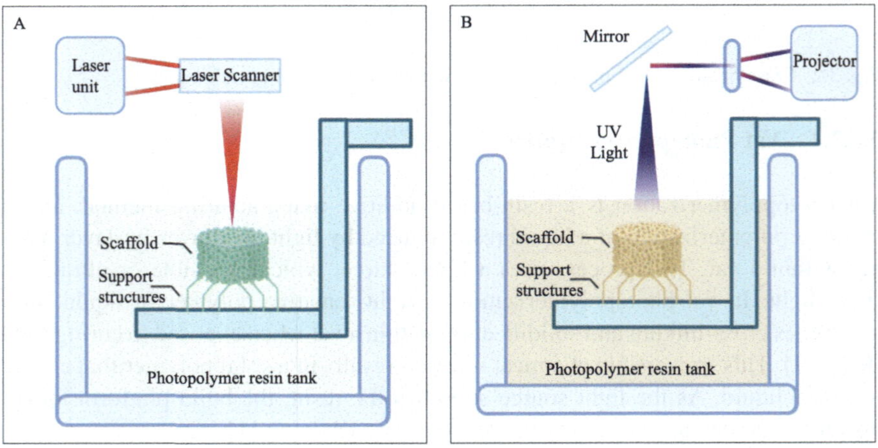

Fig. 3.3 Schematic representation of 3D printing systems based on the photopolymerization. (**a**) SLA, (**b**) DLP

Table 3.1 Common production parameters for SLA and DLP devices

SLA		DLP	
Laser power	250 mW	Light source	UV LED 405 nm
Layer thickness	25–300 μm	Layer thickness	200 μm
XY resolution	25 μm	*XY* resolution	65 μm
Laser diameter	85 μm	Projector	1920 × 1080 px

interfaces and cardiac devices [7–9]. Although technical and regulatory challenges persist, SLA's capacity for patient-specific customization and precision has positioned it as a cornerstone in personalized medicine and tissue engineering innovation.

3.2.1.1.2 Digital Light Processing (DLP)

The curing method used by DLP is different from that of SLA. Instead of using a laser as in SLA, DLP uses light projection for polymerization (Fig. 3.3b). Since the light projection solidifies the entire layer simultaneously, DLP achieves faster results than SLA. In DLP, print quality is directly related to the pixel size of the projected light. Certain features of a DLP device are provided in Table 3.2 [14].

DLP has emerged as a prominent technique in 3D bioprinting, offering high-resolution fabrication capabilities. A droplet bioprinting method utilizing DLP technology was introduced that eliminates the need for the use of a vat, which is free of waste at the same time. This method makes it possible to produce high-resolution 3D structures rapidly [15]. This approach addresses common challenges in traditional DLP setups, such as the need for large volumes of bioink and challenges related to cell settling and contamination. By eliminating the vat, the method reduces bioink waste and minimizes the risk of contamination, making it more cost-effective and efficient. The formation of an endothelial monolayer within the lumens produced via DLP-based droplet bioprinting was shown, which demonstrates the effectiveness of the system for intricate and complex 3D bioprinting approaches [15]. Therefore, DLP is seen as a promising technique for 3D bioprinting applications.

3.2.2 Powder-Based 3D Printing Methods

Powder-based 3D printing methods can be classified into four groups: (i) powder bed fusion (PBF), (ii) selective laser sintering (SLS), (iii) selective laser melting (SLM), and (iv) electron beam melting (EBM). The intended qualities of the finished product determine which application, material, and particular technique are used.

Table 3.2 Examples for the use of extrusion-based 3D printing technique in tissue engineering research

Method	Application	Result	Ref
Thermal extrusion-based 3D printing	Bone tissue engineering scaffolds were produced using composite filaments made of polylactic acid (PLA) and bioactive glass (BG). The viability and spreading of the MC3T3E1 cell line on the 3D scaffold, as well as the differentiation of adipose-derived stem cells (ASC) were examined	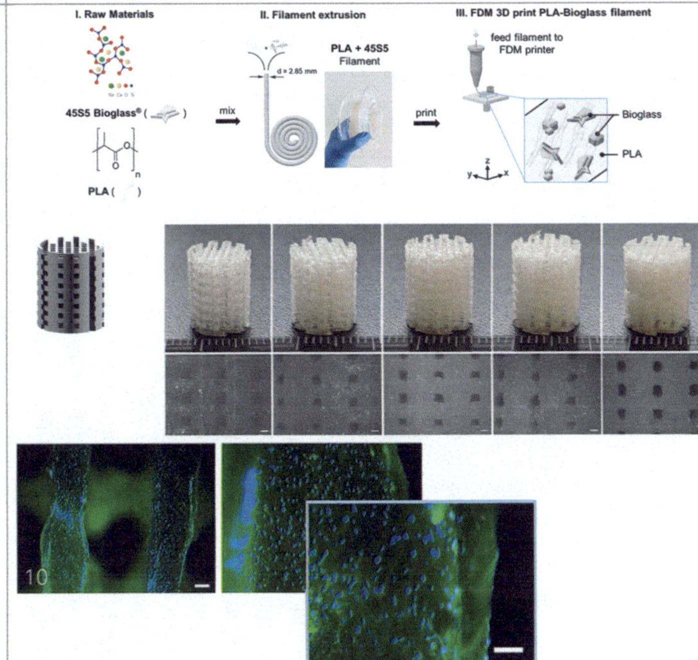	[10]

Thermal extrusion-based 3D printing	3D scaffolds were produced using composite filaments made of PLA, polycaprolactone (PCL), and halloysite nanotubes, and the viability and spreading of human dermal fibroblasts were examined	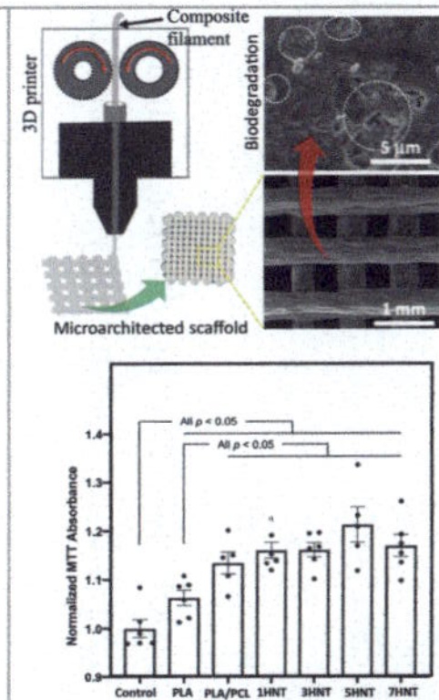	[11]
Thermal extrusion-based 3D printing	A radial 3D model was developed using melt electrospinning-based 3D printing to understand the migration and spreading pathways of glioblastoma cells	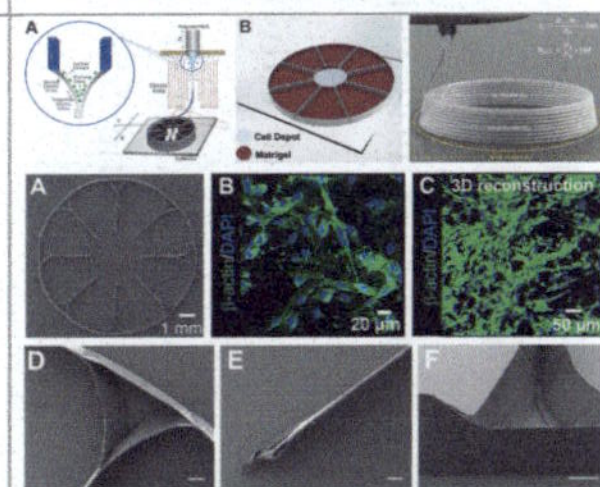	[12]
Solvent-based extrusion 3D printing	The convoluted 3D structure of the proximal tubule was produced using extrusion-based 3D printing, and the spreading and functional properties of epithelial cells seeded along the channel under perfusion were examined	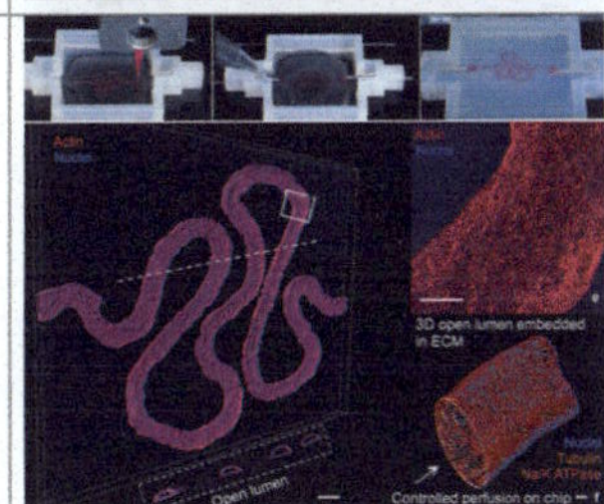	[13]

3.2.2.1 Powder Bed Fusion-Based Manufacturing

PBF is a 3D printing process that operates on the basic principle of adding materials in a layer-by-layer principle. A scan path is calculated for each layer, defining both the boundary contour and a filling strategy [16]. The scan path is typically chosen in a grid pattern due to the nature of the heat source (e.g., laser). After the powder particles in the first layer are melted and bonded, subsequent layers are added sequentially to the top. PBF prepares to form each new layer by spreading powdered material over the previously bonded layer. The powdered material is supplied by a hopper and then uniformly distributed using a roller or blade over the powder bed build platform surface. Each powder layer's ideal thickness, which typically ranges from 25 to 100 μm, depends on the material and processing parameters [17].

3.2.2.2 Selective Laser Sintering (SLS)

SLS is the most used technique for the powder-based 3D printing of biomaterials. As illustrated in Fig. 3.4a, this technique employs a high-energy laser beam to scan a powder layer. The sintered material layers fuse to form complex 3D structures. The process sequence is as follows [18]:

(a) Leveling the powder bed to match the layer reservoir
(b) Spreading powder over the powder bed
(c) Selectively scanning the powder bed with a laser beam

Materials used in SLS include glass, ceramics, polymers, and polymer composites. According to published reports, the SLS 3D printing technology can create features that are about 200 μm in size [17, 18].

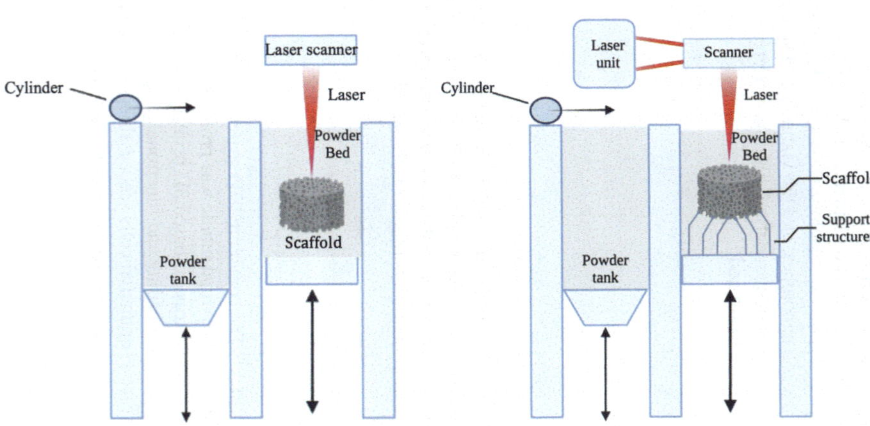

Fig. 3.4 Schematic representation of powder-based 3D printing systems. (**a**) SLS, (**b**) SLM

Shuai et al. [18] investigated the fabrication of electro-actuated shape memory polymer scaffolds using SLS, integrating multiwall carbon nanotubes (MWCNTs) with zinc oxide (ZnO) template to construct a 3D conductive network. They observed that the scaffold incorporating the t-ZnO@MWCNT assembly demonstrated superior shape memory performance actuated by the electrical field. The osteogenic inductivity of these scaffolds was significantly demonstrated by the high alkaline phosphatase (ALP) activity and the expression of osteogenic proteins, facilitated by the combination of electrical stimulation and the osteoconductive properties of ZnO. These findings underline the potential of SLS in creating responsive biomaterials for bone defect repair, where precision and bioactivity are critical. The ability of SLS to integrate complex conductive networks while preserving the scaffold's structural integrity highlights its versatility in producing functional biomaterials with tailored properties [19].

3.2.2.3 Selective Laser Melting (SLM)

Despite the apparent similarities between SLM and SLS in terms of using a laser energy source and powder bed feeding, SLM completely melts the powder particles during the sintering process at a much greater laser power (Fig. 3.4b). This makes SLM more appropriate for producing dense metal parts [20]. The surface roughness of samples produced with SLM is generally higher than those produced with SLS. Commercial SLM 3D printing processes typically use metal powder particles with diameters between 20 and 50 μm to print metal layers with thicknesses ranging from 20 to 100 μm [21]. SLM is capable of printing features with minimum sizes commonly cited as being between 40 and 200 μm. It is also possible to use SLM to produce glass-like materials through 3D printing [22].

The effects of chemical etching were investigated on Ti-6Al-4V biomaterials produced via SLM. Their study demonstrated that chemical etching effectively reduces surface roughness, thereby enhancing the mineralization of BMSCs. This finding underscores the importance of post-processing treatments in optimizing the surface characteristics of SLM-fabricated implants to promote better cell integration and tissue regeneration [23].

In another study, the incorporation of magnesium oxide (MgO) nanoparticles into 3D-printed zinc (Zn) scaffolds was investigated using SLM. The addition of MgO nanoparticles was found to improve the mechanical properties, degradation rates, and antibacterial performance of the Zn scaffolds. Moreover, both in vitro and in vivo assessments indicated enhanced biological properties, suggesting that such composite materials hold promise for biomedical applications where controlled degradation and antibacterial activity are crucial [24].

These investigations demonstrate SLM's adaptability in creating intricate biomaterials with specific characteristics. By adjusting processing parameters and incorporating specific nanoparticles, researchers can engineer materials that meet the stringent requirements of biomedical applications, paving the way for advanced therapeutic solutions.

3.2.2.4 Electron Beam Melting (EBM)

An electron beam acts as the heat source in the EBM process, melting the powder entirely in a vacuum, as shown in Fig. 3.5. EBM 3D printing is exclusively suited for producing metal components. Before melting, the metal powder is preheated by the electron beam. The EBM process eliminates the need for support structures, making it ideal for manufacturing implants and aerospace components, frequently using titanium alloys. Compared to laser beams, electron beams have a somewhat larger spot diameter. The EBM process can achieve structure sizes around 100–200 μm [25]. EBM is considered an alternative to the SLM process. Since EBM employs a high-density energy source, it enables the production of fully dense, void-free components.

Recent advancements in EBM technology have significantly contributed to the development of biomimetic scaffolds for orthopedic applications. Research by Mofazali et al. has highlighted the use of self-assembly systems for antimicrobial properties in porous Ti-6Al-4V scaffolds fabricated using EBM. These systems, incorporating insulin-like growth factor-1 (IGF-1) and cefazolin in polymeric carriers, demonstrated enhanced bone regeneration and bacterial inhibition. Notably, multilayer carriers exhibited superior drug release and osteoblast cell differentiation compared to monolayers, showcasing EBM's potential to produce functionalized implants tailored for orthopedic applications [26].

Similarly, the integration of thermosensitive chitosan hydrogels enriched with icariin (ICA) into EBM-printed titanium scaffolds was investigated. This innovative composite system promoted osteogenesis of BMSCs and facilitated gradual ICA

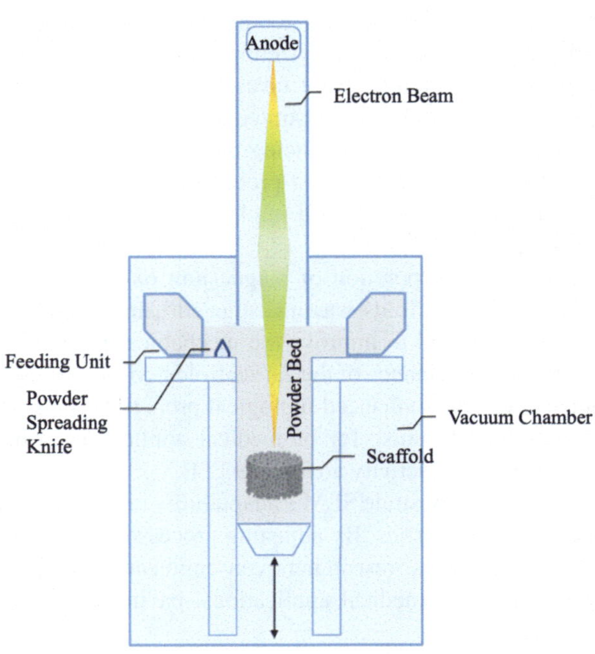

Fig. 3.5 Schematic representation of EBM 3D printing working principle

release over 15 days, significantly enhancing the bone regeneration and osseointegration capacity in vivo. The research underscores EBM's capability to fabricate implants with optimized porosity and mechanical properties, aligning with physiological bone parameters to improve clinical outcomes [27].

Both studies underscore the versatility of EBM in creating advanced biomedical devices, leveraging functional coatings and drug delivery systems to overcome the limitations of traditional implants, such as poor bioactivity and infection risks. These findings emphasize the transformative potential of EBM technology in regenerative medicine and orthopedic solutions.

3.2.3 Extrusion-Based 3D Printing Methods

3D printing done through extrusion is the most widely used technique. It is more practical and less expensive than other solid freeform production methods. It is based on fabrication of successive layers on top of each other through the three-axis (x-y-z) movement of a computer-controlled dispensing head [28, 29]. However, this technique provides a relatively low production resolution (~250 μm). Pneumatic (piston) or mechanical (screw) driving forces are used to place the substance in the dispensing head (Fig. 3.6). The extrusion-based 3D printing technique can be categorized into two main groups: (i) thermal processing-based, and (ii) solvent-based [28–30].

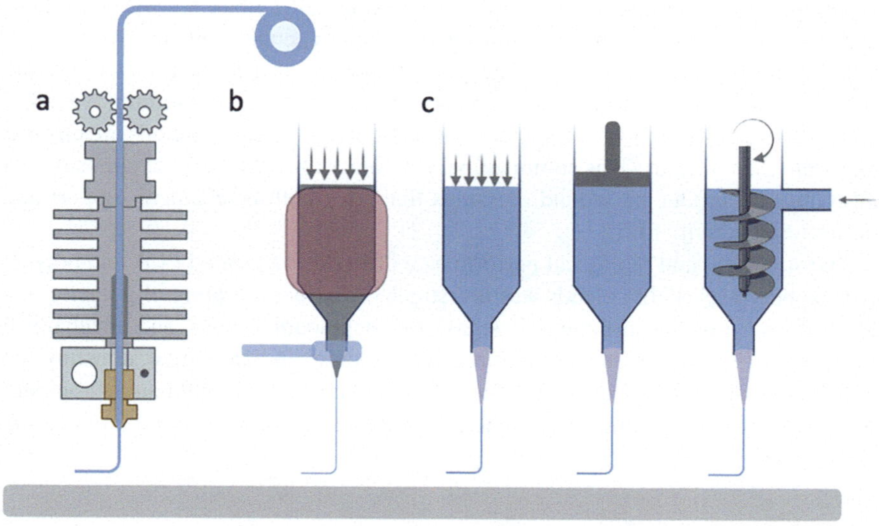

Fig. 3.6 Schematic representation of extrusion-based 3D printing systems. (**a**) Fused Deposition Modeling (FDM). (**b**) Schematic representation of melt electrospinning-based 3D printing technology. (**c**) Schematic representation of extrusion-based 3D printing technology using pneumatic, piston, and screw-driven deposition

3.2.3.1 Thermal Extrusion-Based 3D Printing

The material that is fluidized by heat treatment is fed to the dispensing head by pneumatic (piston) or mechanical (screw) force (Fig. 3.6a) in thermal extrusion-based 3D printing. The three-axis (*x*-*y*-*z*) movement of a computer-controlled dispensing head creates the 3D model layer by layer. The most important advantage of the heat treatment-based extrusion 3D printing technique is that it can deposit material without the need of a solvent, which could potentially be cytotoxic. Scaffolds fabricated through melt extrusion-based 3D printing of thermoplastic polymers are widely studied in tissue engineering.

Fused deposition modeling (FDM) is the most widely used thermal processing-based extrusion 3D printing technique (Fig. 3.6a). The growing range of newly created materials in filament form is driving the continuous expansion of the FDM 3D printing technique's application area. Polylactic acid (PLA), acrylonitrile butadiene styrene (ABS), and polycarbonate (PC) filaments are generally used in the 3D printing by FDM [30]. Additionally, filaments can be created and processed with metal/polymer composite materials made from metal alloys with low melting points.

Another thermal processing-based 3D printing technique is melt electrospinning-based 3D printing technology. In this method, the fluidized polymer is extruded from a nozzle connected to a high-voltage source using pneumatic, piston, or screw-driven force and deposited on a grounded collector plate, moving according to a predefined pattern (Fig. 3.6b, c). This technique provides a unique capability to mimic anatomical systems with fibrous structures, as it enables the production of fibers at the nano- and micrometer scales [12, 31].

Unalan et al. introduced a novel approach by integrating 3D printing with electrospinning to create hierarchical polycaprolactone (PCL)-based scaffolds. These scaffolds incorporate phytotherapeutic agents, such as peppermint and clove oils, to enhance antibacterial properties. The process begins with fabricating 3D-printed PCL struts using FDM, followed by an alkaline surface treatment to improve adhesion. Subsequently, PCL-gelatin nanofibers containing the essential oils are electrospun onto the scaffolds. This combination overcomes the drawbacks of conventional antibiotics by producing wound dressings that have both mechanical support and antibacterial activity [32].

The structural and biological performance of FDM 3D-printed PLA and poly(3-hydroxybutyrate) (PHB) blends was investigated. By incorporating oligomeric lactic acid and varying structural designs, the degradation rates and mechanical properties of the scaffolds were tailored. The findings indicated that adjusting the blend composition and scaffold architecture could effectively modulate the biodegradation behavior, making these materials suitable for various tissue engineering purposes [33].

3.2.3.2 Solvent-Based Extrusion 3D Printing

The material, fluidized with the assistance of a solvent, is fed into the dispensing head by a force, and the cross-sectional layer of the 3D model is created through the horizontal axis movement of the dispensing head in solvent-based extrusion 3D printing (Fig. 3.6c). This technique has a wide range of material options, allowing it to produce complex structures with diverse mechanical properties. Physical, chemical, or photo-crosslinking methods are used for fixation and to stabilize the 3D structures produced from materials fluidized by solvents. One of the most important advantages offered by this method is that cell-laden 3D scaffolds can be fabricated via cells distributed in hydrogel materials, called bioinks. Equipment that performs 3D printing of cell-laden biomaterials is called 3D bioprinters, which is the most widely used 3D printing technique for tissue engineering applications. The drawbacks of this approach include low resolution, reduced cell viability due to high shear stress, and the challenges associated with homogeneous distribution of cells within the scaffolds during fabrication [29, 34].

The recent literature that includes the use of bioprinting includes the following examples: Bharadwaj et al. [34] presented a novel thermogelling bioink for bone tissue engineering, combining polyelectrolyte complexes (PECs) of chitosan and polygalacturonic acid with additional biomaterials such as gelatin, nano-hydroxyapatite, and laponite. They demonstrated that the bioink's thermo-reversible sol-gel transition at physiological conditions enabled direct in situ printing, bypassing additional cross-linking steps. Their findings showed that optimizing gelatin concentration within the bioink significantly improved its print fidelity and mechanical properties, making it a promising candidate for addressing critical-sized bone defects [35].

In another study, Arıcı et al. [35] developed extrusion-based 3D-printed cardiac scaffolds incorporating carboxymethyl cellulose (CMC), gelatin (Gel), and graphene oxide (GO) for cardiac tissue engineering applications. Their research demonstrated that CMC improved scaffold viscosity and printability, while GO enhanced flexibility and electrical conductivity, crucial for mimicking the myocardium's properties. This study highlights the potential of CMC/Gel/GO-based scaffolds in advancing cardiac tissue repair and regeneration through the optimization of material composition and cross-linking parameters [36]. A summary of extrusion-based 3D printing studies using various fabrication techniques is presented in Table 3.2.

In addition to tissue engineering, extrusion-based 3D printing is also frequently used in various healthcare applications, including the production of surgical guides [37], in vitro disease models and related bioactive agent screening applications [38], and production of controlled release systems and pharmacy-on-a-pill systems [39, 40].

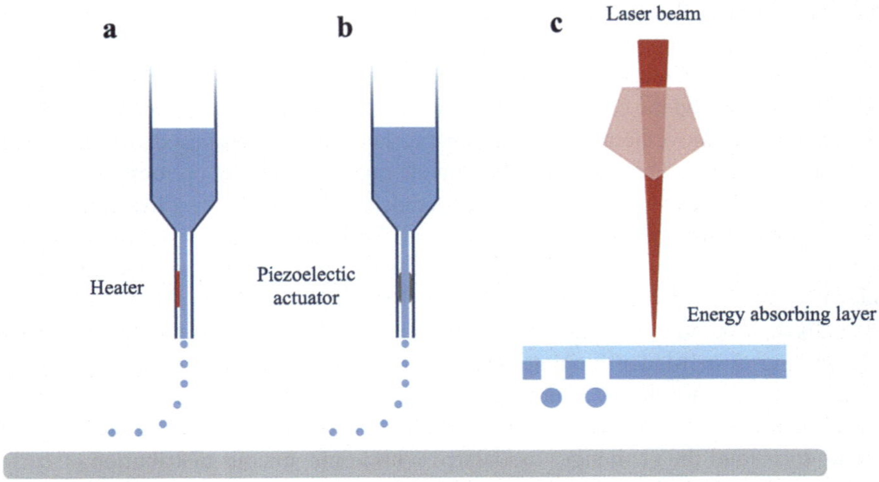

Fig. 3.7 Schematic representation of droplet-based 3D printing techniques. (**a**) Thermal-driven, (**b**) Piezoelectric-driven, (**c**) Laser-driven fabrication systems

3.2.3.3 Droplet-Based 3D Printing Methods

One of the earliest printing techniques utilized in tissue and organ biofabrication is droplet-based 3D printing. This approach may also be used as a non-contact fabrication strategy where cells and non-living materials encapsulated in a droplet can be deposited and patterned layer-by-layer on a surface (Fig. 3.7).

The droplet-based 3D printing technique can be divided into three categories based on the mechanisms used to produce droplets: (i) thermal, (ii) piezoelectric, and (iii) laser. Thermal-sourced droplet-based 3D printing technique is based on depositing the ink heated by the heating element in the dispensing chamber into picoliter volume droplets by spraying them under pressure (Fig. 3.7a). In the piezoelectric-jet modulated droplet-based 3D printing technique, the ink is deposited as a droplet by the effect of acoustic waves created by piezoelectric crystal actuators in the dispensing chamber (Fig. 3.7b). In the laser-based droplet-based 3D printing technique, the focal point made by the laser source causes vaporization to form droplets locally, transferring the droplet to the collector platform with high pressure and thus providing 3D printing (Fig. 3.7c). Although this technique allows for producing higher-resolution models compared to the other two droplet-based printing techniques, cell viability is lower than other dispensing mechanisms [28–30, 34].

Jian et al. [39] utilized a multicellular droplet-based 3D bioprinting strategy to fabricate liver organoids with biomimetic lobule structures. This method involved optimizing bioink formulations to ensure structural integrity and design flexibility, employing sodium alginate-based and dipeptide-based bioinks to support the encapsulation and co-culture of HepaRG cells, HUVECs, and LX-2 cells. The resultant liver organoids retained their multicellular distribution and structural integrity within the printed lobule-like architecture after 7 days of incubation. Compared to traditional 2D monolayer cultures, these 3D organoids exhibited enhanced cell viability, albumin secretion, and urea synthesis, indicating improved hepatic functionality. This study underscores the potential of droplet-based 3D bioprinting in replicating the intricate cellular heterogeneity and spatial organization of liver tissue [40], offering valuable insights for applications in drug development, disease modeling, and tissue regeneration.

Physical, chemical, or photo-crosslinking methods are used to ensure the post-production stability of the 3D structures produced via the droplet-based 3D printing. The advantages of the droplet-based 3D printing technique include high production speed and the ability to eject picoliter-volume droplets, enabling high-resolution 3D products. Since droplet-based 3D printing technology operates faster than the FDM, materials that solidify quickly are required to ensure smooth progress in the layer-by-layer production process. This requirement also limits the selection of bioinks used in droplet-based 3D bioprinting processes. On the other hand, the high operating speed of the droplet-based 3D printing technique and the very short exposure of cells to high temperatures (~2 μs) during thermal-based production allow cell viability to be preserved. Additionally, since each droplet volume produced is very small, the process progresses quickly, but it may take a long time to complete the 3D structure [28, 29].

In droplet-based 3D printing, cell viability is adversely affected by high surface tension. When low-viscosity solutions and high cell concentrations are used in bioink preparation, cells absorb at the liquid-gas interface, reducing free energy. This results in lower surface tension and supports cell viability. For this reason, low-viscosity materials such as fibrin, gelatin, and collagen are preferred as bioinks in droplet-based 3D bioprinting [34, 41].

3.3 Bioprinting

3D bioprinting has become a prominent technology enabling the production of 3D, functional biological tissues by utilizing biocompatible materials, cells, and supporting components in unison. This process generally follows several steps: computerized modeling, bioink preparation, 3D printing, and enabling the printed products to become functional within a cell culture environment [42, 43]. With advancements in tissue engineering, bioprinting technology contributes to research in multiple areas, including bone fracture repair, cartilage regeneration, restoration of large-volume muscle loss, skin tissue repair for dermatological conditions and burn treatment, functional heart valve and cardiac tissue production, and vascular structure development. Beyond the production of functional tissues and organs, 3D printing technologies also support regenerative medicine by improving drug and therapeutic development processes through the accurate modeling of disease. Various examples of bioprinting applications in tissue engineering are presented in Table 3.3.

Table 3.3 Examples of tissue engineering studies using bioprinting

Material	Cell	Application	Ref
Polyethylene glycol diacrylate (PEG-DA)	Porcine aortic valve interstitial cells (PAVICs)	Aortic valve	[44]
Poly(e-caprolactone) (PCL)	Bladder smooth muscle cells	Urethra	[45]
Hyaluronic acid (HA)/alginate	Chondrocytes	Cartilage	[46]
Polyurethane (PU)/PCL, HA/gelatin/fibrinogen	C2C12-NIH/3T3	Muscle—tendon	[47]
GelMA (gelatin methacryloyl)/PCL	HiPSC-derived cardiomyocytes	Cardiac	[48]
Silk fibroin/pectin	Neural stem/progenitor cells (NSPCs)	Neural	[49]
Gelatin/decellularized bone particles	Pre-osteoblast MC3T3-E1	Bone	[50]

3.3.1 Bioinks

3D bioprinters require specialized biomaterials to carry out bioprinting technologies. When these biomaterials are combined with cells and other elements necessary for cellular function, the resulting mixture is referred to as "bioink" [51]. The printability and biocompatibility of bioinks are the most critical parameters. In 3D bioprinting, the shape and durability of the printed product depend on the biomaterial's printability, while cell activities like proliferation and differentiation rely on its biocompatibility [52]. The choice of bioink is determined by the specific characteristics of the target tissue, the bioprinter technology, and the selected bioprinting method. Numerous criteria, including tissue vascularization, gas and nutrient exchange, and the material's biodegradability, are intended to be satisfied by this selection procedure [53]. To this end, natural and synthetic polymers—silk fibroin, alginate, pectin, gelatin, collagen, fibrin, polyethylene glycol (PEG), and hydroxyapatite—are suitable as bioink components due to their biocompatibility and tunable physicochemical properties, making them well-suited for extracellular matrix (ECM) structure and formation.

3.3.2 Cells

Over the past 10 years, somatic cells—fibroblasts, myoblasts, chondrocytes, and cardiac myocytes—have been widely used in 3D bioprinting processes. Yet, there is a further growing interest in using stem cells in bioprinting applications. Pluripotent and multipotent stem cells were used within bioink materials to facilitate and provide the regeneration of the target tissue through 3D bioprinting [43, 54, 55].

Pluripotent stem cells can self-renew and are able to transform into approximately 200 different cell types. They can be classified as: (i) embryonic stem cells (ESCs) and (ii) induced pluripotent stem cells (iPSCs). ESCs are isolated from blastocysts, bringing in various ethical concerns for its use; moreover, they possess a risk of tumor formation [43, 56, 57]. iPSCs, on the other hand, are derived from adult somatic cells through genetic modification; therefore, the ethical concerns related to their use are not as severe. However, reprogramming and clinical use of these cells present certain challenges that are yet to be solved [58, 59].

Adult stem cells that can differentiate into a limited number of different cell types, specific to the tissue or organ they originate in, are known as multipotent stem cells. They possess limited potency compared to the pluripotent stem cells. These cells are essential for tissue regeneration and repair. As a result, they are regarded as viable options for 3D bioprinting and other tissue engineering and regenerative medicine applications [60].

3.3.3 3D Bioprinting Approaches

To engineer a functional tissue using 3D bioprinting, a supporting infrastructure is required to enable cell cohabitation and communication [61]. There are two primary techniques for creating tissue architecture: scaffold-free bioprinting, in which cells develop the architecture on their own, and scaffold-based bioprinting, which employs a structural scaffold for mechanical support. Due to the distinct characteristics of these approaches, each requires specific bioinks for effective tissue production.

3.3.3.1 Scaffold-Based Bioprinting

Biomaterials are used in scaffold-based techniques to provide a transient framework that promotes tissue development and cell adhesion. By mimicking natural tissue structures, this technique creates a 3D replica that can be more cost-effective and scalable due to lower cell density needs, and it often offers higher resolution in 3D printing compared to scaffold-free methods. However, the scaffold can limit cell-to-cell interactions and may lead to the inability to maintain structural integrity during the biodegradation process [51].

Scaffold-based bioprinting often uses bioinks made from hydrogels or decellularized extracellular matrix (dECM), with hydrogel-based 3D printing being the most common method. Hydrogels can be made into a variety of forms and sizes, absorb water drastically, and fuse cells with biological components to make an appropriate bioink. These qualities make hydrogels effective as cell carriers while also offering flexibility during the production process [42, 62].

Hydrogels used in bioprinting may be derived from natural sources, including alginate, gelatin, fibrin, or collagen, or they can be prepared from synthetic polymers including polyethylene glycol (PEG) and polyvinylpyrrolidone (PVP), which are often preferred for more precise control over degradation rates and mechanical strength. Once printed, hydrogels are cross-linked through various methods—thermal, chemical, or physical—to stabilize their shape and structure [43].

Decellularization is a process used to remove all cellular material from a tissue or organ, leaving behind only the ECM. This matrix, known as decellularized ECM (dECM), maintains the structural and biochemical properties of the original tissue, making it an ideal scaffold for tissue engineering and regenerative medicine [63]. During decellularization, chemical, enzymatic, or mechanical methods are used to eliminate cells and cellular components, while preserving important proteins, structural fibers, and other molecules in the ECM. This acellular scaffold can then be repopulated with new cells, which attach to and proliferate within the matrix, eventually forming functional tissue [64]. Such dECM has been utilized to create gel-like materials that can be printed with 3D bioprinters. On the other hand, mechanical strength of dECM bioinks is low which limits their use as standalone tissue scaffolds. Additionally, the requirement for organ donation for dECM isolation may render their use unsuitable for large-scale production [65].

3.3.3.2 Scaffold-Free Bioprinting

Scaffold-free bioprinting is based on the principle of self-assembly, where cellular components spontaneously organize and come together as the tissue develops. This approach aims to produce complex tissue architecture in a culture environment by enabling cells to aggregate, like the effects of processes active from the embryonic stage on tissue development, both environmentally and structurally [66]. In scaffold-free bioprinting, cell pellets are printed at high cell densities and designed to form the necessary ECM components of the tissue by creating 3D structures [67]. In this cell-based approach, no materials other than the cellular structure are used, thereby reducing toxicity, enhancing cell viability, and increasing cell-to-cell interactions. Compared to scaffold-based bioprinting, this method shortens the maturation process of the produced tissue. However, this method requires very high cell densities for successful bioprinting, which may require specialized bioprinters, thus offering lower scalability and additionally lacking the mechanical support provided by scaffolds [45, 51].

3.3.3.3 Functionalization of Bioprinted Tissues

The development of bioprinted structures takes place in a bioreactor or other strictly controlled environment. Bioreactors provide the necessary biomimetic cues to guide in vitro tissue development. Numerous bioreactor systems have been created, including perfusion systems for dynamic culture, incubators for static culture, and specialized bioreactors that use electrical stimulation or mechanical stretching [68].

According to the requirements of the simulated environment, bioreactor systems can be made up of one or more devices that can control variables like temperature, pressure, pH, the necessary speed and frequency for tensile-compressive stress, and voltage for electrical applications for the desired amount of time [68]. This process enables a biomimetic approach to structure the bioprinted tissue to suit the microenvironment intended for transplantation [68]. For example, daily activities expose cartilage tissues to tension and compression, leading to temporary increases in hydrostatic pressure. This pressure helps cartilage cells retain synovial fluid, enhancing proteoglycan and collagen synthesis, thereby strengthening the surrounding architecture and supporting further tissue development [62, 68]. Another example is skeletal muscle tissue, which is continuously affected by forces such as stretching. Simulating these activities in vitro is essential for muscle tissue engineering, as these forces are critical for muscle development.

In recent years, in situ bioprinting has emerged as an alternative approach, involving the direct 3D printing of tissues within a living host. This technique enables the use of the tissue's natural microenvironment rather than relying on a bioreactor. Instead of traditional 3D printers, this method employs robotic arms and biopens. This process aims to utilize the natural tissue environment as an "in vivo bioreactor," leveraging signals from adjacent tissue. By relying on natural cues, in situ bioprinting presents a potential solution to the weak integration typically seen between in vitro structures and the implantation site [69, 70].

3.3.3.4 Emerging Trends in 3D Bioprinting

The aging global population is leading to a continuous increase in medical needs and demand for organ donation. Meanwhile, the use of animals in therapeutic testing raises ethical concerns, highlighting the need to reduce this practice. At this point, 3D bioprinting stands out as an application with significant potential across various areas in healthcare [70]. The influence of 3D bioprinting and biofabrication is increasingly significant, providing cost-effective advancements in medical and clinical research through rapid prototyping. A rapidly advancing area within this field is cancer research, where bioprinted tissues and organs are instrumental. Through 3D bioprinted products, more insight into drug resistance and metastasis can be gained. This can help reduce the need for animal testing before human trials. Additionally, treatments for infectious diseases could be produced more efficiently, particularly in emergency situations where minimizing human life loss is critical. However, several challenges must be addressed to achieve future goals. Ensuring the proper maturation of cells to create biological structures remains a significant obstacle. Furthermore, designing precise sequences of stimuli to induce specific tissue differentiation is seen as one of the greatest challenges in this field [62, 70].

3.4 Conclusion

The 1990s saw the emergence of 3D printing technology in the medical field, which allowed for the creation of prosthetics, orthoses, and implants to replace or enhance the function of tissues and organs. Additionally, customized, living, functional tissues and organs can now be produced using 3D bioprinting with cell-laden bioinks.

Various 3D printers and bioprinters are evolving to meet emerging needs, and the research and production of materials suitable for these devices are also rapidly advancing. Studies highlight the necessity of enhancing the structural and mechanical properties of bioinks, alongside characteristics that support cell viability, particularly in terms of cross-linking mechanisms and rheological attributes. Currently, there is a limited range of materials available to be used as bioinks, including hyaluronic acid, gelatin, collagen, alginate, photocurable acrylates, and modified copolymers. Thus, the development of bioink materials with required rheological properties that support cell viability and sprouting is a global need. In conclusion, the diagnostic, therapeutic, and procedural approaches of the future are still being shaped by 3D modeling and printing technology, which has impacted many facets of healthcare.

References

1. Pagac, M., Hajnys, J., Ma, Q.P., Jancar, L., Jansa, J., Stefek, P., Mesicek, J.: A review of vat photopolymerization technology: materials, applications, challenges, and future trends of 3d printing. Polymers (Basel). **13**, 1–20 (2021). https://doi.org/10.3390/polym13040598
2. Torras, N., Zabalo, J., Abril, E., Carré, A., García-Díaz, M., Martinez, E.: A bioprinted 3D gut model with crypt-villus structures to mimic the intestinal epithelial-stromal microenvironment. Biomater. Adv. **153**, 213534 (2023). https://doi.org/10.1016/j.bioadv.2023.213534
3. Liang, R., Gu, Y., Wu, Y., Bunpetch, V., Zhang, S.: Lithography-based 3D bioprinting and bioinks for bone repair and regeneration. ACS Biomater Sci. Eng. **7**(3), 806–816 (2021). https://doi.org/10.1021/acsbiomaterials.9b01818
4. Shopperly, L.K., Spinnen, J., Krüger, J.P., Endres, M., Sittinger, M., Lam, T., Kloke, L., Dehne, T.: Blends of gelatin and hyaluronic acid stratified by stereolithographic bioprinting approximate cartilaginous matrix gradients. J. Biomed. Mater. Res. **110**(10), 2310–2322 (2022). https://doi.org/10.1002/jbm.b.35079
5. Momin, M., Feng, L., Ahmed, S., Ren, J., Hossain, A., Zhang, S., Zhou, T.: 3D-printed flexible neural probes for recordings at single-neuron level. Dev. Dent., 100519 (2024). https://doi.org/10.1016/j.device.2024.100519
6. Baldock, S.J., Kevin, P., Harper, G.R., Griffin, R., Genedy, H.H., Fong, M.J., Zhao, Z., Zhang, Z., Shen, Y., Lin, H., Au, C., Martin, J.R., Ashton, M.D., Haskew, M.J., Stewart, B., Efremova, O., Esfahani, R.N., Emsley, H.C.A., Appleby, J.B., Cheneler, D., Cummings, D.M., Benedetto, A., Hardy, J.G.: Creating 3D objects with integrated electronics via multiphoton fabrication in vitro and in vivo. Adv. Mater. Technol. **8**, 2201274 (2023). https://doi.org/10.1002/admt.202201274
7. Duan, B., Kapetanovic, E., Hockaday, L.A., Butcher, J.T.: Three-dimensional printed trileaflet valve conduits using biological hydrogels and human valve interstitial cells. Acta Biomater. **10**(5), 1836–1846 (2014). https://doi.org/10.1016/j.actbio.2013.12.005
8. Sun, Z.: Patient-specific 3D-printed models in pediatric congenital heart disease. Children (Basel). **10**(2), 319 (2023). https://doi.org/10.3390/children10020319
9. Janusziewicz, R., Tumbleston, J.R., Quintanilla, A.L., Mecham, S.J., DeSimone, J.M.: Layerless fabrication with continuous liquid interface production. Proc. Natl. Acad. Sci. U. S. A. **113**, 11703–11708 (2016). https://doi.org/10.1073/pnas.1605271113
10. Alam, F., Verma, P., Mohammad, W., Teo, J., Varadarajan, K.M., Kumar, S.: Architected poly(lactic acid)/poly(ε-caprolactone)/halloysite nanotube composite scaffolds enabled by 3D printing for biomedical applications. J. Mater. Sci. **56**, 14070–14083 (2021). https://doi.org/10.1007/s10853-021-06145-0

11. Homan, K.A., Kolesky, D.B., Skylar-Scott, M.A., Herrmann, J., Obuobi, H., Moisan, A., Lewis, J.A.: Bioprinting of 3D convoluted renal proximal tubules on perfusable chips. Sci. Rep. **6**, 1–13 (2016). https://doi.org/10.1038/srep34845
12. Unalan, I., Slavik, B., Buettner, A., Boccaccini, A.R.: Phytotherapeutic hierarchical PCL-based scaffolds as a multifunctional wound dressing: combining 3D printing and electrospinning. Macromol. Biosci. **24**, 2400253 (2024). https://doi.org/10.1002/mabi.202400253
13. Skardal, A., Mack, D., Kapetanovic, E., Atala, A., Jackson, J.D., Yoo, J., Soker, S.: Bioprinted amniotic fluid-derived stem cells accelerate healing of large skin wounds. Stem Cells Transl. Med. **1**, 792–802 (2012). https://doi.org/10.5966/sctm.2012-0088
14. Kunwar, P., Aryal, U., Poudel, A., Fougnier, D., Geffert, Z.J., Xie, R., Li, Z., Soman, P.: Droplet bioprinting of acellular and cell-laden structures at high-resolutions. Biofabrication. **16**, 35019 (2024). https://doi.org/10.1088/1758-5090/ad4c09
15. Lee, Y.S., Kirka, M.M., Ferguson, J., Paquit, V.C.: Correlations of cracking with scan strategy and build geometry in electron beam powder bed additive manufacturing. Addit. Manuf. **32**, 101031 (2020). https://doi.org/10.1016/j.addma.2019.101031
16. Wischeropp, T.M., Emmelmann, C., Brandt, M., Pateras, A.: Measurement of actual powder layer height and packing density in a single layer in selective laser melting. Addit. Manuf. **28**, 176–183 (2019). https://doi.org/10.1016/j.addma.2019.04.019
17. Shirazi, S.F.S., Gharehkhani, S., Mehrali, M., Yarmand, H., Metselaar, H.S.C., Adib Kadri, N., Osman, N.A.A.: A review on powder-based additive manufacturing for tissue engineering: selective laser sintering and inkjet 3D printing. Sci. Technol. Adv. Mater. **16**, 1–20 (2015). https://doi.org/10.1088/1468-6996/16/3/033502
18. Shuai, C., Wang, Z., Yang, F., Zhang, H., Liu, J., Feng, P.: Laser additive manufacturing of shape memory biopolymer bone scaffold: 3D conductive network construction and electrically driven mechanism. J. Adv. Res. **65**, 167–181 (2024). https://doi.org/10.1016/j.jare.2023.11.031
19. Jang, T.S., Kim, D.E., Han, G., Yoon, C.B., Do Jung, H.: Powder based additive manufacturing for biomedical application of titanium and its alloys: a review. Biomed. Eng. Lett. **10**, 505–516 (2020). https://doi.org/10.1007/s13534-020-00177-2
20. Spears, T.G., Gold, S.A.: In-process sensing in selective laser melting (SLM) additive manufacturing. Integr. Mater. Manuf. Innov. **5**, 16–40 (2016). https://doi.org/10.1186/s40192-016-0045-4
21. Zhang, P., Tan, J., Tian, Y., Yan, H., Yu, Z.: Research progress on selective laser melting (SLM) of bulk metallic glasses (BMGs): a review. Int. J. Adv. Manuf. Technol. **118**, 2017–2057 (2022). https://doi.org/10.1007/s00170-021-07990-8
22. O'Keeffe, C., Kotlarz, M., Gonçalves, I.F., Lally, C., Kelly, D.J.: Chemical etching of Ti-6Al-4V biomaterials fabricated by selective laser melting enhances mesenchymal stromal cell mineralization. J. Biomed. Mater. Res. A. **112**, 1548–1564 (2024). https://doi.org/10.1002/jbm.a.37709
23. Yu, L., Sun, F., Wang, Y., Li, W., Zheng, Y., Shen, G., Wang, Y., Chen, M.: Effects of MgO nanoparticle addition on the mechanical properties, degradation properties, antibacterial properties and in vitro and in vivo biological properties of 3D-printed Zn scaffolds. Bioact. Mater. **37**, 72–85 (2024). https://doi.org/10.1016/j.bioactmat.2024.03.016
24. Sing, S.L., An, J., Yeong, W.Y., Wiria, F.E.: Laser and electron-beam powder-bed additive manufacturing of metallic implants: a review on processes, materials and designs. J. Orthop. Res. **34**, 369–385 (2016). https://doi.org/10.1002/jor.23075
25. Mofazali, P., Atapour, M., Nakamura, M., Galati, M., Saboori, A.: Evaluation of layer-by-layer assembly systems for drug delivery and antimicrobial properties in orthopaedic application. Int. J. Pharm. **657** (2024). https://doi.org/10.1016/j.ijpharm.2024.124148
26. Zhu, C., Jia, Y., Tang, Y., Guo, C., Xi, J., Sun, C., Li, H., Wang, W., Zhai, Y., Zhu, Y., Liu, Y.: Functionalized chitosan hydrogel promotes osseointegration at the interface of3D printed titanium alloy scaffolds. Int. J. Biol. Macromol. **266** (2024). https://doi.org/10.1016/j.ijbiomac.2024.131169
27. Hölzl, K., Lin, S., Tytgat, L., Van Vlierberghe, S., Gu, L., Ovsianikov, A.: Bioink properties before, during and after 3D bioprinting. Biofabrication. **8** (2016). https://doi.org/10.1088/1758-5090/8/3/032002

28. Arslan-Yildiz, A., El Assal, R., Chen, P., Guven, S., Inci, F., Demirci, U.: Towards artificial tissue models: past, present, and future of 3D bioprinting. Biofabrication. **8** (2016). https://doi.org/10.1088/1758-5090/8/1/014103
29. E. Provaggi, K. M. Deepak. 3D Printin in Medicine. (K. M. Deepak, Ed.) Woodhead Publishing
30. Kade, J.C., Dalton, P.D.: Polymers for melt electrowriting. Adv. Healthc. Mater. **10** (2021). https://doi.org/10.1002/adhm.202001232
31. Bakirci, E., Schaefer, N., Dahri, O., Hrynevich, A., Strissel, P., Strick, R., Dalton, P.D., Villmann, C.: Melt electrowritten in vitro radial device to study cell growth and migration. Adv. Biosyst. **4**, 1–8 (2020). https://doi.org/10.1002/adbi.202000077
32. Čajková, J., Trebuňová, M., Modrák, M., Ižaríková, G., Bačenková, D., Balint, T., Živčák, J.: Influence of oligomeric lactic acid and structural design on biodegradation and absorption of PLA-PHB blends for tissue engineering. Polymers (Basel). **16** (2024). https://doi.org/10.3390/polym16212969
33. Seol, Y.J., Kang, H.W., Lee, S.J., Atala, A., Yoo, J.J.: Bioprinting technology and its applications. Eur. J. Cardiothorac. Surg. **46**, 342–348 (2014). https://doi.org/10.1093/ejcts/ezu148
34. Bharadwaj, T., Chrungoo, S., Verma, D.: Development of a novel thermogelling PEC-based ECM mimicking nanocomposite bioink for bone tissue engineering. J. Biomater. Sci. Polym. Ed. **34**, 2516–2536 (2023). https://doi.org/10.1080/09205063.2023.2265143
35. Arıcı, S., Kamali, A.R., Ege, D.: CMC/Gel/GO 3D-printed cardiac patches: GO and CMC improve flexibility and promote H9C2 cell proliferation, while EDC/NHS enhances stability. Biofabrication. **17**, 015025 (2025). https://doi.org/10.1088/1758-5090/ad8e87
36. Naddeo, F., Fontana, C., Naddeo, A., Cataldo, E., Cappetti, N., Narciso, N.: Novel design for a customized, 3D-printed surgical template for thoracic spinal arthrodesis. Int. J. Med. Robot. Comput. Assist. Surg. **15**, e2005 (2019). https://doi.org/10.1002/rcs.2005
37. Knoedler, M., Feibus, A.H., Lange, A., Maddox, M.M., Ledet, E., Thomas, R., Silberstein, J.L.: Individualized physical 3-dimensional kidney tumor models constructed from 3-dimensional printers result in improved trainee anatomic understanding. Urology. **85**, 1257–1262 (2015). https://doi.org/10.1016/j.urology.2015.02.053
38. Khaled, S.A., Burley, J.C., Alexander, M.R., Yang, J., Roberts, C.J.: 3D printing of five-in-one dose combination polypill with defined immediate and sustained release profiles. J. Control. Release. **217**, 308–314 (2015). https://doi.org/10.1016/j.jconrel.2015.09.028
39. Jian, H., Li, X., Dong, Q., Tian, S., Bai, S.: In vitro construction of liver organoids with biomimetic lobule structure by a multicellular 3D bioprinting strategy. Cell Prolif. **56**, e13465 (2023). https://doi.org/10.1111/cpr.13465
40. Distler, T., Fournier, N., Grünewald, A., Polley, C., Seitz, H., Detsch, R., Boccaccini, A.R.: Polymer-bioactive glass composite filaments for 3D scaffold manufacturing by fused deposition modeling: fabrication and characterization. Front. Bioeng. Biotechnol. **8**, 1–17 (2020). https://doi.org/10.3389/fbioe.2020.00552
41. Murphy, S.V., Atala, A.: 3D bioprinting of tissues and organs. Nat. Biotechnol. **32**, 773–785 (2014). https://doi.org/10.1038/nbt.2958
42. Jamieson, C., Keenan, P., Kirkwood, D., Oji, S., Webster, C., Russell, K.A., Koch, T.G.: A review of recent advances in 3D bioprinting with an eye on future regenerative therapies in veterinary medicine. Front. Vet. Sci. **7** (2021). https://doi.org/10.3389/fvets.2020.584193
43. Hockaday, L.A., Kang, K.H., Colangelo, N.W., Cheung, P.Y.C., Duan, B., Malone, E., Wu, J.W., Girardi, L.N., Bonassar, L.J., Lipson, H., Chu, C.C., Butcher, J.T.: Rapid 3D printing of anatomically accurate and mechanically heterogeneous aortic valve hydrogel scaffolds. Biofabrication. **4**(3), 35005 (2012)
44. Zhang, K., Fu, Q., Yoo, J., Chen, X., Chandra, P., Mo, X., Song, L., Atala, A., Zhao, W.: 3D bioprinting of urethra with PCL/PLCL blend and dual autologous cells in fibrin hydrogel: an in vitro evaluation of biomimetic mechanical property and cell growth environment. Acta Biomater. **50**, 154–164 (2017). https://doi.org/10.1016/j.actbio.2016.12.008
45. Antich, C., de Vicente, J., Jiménez, G., Chocarro, C., Carrillo, E., Montañez, E., Gálvez-Martín, P., Marchal, J.A.: Bio-inspired hydrogel composed of hyaluronic acid and alginate

as a potential bioink for 3D bioprinting of articular cartilage engineering constructs. Acta Biomater. **106**, 114–123 (2020). https://doi.org/10.1016/j.actbio.2020.01.046
46. Merceron, T.K., Burt, M., Seol, Y.-J., Kang, H.-W., Lee, S.J., Yoo, J.J., Atala, A.: A 3D bioprinted complex structure for engineering the muscle–tendon unit. Biofabrication. **7**, 35003 (2015). https://doi.org/10.1088/1758-5090/7/3/035003
47. Wu, H., Xu, F., Jin, H., Xue, M., Zhang, W., Yang, J., Huang, J., Jiang, Y., Qiu, B., Lin, B., Gao, Q., Chen, S., Sun, D.: 3D nanofiber-assisted embedded extrusion bioprinting for oriented cardiac tissue fabrication. ACS Biomater Sci. Eng. (2024). https://doi.org/10.1021/ACSBIOMATERIALS.4C01611/ASSET/IMAGES/LARGE/AB4C01611_0006.JPEG
48. Lee, H.W., Chen, K.T., Li, Y.C.E., Yeh, Y.C., Chiang, C.Y., Lee, I.C.: Dual crosslinking silk fibroin/pectin-based bioink development and the application on neural stem/progenitor cells spheroid laden 3D bioprinting. Int. J. Biol. Macromol. **269**, 131720 (2024). https://doi.org/10.1016/J.IJBIOMAC.2024.131720
49. Kara Özenler, A., Distler, T., Akkineni, A.R., Tihminlioglu, F., Gelinsky, M., Boccaccini, A.R.: 3D bioprinting of mouse pre-osteoblasts and human MSCs using bioinks consisting of gelatin and decellularized bone particles. Biofabrication. **16**, 025027 (2024). https://doi.org/10.1088/1758-5090/AD2C98
50. Ozbolat, I.T.: Scaffold-based or scaffold-free bioprinting: competing or complementing approaches? J. Nanotechnol. Eng. Med. **6** (2015). https://doi.org/10.1115/1.4030414
51. Yu, K., Zhang, X., Sun, Y., Gao, Q., Fu, J., Cai, X., He, Y.: Printability during projection-based 3D bioprinting. Bioact. Mater. **11**, 254–267 (2022). https://doi.org/10.1016/j.bioactmat.2021.09.021
52. Xu, C., Chai, W., Huang, Y., Markwald, R.R.: Scaffold-free inkjet printing of three-dimensional zigzag cellular tubes. Biotechnol. Bioeng. **109**, 3152–3160 (2012). https://doi.org/10.1002/bit.24591
53. Albanna, M., Binder, K.W., Murphy, S.V., Kim, J., Qasem, S.A., Zhao, W., Tan, J., El-Amin, I.B., Dice, D.D., Marco, J., Green, J., Xu, T., Skardal, A., Holmes, J.H., Jackson, J.D., Atala, A., Yoo, J.J.: In situ bioprinting of autologous skin cells accelerates wound healing of extensive excisional full-thickness wounds. Sci. Rep. **9**(2019) (1856). https://doi.org/10.1038/s41598-018-38366-w
54. Koti, P., Muselimyan, N., Mirdamadi, E., Asfour, H., Sarvazyan, N.A.: Use of GelMA for 3D printing of cardiac myocytes and fibroblasts. J. 3D Print. Med. **3**, 11–22 (2019). https://doi.org/10.2217/3dp-2018-0017
55. Volarevic, V., Markovic, B.S., Gazdic, M., Volarevic, A., Jovicic, N., Arsenijevic, N., Armstrong, L., Djonov, V., Lako, M., Stojkovic, M.: Ethical and safety issues of stem cell-based therapy. Int. J. Med. Sci. **15**, 36–45 (2018). https://doi.org/10.7150/ijms.21666
56. Singh, V.K., Saini, A., Kalsan, M., Kumar, N., Chandra, R.: Describing the stem cell potency: the various methods of functional assessment and in silico diagnostics. Front. Cell Dev. Biol. **4**, 134 (2016). https://doi.org/10.3389/fcell.2016.00134
57. Roberts, C.L., Chen, S.S., Murchison, A.C., Ogle, R.A., Francis, M.P., Ogle, R.C., Sachs, P.C.: Preferential lineage-specific differentiation of osteoblast-derived induced pluripotent stem cells into osteoprogenitors. Stem Cells Int. **2017**, 1513281 (2017). https://doi.org/10.1155/2017/1513281
58. Moradi, S., Mahdizadeh, H., Šarić, T., Kim, J., Harati, J., Shahsavarani, H., Greber, B., Moore 4th, J.B.: Research and therapy with induced pluripotent stem cells (iPSCs): social, legal, and ethical considerations. Stem Cell Res Ther. **10**, 341 (2019). https://doi.org/10.1186/s13287-019-1455-y
59. Tasnim, N., De la Vega, L., Anil Kumar, S., Abelseth, L., Alonzo, M., Amereh, M., Joddar, B., Willerth, S.M.: 3D bioprinting stem cell derived tissues. Cell. Mol. Bioeng. **11**, 219–240 (2018). https://doi.org/10.1007/s12195-018-0530-2
60. Yilgor, P., Tuzlakoglu, K., Reis, R.L., Hasirci, N., Hasirci, V.: Incorporation of a sequential BMP-2/BMP-7 delivery system into chitosan-based scaffolds for bone tissue engineering. Biomaterials. **30**, 3551–3559 (2009)

61. Ovsianikov, A., Khademhosseini, A., Mironov, V.: The synergy of scaffold-based and scaffold-free tissue engineering strategies. Trends Biotechnol. **36**, 348–357 (2018). https://doi.org/10.1016/j.tibtech.2018.01.005
62. Park, S.M., Yang, S., Rye, S.-M., Choi, S.W.: Effect of pulsatile flow perfusion on decellularization. Biomed. Eng. Online. **17**, 15 (2018). https://doi.org/10.1186/s12938-018-0445-0
63. Garreta, E., Oria, R., Tarantino, C., Pla-Roca, M., Prado, P., Fernández-Avilés, F., Campistol, J.M., Samitier, J., Montserrat, N.: Tissue engineering by decellularization and 3D bioprinting. Mater. Today. **20**, 166–178 (2017). https://doi.org/10.1016/j.mattod.2016.12.005
64. Dzobo, K., Motaung, K.S.C.M., Adesida, A.: Recent trends in decellularized extracellular matrix bioinks for 3D printing: an updated review. Int. J. Mol. Sci. **20** (2019). https://doi.org/10.3390/ijms20184628
65. Bishop, E.S., Mostafa, S., Pakvasa, M., Luu, H.H., Lee, M.J., Wolf, J.M., Ameer, G.A., He, T.-C., Reid, R.R.: 3-D bioprinting technologies in tissue engineering and regenerative medicine: current and future trends. Genes Dis. **4**, 185–195 (2017). https://doi.org/10.1016/j.gendis.2017.10.002
66. J. Rosser and D. J. Thomas, 10—Bioreactor processes for maturation of 3D bioprinted tissue, in D. J. Thomas, Z. M. Jessop, & I. S. Whitaker (Eds.), 3D bioprinting for reconstructive surgery Woodhead Publishing, (pp. 191–215); DOI:https://doi.org/10.1016/B978-0-08-101103-4.00010-7
67. Unagolla, J.M., Jayasuriya, A.C.: Hydrogel-based 3D bioprinting: a comprehensive review on cell-laden hydrogels, bioink formulations, and future perspectives. Appl. Mater. Today. **18** (2020). https://doi.org/10.1016/j.apmt.2019.100479
68. Blevins, K.M., Danilkowicz, R.M., Fletcher, A.N., Allen, N.B., Johnson, L.G., Adams, S.B.: In situ 3D bioprinting of musculoskeletal tissues in orthopedic surgery. J. 3D Print. Med. **6**, 25–36 (2022). https://doi.org/10.2217/3dp-2021-0022
69. Bejoy, A.M., Makkithaya, K.N., Hunakunti, B.B., Hegde, A., Krishnamurthy, K., Sarkar, A., Lobo, C.F., Keshav, D.V.S., Dharshini, G., Dharshini, D., Mascarenhas, S., Chakrabarti, S., Kalepu, R., Paul, B., Mazumder, N.: An insight on advances and applications of 3d bioprinting: a review. Bioprinting. **24**, e00176 (2021). https://doi.org/10.1016/j.bprint.2021.e00176
70. Agarwal, S., Saha, S., Balla, V.K., Pal, A., Barui, A., Bodhak, S.: Current developments in 3D bioprinting for tissue and organ regeneration—a review. Front. Mech. Eng. **6** (2020). https://doi.org/10.3389/fmech.2020.589171

Chapter 4
Properties and Preparation Techniques of Hydrogels

Neval Sevinc Ozdemir and Halime Kenar

Abstract Hydrogels are solid and network-structured systems consisting of crosslinked hydrophilic macromolecules, which refer to materials that absorb high amounts of water, and turn into a flexible and soft jelly consistency. In aqueous environments, they can absorb water at 20–95% of their own dry weight. Although hydrogels like water, they generally do not dissolve in water due to their crosslinked structure, but they swell and get larger in volume. They are widely used in the medical field, as scaffolds in tissue engineering applications, as carriers in drug delivery systems, as water-retaining components of wound and burn dressings, and in cosmetics. Their first application was in contact lens production. Since their structure is soft, they adapt well to natural tissues, and when used inside the body, they do not disturb the patient and do not give frictional damage to the tissues. Some hydrogels are able to adhere to tissues or mucosal surfaces.

Keywords Hydrogels · Hydrogen bonding · Complex coacervation · Chemical crosslinkers · Free radical polymerization · Enzyme mediated crosslinking · Click reaction

N. Sevinc Ozdemir
Biomaterials A&R Center, Acibadem Mehmet Ali Aydinlar University, Istanbul, Türkiye

Department of Pharmaceutical Basic Sciences, School of Pharmacy, Acibadem Mehmet Ali Aydinlar University, Istanbul, Türkiye

H. Kenar (✉)
Biomaterials A&R Center, Acibadem Mehmet Ali Aydinlar University, Istanbul, Türkiye

Department of Biomedical Engineering, Faculty of Engineering and Natural Sciences, Acibadem Mehmet Ali Aydinlar University, Istanbul, Türkiye
e-mail: halime.kenar@acibadem.edu.tr

N. Hasirci, V. Hasirci (eds.), *Hydrogels and Bioinks in Tissue Engineering*, Biological and Medical Physics, Biomedical Engineering,
https://doi.org/10.1007/978-3-031-90966-5_4

4.1 Properties of Hydrogels

The properties of hydrogels vary depending on the method preferred in their synthesis. Swelling properties in different liquids vary depending on their chemical structure, the chemistry of functional groups, and the type and density of crosslinking bonds. In general, hydrogels can be classified as physically or chemically crosslinked gels, according to the structure of the bond between the polymeric molecules that form the network.

Physical (Reversible) Gel Molecules form the "reversible" or "physical" gels when held together by secondary forces, including entanglements between chains and/or hydrogen bonding, ionic bonding, or hydrophobic interactions. Dissolution is hindered in these gels by the physical interactions between the different polymer chains which are reversible, meaning that they can be disrupted by changes in physical conditions or the application of stress.

Chemical (Permanent) Gel Gels formed by networks that have covalent bonds between molecules (replacing the hydrogen bond with stronger and more stable covalent bonds), which are crosslinking them. These are called "permanent" or "chemical" gels. These gels reach an equilibrium swelling state depending on the polymer-water interaction parameter and crosslink density (Fig. 4.1).

The presence of covalent bonds between polymer chains of chemically crosslinked gels enables stable connections, whereas in the case of physical crosslinking, there are physical interactions between the chains. Due to such weak and non-permanent connections, physical hydrogels, in addition to being in reversible form, can soften and flow or degrade uncontrollably under certain conditions, such as mechanical loading.

When these two techniques are compared, chemical hydrogels have higher mechanical strength, stability, better surface properties, and longer and more controllable degradation time than physically crosslinked gels. Due to these properties, interest in chemically crosslinked hydrogels has been increasing in recent years. On the other hand, most of the compounds used in the preparation of chemically crosslinked hydrogels are toxic and can adversely affect biological tissues. However, biocompatible hydrogels can be obtained by taking appropriate precautions during

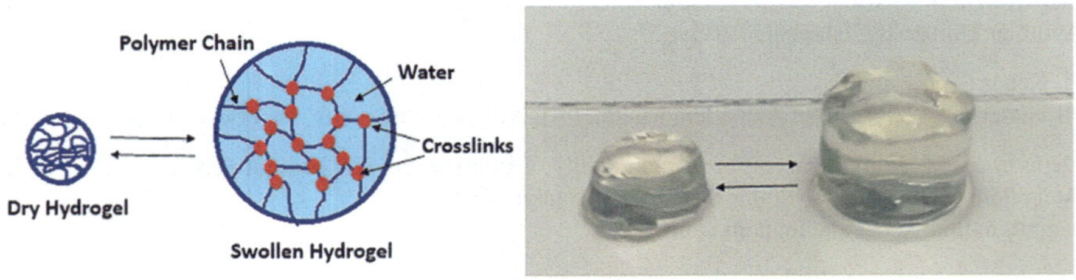

Fig. 4.1 Equilibrium swelling state reached by hydrogels in water

synthesis followed by strict washing procedures [1]. In order to increase the functionality of hydrogels used in biomedical applications, it is also possible to combine the chemical and physical crosslinking methods to obtain gels with the desired properties depending on the purpose and the application area [2, 3].

4.2 Hydrogel Preparation Techniques

Hydrogels are prepared from hydrophilic polymers by crosslinking the polymeric chains. It is possible to divide the hydrogel crosslinking techniques into three groups: (1) Physical crosslinking, (2) Chemical crosslinking, and (3) Radiation crosslinking. Modifications can improve mechanical properties and viscoelasticity for applications in biomedical and pharmaceutical fields. General methods applied for the production of hydrogels are described below.

4.2.1 Physical Crosslinking

There is a growing interest in physical gels due to their advantages, such as relative ease of production and lack of use of crosslinkers which are mostly toxic chemicals. Crosslinking chemicals affect the integrity of the agents (e.g. cells, proteins, drug, etc.) to be encapsulated within the hydrogel and may need to be removed from the system prior to application. Careful selection of the type, concentration, and pH of a hydrocolloid, defined as a high molecular weight, water-soluble macromolecule that can bind large amounts of water and modify the flow properties of aqueous systems to which it is added, can lead to the formation of a wide variety of gel textures. There are various methods reported in the literature to obtain physically crosslinked hydrogels [4].

4.2.1.1 Heating or Cooling of Polymer Solution

It is possible to obtain physical hydrogels by simply heating or cooling some polymer solutions. Gels form when cooling hot gelatin or carrageenan solutions. Gel formation is due to the helical structure, the binding of the helices, and the physical crosslinking of the junction regions [5]. In a hot solution above its melting temperature, carrageenan exists as a random coil structure. When it cools, it turns into hard spiral rods. In the presence of a salt (K^+, Na^+, etc.), the double helices further aggregate to form stable gels due to blocking of the sulfonic group (SO_3^-), which causes repulsion between the helical rods. In other cases, block copolymerization, which occurs as a result of simple heating of different polymer solutions, may also lead to hydrogel formation. Some of the examples are polyethylene oxide-polypropylene oxide [6] and polyethylene glycol-polylactic acid [7] hydrogels. Injectable,

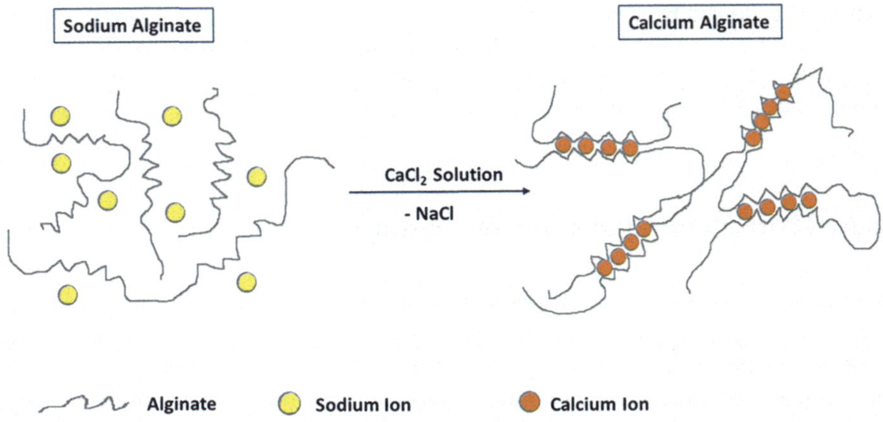

Fig. 4.2 Ionic gelation that occurs through the interaction between anionic groups (COO^-) on the alginate and divalent metal ions (Ca^{2+})

self-healing, and antibacterial adenine-modified chitosan hydrogels can be prepared in aqueous solutions by a simple heating/cooling process [8].

4.2.1.2 Ionic Interaction

Ionic polymers can be crosslinked by adding divalent or trivalent counterions. This method is based on the gelation of a polyelectrolyte solution (e.g. sodium alginate: Na^+ alginate$^-$) with a multivalent counterion (e.g. calcium chloride: $Ca^{2+} + 2Cl^-$) (Fig. 4.2). In an acidic environment, the amino groups of chitosan are protonated, i.e., positively charged, and can form ionic bonds with anionic molecules. It is possible to create chitosan microparticles using tripolyphosphate (TPP) with this method called ionic gelation [9]. Some other examples are chitosan-polylysine [10], chitosan-glycerol phosphate [11], and alginate-pectin [12] hydrogels.

4.2.1.3 Complex Coacervation

The principle underlying this method is that oppositely charged polymers, i.e., a polyanion and a polycation, stick together and depending on their concentration and pH of the respective solutions can form soluble and insoluble complexes (Fig. 4.3). One such example is polycationic chitosan, which forms coacervates with polyanionic xanthan [13]. In a study by Lalevee et al. [14], hydrogels with controllable properties in terms of elasticity and strength were obtained by the controlled complex coacervation of oppositely charged hyaluronic acid and chitosan. A synthetic and in situ curing adhesive based on complex coacervation consisting of oppositely charged polyelectrolytes grafted with thermosensitive poly(N-isopropylacrylamide) (PNIPAM) chains has been reported [15]. The adhesive can be injected at room

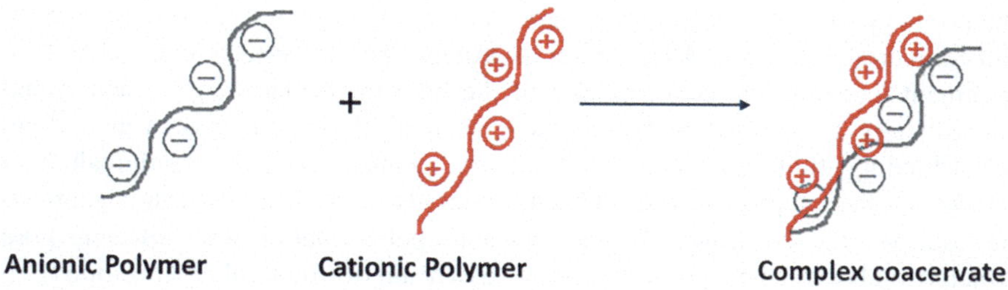

Fig. 4.3 Complex coacervation between a polyanion and a polycation

temperature and when the temperature is increased above the low critical solution temperature of PNIPAM, the complex coacervate transitions to a non-flowing hydrogel structure, while maintaining its volume and water content. The adhesive functions in the presence of water and has the ability to adhere to different surfaces regardless of their charge.

4.2.1.4 Hydrogen Bonding

H-bonded hydrogels can be obtained in some applications by lowering the pH of the aqueous solution of polymers bearing carboxyl groups. The H-bonded carboxymethyl cellulose (CMC) network formed by dispersing CMC in 0.1 M HCl is an example of such hydrogels [16]. Sodium in the CMC is replaced with hydrogen in the acid solution to promote hydrogen bond formation between the chains that cause a decrease in the solubility of CMC in water, resulting in the formation of an elastic hydrogel. Another example is polyethylene oxide and polyacrylic acid (PEO-PAAc)-based hydrogels prepared by lowering the pH in their aqueous solution to form the H-bonds [6]. Similarly, when mixed at a high temperature, the multiple H bonds may form between poly(vinyl alcohol) (PVA) and tannic acid (TA), which leads to their coagulation and gelation at room temperature. The easy production of a new type of shape memory hydrogel with this method has been reported [17]. The reversible breaking and reformation of the weak H-bond between PVA chains in PVA-TA hydrogels gives these hydrogels temperature-sensitive shape memory property. Deformed wet PVA-TA hydrogel samples can return to their original shape within a few seconds when immersed in water at 60 °C. Another example of shape memory polymers is physically crosslinked poly(methacrylic acid)/gelatin hydrogels, which have very good mechanical properties [18].

4.2.1.5 Maturation (Heat-Induced Aggregation)

Gelation of polymer chains may occur via application of heat. One example is gum arabic (Acacia gums), which is predominantly a carbohydrate but contains 2–3% protein as an integral part of its structure [19]. The maturation of gum arabic causes the protein part of the molecules to aggregate, leading to the formation of a crosslinked hydrogel network. Aggregation of protein-like components induced by heat treatment increases the concentration of the high molecular weight fraction and subsequently a hydrogel with improved mechanical properties as well as high water-binding ability is formed [20]. Similarly, incubation of bovine serum albumin (BSA), a spherical protein, at 60 °C at different pH values triggers protein aggregation due to conformational change, and hydrogel formation with different pore structure and mechanical properties occurs as a result of structural organization [21].

4.2.1.6 Freeze-Thaw

Freeze-thaw cycles can also lead to physical crosslinking of a polymer. Microcrystals are formed in the structure due to freeze-thawing. Polyvinyl alcohol and xanthan gels are good examples for this type of gelation [6, 22]. During freezing, due to conversion of the solvent molecules (e.g. water) into ice crystals, xanthan chains are forced to align and interact. The resulting alignment and association survive during thawing, thus forming the cryogel network. It is also possible to produce chitin-glucan hydrogels as drug release systems by the freeze-thaw method [23]. Porous cryogel tissue scaffolds can also be produced by the freeze-thaw method from bean gum, xanthan gum (xanthan gum), and mastic gum (gum gum) blended in different proportions [24].

4.2.2 Chemical Crosslinking

Different approaches are used to synthesize the chemically crosslinked hydrogels. Choosing the appropriate method is associated with the target application area and purpose. Chemical crosslinking methods are preferred in applications where the produced hydrogels are required to have longer-term stability and higher mechanical strength than that of physically crosslinked ones [3]. For example, the free radical polymerization method is often used to keep protein molecules in their three-dimensional structure in the formed protein-based hydrogels. The basic chemical crosslinking methods used in hydrogel preparation are summarized below.

4.2.2.1 Crosslinking Through Functional Groups

Hydrophilic groups such as $-NH_2$ (amine), –COOH (carboxyl), and –OH (hydroxyl) present in the polymers are used in hydrogel synthesis. Polymer chains that do not possess these functional groups can also be chemically modified and crosslinked. Hydrogel synthesis using functional groups comprises some subgroups such as Schiff-base reaction, epoxide coupling, addition reaction, and condensation reaction. In hydrogel synthesis by Schiff-base reaction, two polymers containing amine

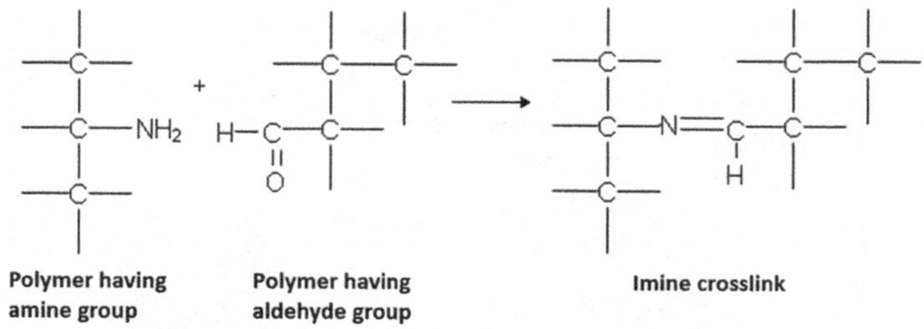

Fig. 4.4 Simple demonstration of the formation of an imine bond by a Schiff base reaction

and aldehyde functional groups or a polymer containing an amino group and a crosslinker with a suitable function are needed for an interaction that leads to the formation of imine bonds (Fig. 4.4). The imine crosslinks formed as a result of this interaction are pH dependent and are called pseudo-covalent bonds. The Schiff-base reaction takes advantage of the dynamic equilibrium in the breakdown and re-formation of imine bonds in hydrogel networks. This gives the hydrogel the ability to heal itself [3] (Fig. 4.5).

4.2.2.2 Chemical Crosslinkers

Covalent crosslinking of the polymeric chains may be obtained by using cross-linkers. Glutaraldehyde, genipin, and vanillin can be considered as the most common crosslinkers used to form the network structures of hydrogels for medical applications.

Glutaraldehyde Although glutaraldehyde is a frequently preferred crosslinking agent due to its high water solubility and low cost, it can be toxic if used above a certain concentration. Therefore, when used as a crosslinker, its low concentration should be preferred [25]. Crosslinked networks of proteins such as gelatin and albumin, and polysaccharide-based hydrogels containing amine groups in their structure, such as chitosan, can be synthesized by Schiff base reaction using dialdehyde or formaldehyde as crosslinking agents [1, 26].

Genipin Genipin, which is water soluble and has low cytotoxicity, is obtained from geniposide. It is used with polymers such as chitosan, hyaluronic acid, collagen, and its derivatives.

Vanillin Vanillin is the naturally occurring aromatic chemical compound. Since vanillin can participate in hydrogen bonding with the hydroxyl groups in its structure, it is preferred in both chemical and physical crosslinking in hydrogel synthesis [27]. Vanilla, which is the extract of vanillin, generally used as a flavoring agent in cosmetics and food, can also be used as a crosslinking agent owing to the hydroxyl and aldehyde groups in its structure.

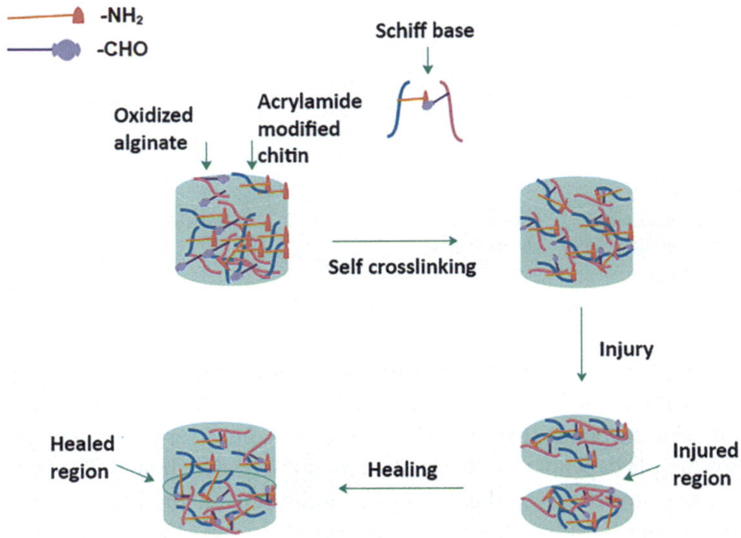

Fig. 4.5 Schematic representation of the Schiff base formation and the self-healing process. Briefly, the hydrogel is synthesized via Schiff base linkage formed between the amino group of acrylamide-modified chitin and the dialdehyde group of oxidized alginate

Epoxides In the epoxide-based crosslinking approach, diglycidyl ether (PEGDE), epichlorohydrin, butanediol diglycidyl ether (BDDE), and ethyleneglycol diglycidyl ether (EGDE) are commonly used compounds for hydrogel synthesis. Epoxides are molecules that are soluble in water and have a high affinity for amine and alcohol groups. The environmental conditions, whether it is acidic or basic, affect the epoxides and direct them to the nucleophile group it will react with [28].

The Michael-Type Addition This method is also called conjugate addition or 1,4-nucleophilic addition. In this method, gelation mainly occurs by adding a nucleophile to carbonyl-containing compounds such as acrylate esters, acrylonitrile, acrylamides, maleimides, alkyl methacrylates, cyanoacrylates, and vinyl sulfones. Nucleophiles, owing to the amine or thiol groups they contain, basically interact with the electrophilic groups of polymers such as acrylate, methacrylate, or vinyl sulfone. This technique is highly selective, has mild reaction conditions, does not form toxic intermediates during the reaction, and has a short duration. In hydrogel synthesis, Michael-type addition mostly involves conjugating a thiol-containing polymer to an activated α,β-unsaturated carbonyl polymer under basic conditions. In contrast to the mentioned positive aspects, this method has disadvantages such as the multi-stage synthesis and the need for purification.

The Condensation Method This is mostly used for the hydrogel synthesis of polyamides and polyesters. The reaction occurs with the formation of ester or amide bonds between carboxyl groups (or derivatives) and hydroxyl or amino groups.

N,N-(3-dimethylaminopropyl)-*N*-ethyl carbodiimide (EDC) is generally used as the crosslinking agent. *N*-hydroxysuccinimide (NHS) is added during the reaction to prevent possible side reactions and to ensure effective crosslinking. Polysaccharide-based hydrogels can be synthesized by condensation reactions [3, 26].

4.2.2.3 Free Radical Polymerization

With this method, crosslinked hydrogels are obtained by polymerization of vinyl monomers such as acrylic acid, acrylamide, vinyl chloride, styrene, epoxide, *N*-vinyl-2-pyrrolidone, and 2-hydroxyethyl methacrylate in the presence of a crosslinking agent. With this frequently preferred method, hydrogel can be synthesized in a short time, with high efficiency and under mild conditions. Initiators such as ammonium persulfate (APS), potassium persulfate (KPS), 2,2-azobisisobutyronitrile (AIBN), ceric ammonium nitrate, and benzoyl peroxide are used to start the reaction. Ethylene glycol dimethacrylate (EGDMA), melamine triacrylamide, *N,N'*-methylene bisacrylamide (MBA), and melamine trimethylacrylamide are the main crosslinking agents used in the free radical mechanism [29]. In the early stages of radical polymerization, the initiator, which forms radicals by being degraded by thermal, chemical, or photochemical triggers, interacts with monomer types to form new radicals. The monomers involved in this polymerization contain carbon double bonds like vinyl, acryl, methacryl, or allyl groups. Using this approach to synthesize natural polymer-based hydrogels, the polymers need to have groups with double carbon bonds. Free radical polymerization is also used in the synthesis of gelatin methacrylate (GelMA)-based hydrogels, which are frequently used in tissue engineering and regenerative medicine applications. After adding methacrylate groups to gelatin, which is a hydrophilic protein obtained from denatured collagen, it can be chemically crosslinked by applying UV (at 365 nm wavelength) or visible light photocuring, in the presence of photoinitiators such as Irgacure or lithium phenyl-2,4,6-trimethyl-benzoylphosphinate (LAP), respectively [3] (Fig. 4.6).

Fig. 4.6 GelMA hydrogel obtained by applying light at a wavelength of 405 nm to the pre-hydrogel in the presence of a photoinitiator

4.2.2.4 Enzyme Mediated Crosslinking

There is great interest in enzyme-catalyzed crosslinking methods due to the substrate specificity of the enzymes, absence of side reactions, and non-harsh reaction conditions (e.g. neutral pH and room temperature). This method is preferred in producing natural polymer-based hydrogels to preserve their bioactivity. Transglutaminase, tyrosinase, lysyl oxidase, and horseradish peroxidase are among the enzymes frequently used for hydrogel synthesis. Transglutaminase enzyme is the most commonly used enzyme as a transfer enzyme that catalyzes the transamidation reaction. The oxidation enzyme tyrosinase catalyzes the oxidation of activated quinine from phenol. Tyrosinase enzyme, like transglutaminase, can catalyze the crosslinking of chitosan and gelatin or chitosan and silk protein. Lysyl oxidase, which is mostly used in the preparation of peptide-based hydrogels, can crosslink polymers that possess amine groups through Schiff-base formation. Peroxidase enzymes require hydrogen peroxide as their substrate for crosslinking, which is catalyzed through the conjugation of phenol and aniline groups. Although this method has some positive aspects compared to other crosslinking methods, the instability of the specific enzymes and poor mechanical strength of the hydrogels can be considered as its drawbacks [3].

4.2.2.5 Click Reaction

Click reactions are chemical reactions that have advantages such as high yield, highly pure product, and high specificity and selectivity under a wide range of mild reaction conditions. With this method, molecules containing more than one functional group can be used for the production of complex polymeric materials. Click reactions such as cycloaddition of azide and alkyne, Diels-Alder reaction between furan and maleimide (Fig. 4.7), and thiolene reactions are used in hydrogel preparation.

Since residues of metal ions used in some Click reactions, such as copper-catalyzed azide-alkyne cycloaddition (CuAAC), may cause toxicity, Click reactions

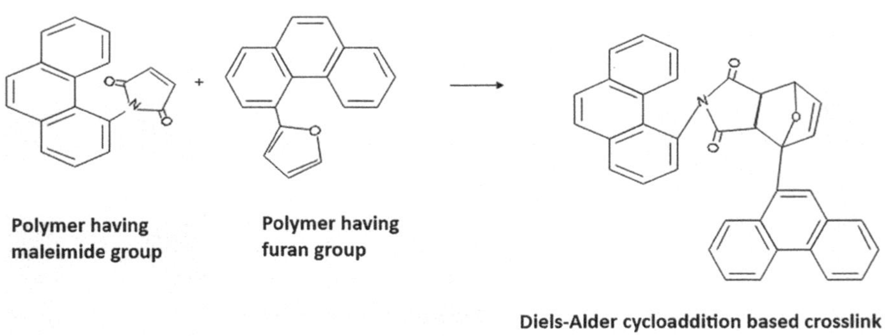

Fig. 4.7 Schematic representation of the Diels-Alder reaction

that do not contain metal ions, such as azide alkyne cycloaddition reaction, oxa norbornadiene cycloaddition reaction, Diels-Alder reaction, thiolene reaction, and oxime reaction, have been developed [3, 26].

4.2.3 Crosslinking via Radiation

Radiation crosslinking is based on the generation of free radicals within a polymer exposed to high-energy sources such as gamma rays, X-rays, or electron beams, and their subsequent reaction to form a gel. Since this technique eliminates the need for initiators, the resulting product consists solely of the polymer. The absence of chemical additives makes this method particularly advantageous for preserving the biocompatibility of biopolymers. Moreover, since the modification and sterilization are carried out in a single step, it is a cost-effective process for the modification of biopolymers, especially those intended for biomedical applications. In addition, as the degree of swelling depends on the degree of crosslinking, it can be controlled by adjusting the exposure to radiation dose. The effect of radiation—whether direct (radical formation on the polymer itself) or indirect (radical formation on intermediate molecules in the surrounding environment)—is influenced by the polymer's physical state and the medium, such as dilute solution, concentrated solution, or solid state.

4.2.3.1 Aqueous Radiation

In aqueous media, radiation is mainly absorbed by water. Radiolysis of water molecules results in the formation of free radicals (Eq. 4.1), which interact with the dissolved polymer. The irradiation of polymers in dilute media induces chemical changes primarily due to the indirect effects of radiation.

$$H_2O \xrightarrow{\text{☢}} {}^{\bullet}OH,\ H^{\bullet},\ H_3O^+,\ H_2O_2,\ H_2,\ e^-_{aq} \qquad 4.1$$

Free radicals generated through the radiolysis of water interact with the polymer to produce macroradicals, whose recombination across different chains leads to the formation of covalent bonds, ultimately resulting in a crosslinked structure. Polysaccharides such as starch, carboxymethylcellulose, carrageenan, chitin, and alginate are frequently used in hydrogel fabrication. In the meantime, they undergo degradation upon irradiation in dilute aqueous solutions regardless of their structure and conformation. This degradation occurs because, at polymer concentrations below the critical overlap concentration, the chain density of the polymer is insufficient to enable chain recombination and the formation of a crosslinked network. In saturated aqueous systems, the two primary radicals—hydroxyl radicals (${}^{\cdot}OH$) and hydrated electrons (e^-_{aq})—react with carbohydrates by abstracting hydrogen atoms bonded to carbon. Among these radicals, the hydroxyl radical is responsible for hydrogen abstraction and is non-specific in its activity, creating radical sites at various positions within a carbohydrate molecule. Hydrated electrons formed during the radiolysis of water react with hydrocolloids only in oxygen-free systems. In oxygenated solutions, hydrated electrons react with oxygen to produce superoxide radicals ($O^{\bullet-2}$).

It has been demonstrated that permanent, covalently crosslinked macroscopic hydrogels can form in dilute, oxygen-depleted solutions of carboxymethylcellulose (CMC) at low pH through $^{\cdot}OH$ radical-induced production and macroradical recombination [30]. Unlike neutral solutions, which require high CMC concentrations for gel formation, gelation was observed at concentrations as low as 0.5% CMC and at doses below 2 kGy under acidic conditions. The swelling equilibrium of the resulting hydrogels decreases with increasing radiation dose. Under the conditions used in this study, the swelling equilibrium value after a typical sterilization dose of 25 kGy ranged between 50 and 200 g water/g dry gel (weight of water absorbed by the hydrogel/dry hydrogel weight). In another study, 1% and 2% κ-carrageenan solutions were exposed to 400–500 kGy gamma radiation, leading to the formation of microgels with particle sizes <5 μm [31]. This crosslinking was attributed to radiation-induced transformations in the alkyl group structure on the side chains of carrageenan. Additionally, it was suggested that the decrease in pH resulted in the release of sulfate groups during irradiation and may further promote structural changes and enhance the crosslinking of κ-carrageenan.

4.2.3.2 Radiation in Paste-Like Systems

Hydrocolloids can undergo crosslinking under aqueous paste-like conditions, where the polymer concentration is sufficiently high to facilitate both the direct effects of radiation in generating free radicals and the radiolysis of available water to produce hydroxyl ($^{\cdot}OH$) radicals and other associated radicals. These conditions result in a high concentration of radicals, closely associated with the radicals of the original polymer and the secondary polymer. Consequently, crosslinking occurs via radical-radical and polymer-polymer radical interactions, leading to the formation of a network structure. If the concentration of the original polymer is insufficient to promote radical-radical reactions, degradation will occur rather than crosslinking. Irradiating polysaccharide derivatives homogeneously dispersed in water at high concentrations, or under paste-like conditions facilitates crosslinking [32]. Here the presence of water contributes to the crosslinking in two ways: First, reaction of $^{\cdot}OH$ radicals (products of water radiolysis) with macromolecules generates high concentration of macroradicals. Second, water increases the mobility of rigid polymer chains, allowing macroradicals to diffuse into proximity, facilitating recombination and network formation. The density of crosslinks between polymer chains can be determined by measuring the swelling degree of the polymer, by comparing the weight of the swollen material in a compatible solvent to its dry weight. Radiation treatment increases the density of crosslinks, reducing the hydrogel's ability to absorb the solvent.

Conversely, when degradation occurs, the hydrogel's swelling capacity increases (Fig. 4.8). An increase in soluble fraction indicates degradation, while an increase in insoluble fraction signifies higher crosslinking density [33].

Factors such as the structure, degree of substitution, distribution of the substituent groups, and the initial molecular weight change the concentration that modification can be achieved. For instance, a higher degree of substitution (the number of carboxymethyl groups per cellulose unit) facilitates the crosslinking of CMC, as intermolecular linkages result from the activity of ether groups [34]. Similar results have been observed for the aqueous irradiation of methylcellulose [33, 35], carboxymethyl starch [36], gum arabic [37], and carboxymethylated chitin and chitosan [38]. In paste-like environments, hydrogels based on carboxymethyl-hyaluronic acid were synthesized through γ-ray-induced crosslinking without addition of initiators or crosslinkers [32]. In such a study, the gel fraction increased with the absorbed dose, and swelling ratios in water ranged from 46 to 168 g water/g dry gel.

In another study, hydrogels composed of hydroxypropyl cellulose (HPC) were produced through radiation, and their potential use in single-use soft contact lenses was investigated [39]. By electron beam irradiation of high concentration solutions containing HPC, polyethylene glycol #1000 dimethacrylate (PEGDMA), and 2-hydroxyethyl methacrylate (HEMA), HPC-based hydrogels with good mechanical properties and transparency were obtained. The addition of PEGDMA contributed to formation of an increased gel fraction, which resulted in improved tensile strength and elongation at break.

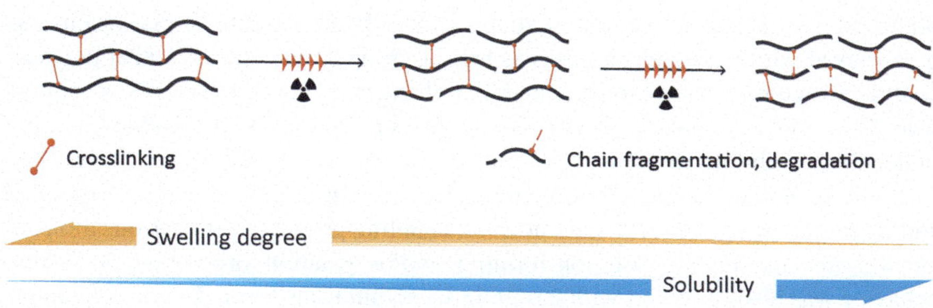

Fig. 4.8 Schematic representation of crosslinking and degradation between polymer chains induced by radiation. In cases where the crosslinking density between polymer chains is high and degradation is minimal, the hydrogel exhibits a lower swelling degree. As degradation increases, the solubility of the hydrogel in a compatible solvent also increases

4.2.3.3 Solid-State Radiation

When hydrocolloids are irradiated in their solid state, radiation directly generates radicals within the molecular chains. The primary reaction occurring during the radiolysis of solid-state hydrocolloids is the breaking of glycosidic bonds, resulting in a decrease in the molecular weight of the macromolecules [40]. The degradation rates are generally dependent on the concentrations of reactants and temperature, as is typical for other chemical reactions. Additionally, these rates are influenced by the purity of the hydrocolloid, the type of substituent groups, and the molecular weight of the polymer. In the degradation process of carbohydrates in the solid state, the main effects include fragmentation, hydrolysis (due to the presence of moisture), and rearrangement, all of which contribute to the formation of low molecular weight products.

The use of radiation processing to induce structural modifications through crosslinking in synthetic polymers has become a well-established industry. Conversely, as previously noted, exposing polysaccharides and other natural polymers to ionizing radiation, whether in solid form or aqueous solution, typically results in degradation instead of crosslinking.

4.2.3.3.1 Natural Polymers

One method for the production of hydrogels from natural polymers can be applied by modifying them using high-energy radiation in the solid state [41]. This technique employs ionizing radiation in the presence of an intermediary alkyne gas, allowing for the controlled modification of polysaccharides and related materials. The method has been applied to a variety of polysaccharides with different types and structures, as well as to proteins such as collagen, gelatin, casein, and combinations of polysaccharides with plant-derived proteins. Irradiation of these polymers in the presence of acetylene gas led to crosslinking, resulting in an increase in the molecular weight of macromolecules and enhanced functionality.

Up to a fourfold increase in molecular weight can be achieved with highly branched polysaccharide structures at doses up to 10 kGy, while similar results can be achieved with hydrogels at doses as high as 50 kGy. In contrast, linear polysaccharide structures can exhibit comparable changes at much lower doses ranging from 1 to 3 kGy. Proteins, however, require doses of up to 25 kGy to achieve similar molecular weight increases.

It has been demonstrated that the molecular weight of gelatin can be controlled and increased using the solid-state process, enabling products with various molecular weights and having solution-forming and/or gelation properties. A similar behavior was also observed with casein in its sodium salt form. The modifications achieved to date can be applied to a broad range of commercially available polysaccharides. These include xanthan, pectin, carrageenan, welan, gellan, alginate, starch, heparin, chitin, chitosan, locust bean gum, and guar gum [42]. A study on the solid-state modification of carrageenan demonstrated that hydrogel formation and an increase in the viscoelasticity of κ-carrageenan were achieved without the use of a gelling agent [43]. The optimal dose range for achieving this modification was determined to be 5–10 kGy, as higher doses led to degradation and a reduction in the gel fraction. The irradiation of carrageenan resulted in approximately 78% hydrogel production, with a fourfold improvement in viscosity compared to the control material.

4.2.3.3.2 Synthetic and Natural Polymer Blends

The radiation technique applied to natural polymers has also been used for various blend systems of water-soluble synthetic and natural polymers, yielding synergistic effects on the functionality of these systems. One example is the irradiation of a solid-state mixture of polyvinylpyrrolidone (PVP) and gum arabic (GA) in a 1:1 ratio. Rheological measurements conducted on a 10% aqueous solution of this system demonstrated a significant improvement in the viscoelasticity of the polymer blend compared to either of the individual components. Another example of crosslinking using an electron beam is the CMC/polyacrylamide hydrogel [44]. It has been shown that both the absorbed dose and the polyacrylamide concentration result in increased gel fraction. Similarly, a crosslinked CMC/polyvinylpyrrolidone (PVP) hydrogel was prepared using ^{60}Co γ-radiation, and the gel fraction increased with higher PVP content and the absorbed dose [45]. In another study, a pH-sensitive sodium alginate-g-poly(acrylic acid) hydrogel was synthesized using γ-radiation. This hydrogel reached swelling equilibrium in water within 6 h, with the swelling ratio sharply increasing as the pH was raised to 6.0, and was followed by a decrease when the pH was further increased to 8.0 [46].

4.3 Conclusion

This section has described the properties of hydrogels, typical synthesis methods, strategies, and specific mechanisms, including physical, chemical, and radiation-induced crosslinking, as well as the conditions for hydrogel formation. Chemical crosslinking is a versatile method for enhancing the mechanical properties of hydrogels. However, crosslinking agents are often toxic and environmentally unfriendly compounds that may interact undesirably with bioactive substances within the hydrogel matrix. The adverse effects of chemical crosslinking can be avoided through radiation or electron beam methods. Radiation offers several advantages, such as precise control over the crosslinking level through the radiation dose and providing a clean process that leaves no unwanted residues in the hydrogel. There is growing interest in physically crosslinked hydrogels, particularly for applications involving cell encapsulation, due to their ease of production and the absence of crosslinking chemicals. Combining chemical and physical crosslinking methods allows for the preparation of hydrogels with diverse properties tailored for specific applications, enhancing their functionality for biomedical uses. The continuous development of hydrogels with desirable mechanical properties and appropriate biocompatibility will be beneficial for tissue engineering and regenerative medicine.

References

1. Rimmer, S.: Biomedical Hydrogels: Biochemistry, Manufacture and Medical Applications. Woodhead Publishing (2011)
2. Bostancı, N.S., Buyuksungur, S., Hasirci, N., Tezcaner, A.: Bioprinted scaffolds assembled as synthetic skin grafts by natural hydrogels containing fibroblasts and bioactive agents. Int. J. Polym. Mater. Polym. Biomater. **73**(11), 927–945 (2024)
3. Hu, W., Wang, Z., Xiao, Y., Zhang, S., Wang, J.: Advances in crosslinking strategies of biomedical hydrogels. Biomater. Sci. **7**(3), 843–855 (2019)
4. Gulrez, S.K.H., Al-Assaf, S., Phillips, G.O.: Hydrogels: methods of preparation, characterisation and applications. In: Carpi, A. (ed.) Progress in Molecular and Environmental Bioengineering—From Analysis and Modeling to Technology Applications. IntechOpen (2011)
5. Funami, T., Hiroe, M., Noda, S., Asai, I., Ikeda, S., Nishimari, K.: Influence of molecular structure imaged with atomic force microscopy on the rheological behavior of carrageenan aqueous systems in the presence or absence of cations. Food Hydrocoll. **21**, 617–629 (2007)
6. Hoffman, A.S.: Hydrogels for biomedical applications. Adv. Drug Deliv. Rev. **43**, 312 (2002)
7. Hennink, W.E., Nostrum, C.F.: Novel crosslinking methods to design hydrogels. Adv. Drug Deliv. Rev. **54**, 13–36 (2002)
8. Deng, P., Yao, L., Chen, J., Tang, Z., Zhou, J.: Chitosan-based hydrogels with injectable, self-healing and antibacterial properties for wound healing. Carbohydr. Polym. **276**, 118718 (2022)
9. Küçükoğlu, V., Uzuner, H., Kenar, H., Karadenizli, A.: In vitro antibacterial activity of ciprofloxacin loaded chitosan microparticles and their effects on human lung epithelial cells. Int. J. Pharm. **569**, 118578 (2019)
10. Bajpai, A.K., Shukla, S.K., Bhanu, S., Kankane, S.: Responsive polymers in controlled drug delivery. Prog. Polym. Sci. **33**, 1088–1118 (2008)
11. Zhao, Q.S., Ji, Q.X., Xing, K., Li, X.Y., Liu, C.S., Chen, X.G.: Preparation and characteristics of novel porous hydrogel films based on chitosan and glycerophosphate. Carbohydr. Polym. **76**, 410–416 (2009)
12. Rezvanian, M., Ahmad, N., Mohd Amin, M.C., Ng, S.F.: Optimization, characterization, and in vitro assessment of alginate-pectin ionic cross-linked hydrogel film for wound dressing applications. Int. J. Biol. Macromol. **97**, 131–140 (2017)
13. Esteban, C., Severian, D.: Polyionic hydrogels based on xanthan and chitosan for stabilising and controlled release of vitamins, Vol. WO0004086 (A1) (ed. U. United States Patent), Kemestrie Inc [CA], USA (2000)
14. Lalevee, G., David, L., Montembault, A., Blanchard, K., Meadows, J., Malaise, S., Crépet, A., Grillo, I., Morfin, I., Delair, T., Sudre, G.: Highly stretchable hydrogels from complex coacervation of natural polyelectrolytes. Soft Matter. **13**(37), 6594–6605 (2017)
15. Dompe, M., Cedano-Serrano, F.J., Heckert, O., van den Heuvel, N., van der Gucht, J., Tran, Y., Hourdet, D., Creton, C., Kamperman, M.: Thermoresponsive complex coacervate-based underwater adhesive. Adv. Mater. **31**(21), e1808179 (2019)
16. Takigami, M., Amada, H., Nagasawa, N., Yagi, T., Kasahara, T., Takigami, S., Tamada, M.: Preparation and properties of CMC gel. Trans. Mater. Res. Soc. Jpn. **32**(3), 713–716 (2007)
17. Chen, Y.N., Peng, L., Liu, T., Wang, Y., Shi, S., Wang, H.: Poly(vinyl alcohol)-tannic acid hydrogels with excellent mechanical properties and shape memory behaviors. ACS Appl. Mater. Interfaces. **8**(40), 27199–27206 (2016)

18. Ugrinovic, V., Markovic, M., Bozic, B., Panic, V., Veljovic, D.: Physically crosslinked poly(methacrylic acid)/gelatin hydrogels with excellent fatigue resistance and shape memory properties. Gels. **10**(7), 444 (2024)
19. Williams, P.A., Phillips, G.O.: Physicochemical characterisation of gum Arabic arabinogalactan protein complex. Food Food Ingred. J. Jpn. **211**, 181–188 (2006)
20. Aoki, H., Al-Assaf, S., Katayama, T., Phillips, G.O.: Characterization and properties of Acacia senegal (L.) Willd. var. senegal with enhanced properties (Acacia (sen) SUPER GUM™): part 2—mechanism of the maturation process. Food Hydrocoll. **21**, 329–337 (2007)
21. Navarra, G., Peres, C., Contardi, M., Picone, P., San Biagio, P.L., Di Carlo, M., Giacomazza, D., Militello, V.: Heat- and pH-induced BSA conformational changes, hydrogel formation and application as 3D cell scaffold. Arch. Biochem. Biophys. **606**, 134–142 (2016)
22. Giannouli, P., Morris, E.R.: Cryogelation of xanthan. Food Hydrocoll. **17**, 495–501 (2003)
23. Araujo, D., Rodrigues, T., Alves, V.D., Freitas, F.: Chitin-glucan complex hydrogels: optimization of gel formation and demonstration of drug loading and release ability. Polymers (Basel). **14**(4), 785 (2022)
24. Bektaş, E.I., Gurel Pekozer, G., Kök, F.N., Torun Kose, G.: Evaluation of natural gum-based cryogels for soft tissue engineering. Carbohydr. Polym. **271**, 118407 (2021)
25. Ulubayram, K., Aksu, E., Gurhan, S.I., Serbetci, K., Hasirci, N.: Cytotoxicity evaluation of gelatin sponges prepared with different cross-linking agents. J. Biomater. Sci. Polym. Ed. **13**(11), 1203–1219 (2002)
26. Nezhad-Mokhtari, P., Ghorbani, M., Roshangar, L., Soleimani Rad, J.: A review on the construction of hydrogel scaffolds by various chemically techniques for tissue engineering. Eur. Polym. J. **117**, 64–76 (2019)
27. Hunger, M., Domalik-Pyzik, P., Reczyńska, K., Chłopek, J.: Double crosslinking of chitosan/vanillin hydrogels as a basis for mechanically strong gradient scaffolds for tissue engineering. Eng. Biomater. **23**(155) (2020)
28. Ghorpade, V.S.: Preparation of hydrogels based on natural polymers via chemical reaction and cross-linking. In: Hydrogels Based on Natural Polymers. Elsevier (2020)
29. Hasirci, V., Hasirci, N.: Polymers as biomaterials. In: Fundamentals of Biomaterials, 2nd edn, pp. 83–100. Springer (2024)
30. Wach, R.A., Rokita, B., Bartoszek, N., Katsumura, Y., Ulanski, P., Rosiak, J.M.: Hydroxyl radical-induced crosslinking and radiation-initiated hydrogel formation in dilute aqueous solutions of carboxymethylcellulose. Carbohydr. Polym. **112**, 412–415 (2014)
31. Relleve, L.S., Lopez, G.E.P., Dela Cruz, R.M.M., Abad, L.V.: High radiation dose studies of kappa-carrageenan in dilute aqueous solution. Radiat. Phys. Chem. **197**, 110165 (2022)
32. Relleve, L.S., Gallardo, A.K.R., Abad, L.V.: Radiation crosslinking of carboxymethyl hyaluronic acid. Radiat. Phys. Chem. **151**, 211–216 (2018)
33. Wach, R.A., Mitomo, H., Nagasawa, N., Yoshii, F.: Radiation crosslinking of methylcellulose and hydroxyethylcellulose in concentrated aqueous solutions. Nucl. Inst. Methods Phys. Res. B. **211**, 533–544 (2003)
34. Shen, X., Kitajyo, Y., Duan, Q., Narumi, A., Kaga, H., Kaneko, N., Satoh, T., Kakuchi, T.: Synthesis and photocrosslinking reaction of N-allylcarbamoylmethyl cellulose leading to hydrogel. Polym. Bull. **56**, 137–143 (2006)
35. Horikawa, Y., Kawachi, S., Honna, T.: Sedimentation behavior of dispersed particles of clay and silt in acidic and alkaline suspensions of inorganic materials from various volcanic ash soils. Soil Sci. Plant Nutr. **50**, 19–25 (2004)
36. Yoshii, F., Zhao, L., Wach, R.A., Nagasawa, N., Mitomo, H., Kume, T.: Hydrogels of polysaccharide derivatives crosslinked with irradiation at paste-like condition. Nucl. Inst. Methods Phys. Res. B. **208**, 320–324 (2003)
37. Katayama, T., Nakauma, M., Todoriki, S., Phillips, G.O., Tada, M.: Radiation-induced polymerization of gum arabic (Acacia sengal) in aqueous solution. Food Hydrocoll. **20**, 983–989 (2006)

38. Wasikiewicz, J.M., Mitomo, H., Nagasawa, N., Yagi, T., Tamada, M., Yoshii, F.: Radiation crosslinking of biodegradable carboxymethylchitin and carboxymethylchitosan. J. Appl. Polym. Sci. **102**, 758–767 (2006)
39. Hiroki, A., Taguchi, M.: Development of environmentally friendly cellulose derivative-based hydrogels for contact lenses using a radiation crosslinking technique. Appl. Sci. **11**(19), 9168 (2021)
40. Şen, M., Hayrabolulu, H., Taşkın, P., Torun, M., Demeter, M., Cutrubinis, M.: Radiation induced degradation of xanthan gum in the solid state. Radiat. Phys. Chem. **124**, 225–229 (2016)
41. Al-Assaf, S., Phillips, G.O., Williams, P.A., Plessis, T.A.: Application of ionizing radiations to produce new polysaccharides and proteins with enhanced functionality. Nucl. Inst. Methods Phys. Res. B. **265**, 37–43 (2007)
42. Phillips, G. O., Plessis, T. A. D., Al-Assaf, S., Williams, P. A.: Biopolymers obtained by solid state irradiation in an unsaturated gaseous atmosphere. Vol. 6,841644, (ed. U. S. Patent), Phillips Hydrocolloid Research Limited, UK, USA (2005)
43. Gulrez, S., Al-Assaf, S., Phillips, G.O.: Characterisation and radiation modification of carrageenan in solid state. In: Khandal, R.K. (ed.) Radaiation Processing Technology Applications, vol. 2, pp. 519–531 (2010)
44. Abd El-Rehim, H.A., El-Sayed, H.A., Diaa, D.A.: Characterization of super-absorbent material based on carboxymethylcellulose sodium salt prepared by electron beam irradiation. J. Macromol. Sci. A. **43**(1), 101–113 (2006)
45. Raafat, A.I., Eid, M., El-Arnaouty, M.B.: Radiation synthesis of superabsorbent CMC based hydrogels for agriculture applications. Nucl. Instrum. Meth. B. **283**, 71–76 (2012)
46. Mohamed, S.F., Mahmoud, G.A., Abou Taleb, M.F.: Synthesis and characterization of poly(acrylic acid)-gsodium alginate hydrogel initiated by gamma irradiation for controlled release of chlortetracycline HCl. Monatsh Chem. **144**(2), 129–137 (2013)

Chapter 5
Natural Hydrogels and Bioinks

Arda Buyuksungur

Abstract Polymers are natural or synthetic materials with a wide range of properties, consisting of repeating structural units that form macromolecular structures. Natural polymers are materials of natural origin that are found in or produced by microorganisms, plants, and animals. In living organisms, natural polymers serve multiple functions. Natural polymers provide structural support for the organism and include macromolecules such as carbohydrates, proteins, and nucleic acids, which are essential for life. The types of natural polymers include proteins, polysaccharides, polynucleotides, polyisoprenes, polyesters, and lignin. Natural polymers are easier and more cost-effective to obtain than synthetic and semi-synthetic polymers. Furthermore, due to their origins, these molecules can be biocompatible, biodegradable, and are readily absorbed by the body without any adverse effects.

Keywords Collagen · Gelatin · Fibrin · Silk fibroin · Elastin · Hyaluronic acid · Alginate · Chitosan · Dextran · Cellulose

5.1 The Importance and Use of Natural Polymers

Natural polymers are materials of natural origin that are either found in the plants, microorganisms, and animals or produced by them. Carbohydrates and proteins, which provide structural support to plants and animals, fall into this category. Compared to synthetic and semi-synthetic polymers, natural polymers are cost-effective, easily accessible, potentially biodegradable, and demonstrate excellent compatibility with biological environments due to their origin, making them biocompatible [1, 2]. Many of these polymers are integral components of daily

A. Buyuksungur (✉)
Department of Basic Sciences, Faculty of Dentistry, Ankara University, Ankara, Türkiye
e-mail: abuyuksungur@ankara.edu.tr

N. Hasirci, V. Hasirci (eds.), *Hydrogels and Bioinks in Tissue Engineering*, Biological and Medical Physics, Biomedical Engineering,
https://doi.org/10.1007/978-3-031-90966-5_5

nutrition and hold significant applications in the pharmaceutical, food, and cosmetic industries.

Polymers composed of crosslinked hydrophilic chains can absorb and retain large amounts of water, creating a gel-like consistency. Hydrogels are such hydrophilic network structures having high water content, flexibility, and strength similar to natural soft tissues. Natural hydrogels, derived from biological polymers such as collagen, gelatin, hyaluronic acid, and many others, serve as biomaterials in medical applications. These hydrogels, obtained from various organisms, can be processed in different ways to have diverse mechanical and physical properties as well as controllable biodegradation and are similar to the extracellular matrix. Due to their biocompatibility and similarity of their mechanical and physical characteristics to those of tissues, they are widely used in the design and manufacture of medical devices such as wound dressings, drug delivery systems, and scaffolds for tissue engineering applications. Some of these natural polymers can also be stimuli responsive and can change their properties (e.g., shape, color, conductivity, etc.) depending on environmental conditions such as pH, temperature, presence of ionic materials, charge or magnetic fields, etc. [3, 4]. In wound care, hydrogels are employed as wound dressings to maintain the moist environment and provide a favorable medium for cell growth, thereby accelerating healing. In drug delivery systems, hydrogels prepared as membranes and micro or nanoparticles enable the targeted delivery of drugs and facilitate their controlled release over a specific period. In tissue engineering, hydrogels are utilized as scaffolds to support cell adhesion, growth, and tissue regeneration [5]. Furthermore, these natural hydrogels are commonly used as bioink materials for 3D printing processes involving cells [6]. Such applications make it possible to print various cell types in specific organizations and arrangements, enabling the production of organoids in laboratory settings.

5.2 Sources of Natural Hydrogels

Many chemicals capable of forming hydrogels can be easily extracted from their natural sources. Collagen, fibrin, and hyaluronic acid are components of the natural extracellular matrix. Hydrogels made from these polymers are known to encourage tissue regeneration and have biofunctional properties including regulation of matrix formation and cell activity. Some natural hydrogels, such as alginate hydrogels derived from algae or nanofibrillar hydrogels prepared from cellulose, consist of plant-based fibrillar glucan chains. These hydrogels are biologically compatible, exhibit low cytotoxicity, and can be enzymatically degraded within the human body. Non-animal derived natural hydrogels are easily sourced from plants or produced by microorganisms and carry no risk of viral contamination. This makes them particularly advantageous for applications where safety and reproducibility are critical [7].

5.3 Classification of Natural Hydrogels

There are two main classes of polymers that constitute natural hydrogels: (1) proteins (e.g. collagen, gelatin, and lysozyme); and (2) polysaccharides (e.g., hyaluronic acid (HA), alginate, and chitosan) [8, 9]. These hydrogels are biocompatible, biodegradable, and non-toxic. However, they have poor mechanical properties and are not easy to reproducibly produce with similar properties [10]. Another type of natural hydrogels is the decellularized native tissues.

5.3.1 Protein-Based Natural Hydrogels

Protein-based polymers are among the least utilized raw materials compared to those produced for advanced industrial applications. However, these serve as efficient building blocks for the creation of different biomaterials such as composites and films. Notable examples of these proteins include collagen and gelatin obtained from animal skin, silk fibroin derived from silkworm or spider webs, keratin extracted from wool/hair, bones and tendons, elastin from elastic tissues and insect tendons, and fibrin from blood clots (Table 5.1).

In order to synthesize protein-based hydrogels with three dimensional interwoven networks, the free radical polymerization method is commonly employed. This process involves grafting vinyl monomers onto the protein molecules in the presence of crosslinkers and initiators, followed by crosslinking to produce hydrogels with special properties. For instance, super absorbent polymers such as the hydrogels based on collagen and cottonseed proteins are synthesized using graft polymerization techniques [11].

5.3.1.1 Collagen

Collagen is a fibrous protein and a primary component of the extracellular matrix (ECM). One of its main functions is to provide mechanical support to resist opposing forces, thus preventing repeatable plastic deformation. The primary structure of collagen is represented as (Gly-X-Y)n, where n shows number of repeating units, and glycine constitutes approximately 30% of the total amino acid content. In this tripeptide structure, X is typically proline and hydroxyproline, and Y is other amino acids. Its secondary structure consists of repeating triple helices, the tertiary structure is a triple helix comprising about 1000 amino acids, and its quaternary structure forms more than one collagen fibers through self-assembly [12].

Collagen exhibits excellent hydrophilicity, in vivo stability, and cell adhesion-promoting properties due to the specific peptide sequences in its structure. Its low immunogenicity makes it widely used in biomedical applications. The most common type of collagen is type I, derived from tissues such as skin, bone, and

Table 5.1 Examples of molecules forming protein-based natural hydrogels

Collagen	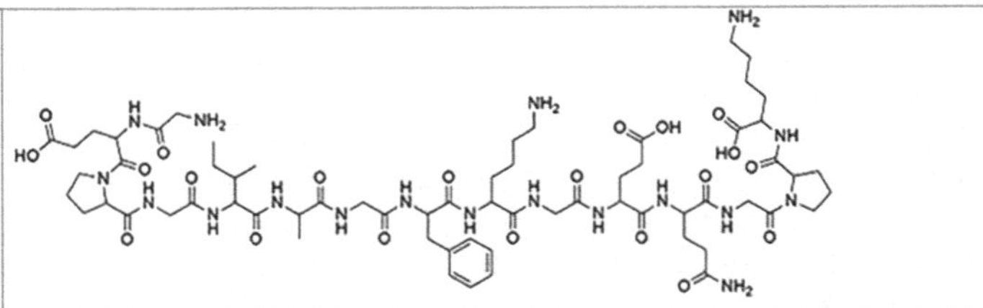
Gelatin	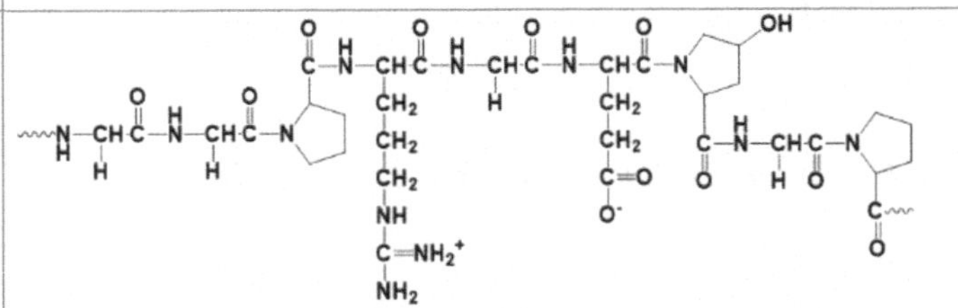
Fibroin	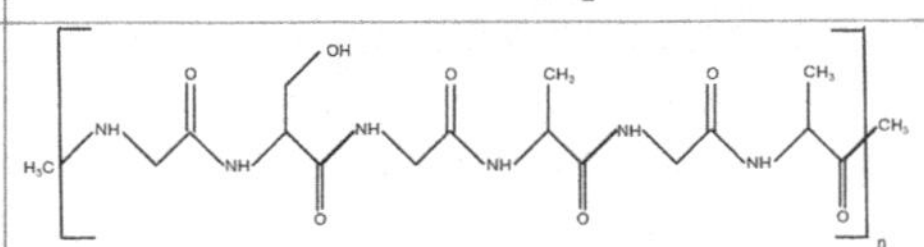
Fibrin	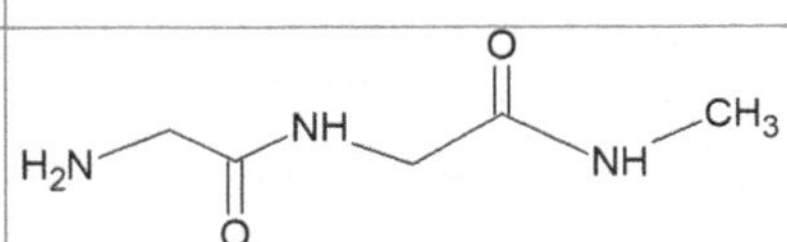
Elastin	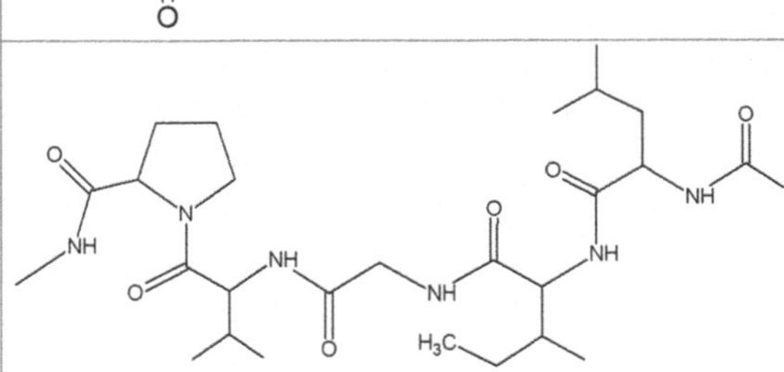

ligaments, and prepared through enzymatic and acidic processing. Collagen is typically dissolved in dilute acid, and when the reaction is neutralized (usually with sodium hydroxide) and slowly heated to body temperature, it forms a hydrogel [13]. Collagen is primarily utilized in tissue regeneration applications, wound dressings, drug delivery systems in tissue engineering, pharmaceuticals, and biomedical products. While collagen's mechanical properties are sufficient for natural tissues, its lack of crosslinking makes its application in biomaterials mechanically challenging. Various crosslinking methods have been tested to address this limitation [13].

Collagen fibers have been specifically used in studies involving cell guidance. By aligning collagen fibers, cell orientation can be controlled. Collagen-based studies can target different tissues and are often used in combination with synthetic materials tailored to specific tissues. However, collagen also has some disadvantages, including weak mechanical properties, challenges in handling, the potential to induce immune reactions, and limited control over biodegradability [14]. On the other hand, it can be used in many different forms for treatments of organs and tissues. For example, it can be prepared as tubular forms for vascular tissue engineering applications [15], can be combined with polysaccharides as skin grafts [16], or structured as scaffold for meniscus treatment [17].

5.3.1.2 Gelatin

Gelatin has very low immunogenicity. It is among the best options used in the field of biomaterials and tissue engineering because it is natural, vascular, low-cost, and degradable. Gelatin is obtained by physically and chemically denaturing the triple helical structure of collagen. These denaturation processes can be done in two different ways. Gelatin A is obtained by denaturing collagen treated with acids (pH 1–3), and gelatin B is obtained by denaturing collagen treated with alkaline solutions. Anionic or cationic gelatin can be synthesized and this depends on the extraction conditions. It is basically in the form of filling the molecules through electrostatic interaction. The resistance of gelatin allows the grafting of synthetic polymers to the gelatin backbone at high temperatures and in a wide pH range [14]. One of these grafting methods is methacryloyl grafted gelatin, also known as Gelatin methacrylate (GelMA). Due to its heat-sensitive nature, gelatin has the ability to enter the sol-gel transition depending on the ambient temperature. Below ambient temperature (30 °C), there is a resolution of the gel transition and heating to physiological temperature reverses this situation. Gelatin also has gel-forming, thickening, emulsifying, and foaming properties. As with many biopolymers, it has disadvantages such as poor mechanical properties and requires physical, chemical, or enzymatic crosslinking to be functional. One very wide application of gelatin is in preparation of scaffolds or dressings for wound healing applications [18–20].

5.3.1.3 Fibrin

Fibrin is a protein formed through the enzymatic polymerization of fibrinogen in the presence of thrombin. It features high water content and plays a critical role in the natural tissue repair process and the coagulation cascade. Fibrin forms a fibrous network that is instrumental in wound tissue regeneration and blood clotting. This network supports the creation of a fibrillar gel scaffold from fibrin clots. The inactive form of fibrin, fibrinogen, can be extracted from blood and used as an autologous scaffold source. This approach significantly reduces or entirely eliminates the risk of immune reactions. Due to its availability and simplicity, fibrin is considered a viable alternative to collagen [21].

Fibrin has disadvantages, such as low mechanical strength and rapid degradation. Its degradation can be controlled through chemical modifications, such as crosslinking and the use of protease inhibitors. Due to its high biocompatibility, fibrin is well-suited for applications in cartilage, bone, and adipose tissue engineering. Additionally, it is used as an adhesive in surgical procedures to prevent leakage and secure skin grafts.

The primary challenge of fibrin-based hydrogels is their mechanical instability. To address this issue, the fibrin matrix is often combined with other materials or molecules to enhance the product's structural integrity. Among the most important of these are protease inhibitors, as previously mentioned. Fibrin gels can be loaded with cells and added to scaffolds to enhance their bioactivity [22].

5.3.1.4 Silk Fibroin

Silk fibroin is a protein derived from the silkworm (*Bombyx mori*) and is also produced by spiders and scorpions. Its primary amino acid structure primarily consists of glycine, alanine, and serine. Fibroins exhibit excellent biocompatibility, flexibility, and mechanical strength. The repetitive amino acid structure leads to a homogeneous secondary structure, resulting in strong yet flexible mechanical properties [23]. These properties make silk fibroin superior to other natural and synthetic polymers in terms of mechanical performance.

Silk fibroin hydrogels exhibit shear-thinning behavior, making them suitable as injectable hydrogels for 3D printing and tissue engineering scaffold fabrication. Lithium bromide, formic acid, ionic liquids, and a ternary solvent system ($CaCl_2$/ethanol/water) are some of the solvents that can dissolve silk fibroin.

Two structural forms of silk exist: silk I, which is water-soluble, and silk II, which has an antiparallel beta-sheet structure and is water-insoluble. Converting silk I into crystalline silk II through annealing improves strain and tensile strength. Due to its superior mechanical properties, low immunogenicity, low thrombogenicity, and controlled degradation rates, silk fibroin is widely used in vascular tissue engineering. It is compatible with various production methods, including electrospinning and gel spinning. Silk hydrogels can also be modified using methacrylate grafting to create 3D printable and photo-crosslinkable hydrogels. To enhance rheological properties, silk fibroin can be combined with other polymers. Also, conductive hydrogels based on fibroin can be prepared to monitor body motions [24].

5.3.1.5 Elastin

Elastin is a structural connective tissue protein found in load-bearing tissues. It forms an insoluble biopolymer through crosslinking of tropoelastin, a water-soluble precursor, creating a network structure. Elastin and tropoelastin-like peptides have the potential for self-assembly under physiological conditions. Due to its elastic nature, elastin is well-suited for use as a scaffold in both skin and vascular tissue engineering. It is also utilized in highly porous hydrogels, 3D cell culture studies, and drug delivery systems. As an extracellular matrix protein, elastin exhibits biocompatibility and biodegradability.

Proteins used in biomaterials must be pure, but purifying elastic fibers during elastin synthesis can be challenging, increasing contamination risks. As a result, elastin is less commonly used than other proteins. Since elastin is insoluble, soluble forms such as tropoelastin, alpha-elastin, elastin-like polypeptides, and genetically recombinant elastin are frequently employed in medical applications. Elastin-based hydrogels can be synthesized using crosslinkers, functional group modifications through click polymerization, or electrospinning. During synthesis, the porosity of the hydrogels can be adjusted using CO_2, achieving optimal conditions for cell proliferation [25]. Incorporating elastin into various hydrogel structures allows for the adjustment of elastic properties, which is crucial for applications such as vascular and skin tissue engineering.

5.3.2 *Polysaccharide-Based Natural Hydrogels*

Polysaccharides are widely used biopolymers for hydrogel development. Their biodegradability and biocompatibility, combined with their non-toxic nature, high swelling capacity, and superior mechanical properties, make them suitable for numerous biomedical applications. These include tissue engineering, regenerative medicine, and gene and drug delivery. Common polysaccharides used in these applications include alginate, chitosan, and cellulose (Table 5.2). Polysaccharide hydrogels can be synthesized using chemical crosslinking methods or physical techniques such as freeze-thaw cycles. Hydrogels prepared through freeze-thawing are reported by some authors to be more biocompatible and biodegradable than those made using chemical methods [11].

5.3.2.1 Hyaluronic Acid

Hyaluronic acid (HA) is a very hydrophilic copolymer of glucuronic acid and N-acetylglucosamine found in mammalian connective tissue. As a primary component of the natural extracellular matrix, it is classified as a glycosaminoglycan (GAG). HA has high water retention capacity, biodegradability, combined with excellent physical flexibility.

HA can be sourced from animals or produced through microbial means to avoid immune reactions [26]. Evidence suggests that HA plays a role in angiogenesis, extracellular matrix organization, and wound healing. High molecular weight HA is highly viscous and difficult to process; therefore, it sometimes is reduced to low molecular weight forms using acids or bases, making it suitable for hydrogel preparation [27].

Chemical modifications can be used to increase HA's hydroxyl and carboxylic groups. Its use in medical applications include cartilage regeneration, chondrocyte growth in vascular diseases, bone and skin tissue regeneration [28], and wound dressings in the form of injectable hydrogels and sponges.

Table 5.2 Examples of polysaccharide-based natural hydrogels

Chitosan	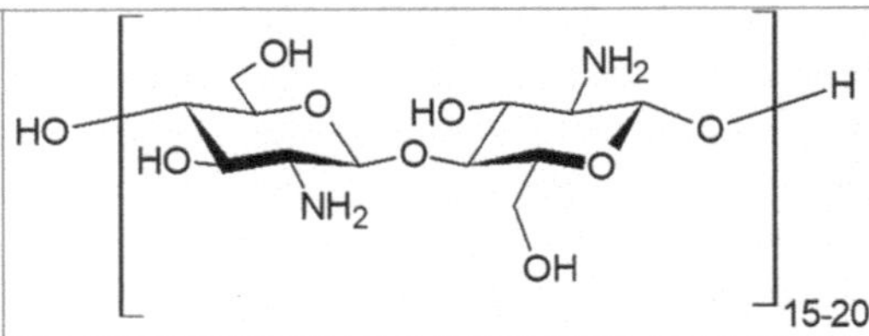
Hyaluronic acid	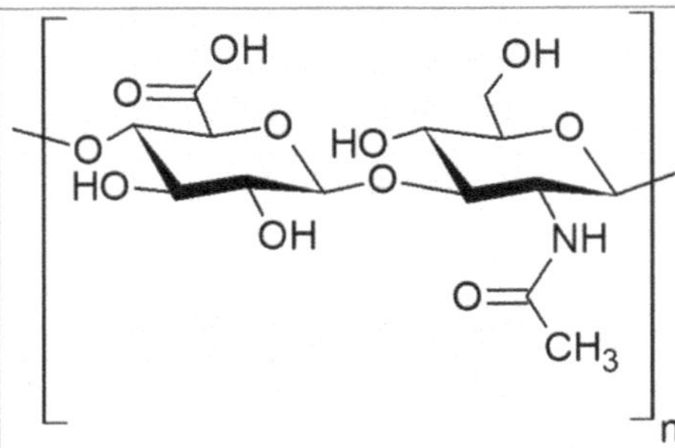
Cellulose	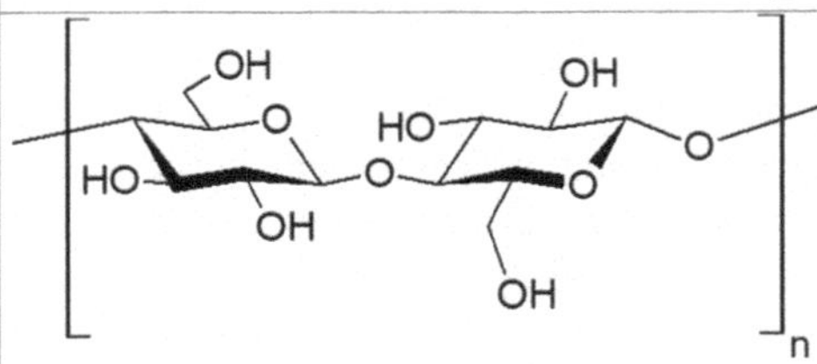
Agarose	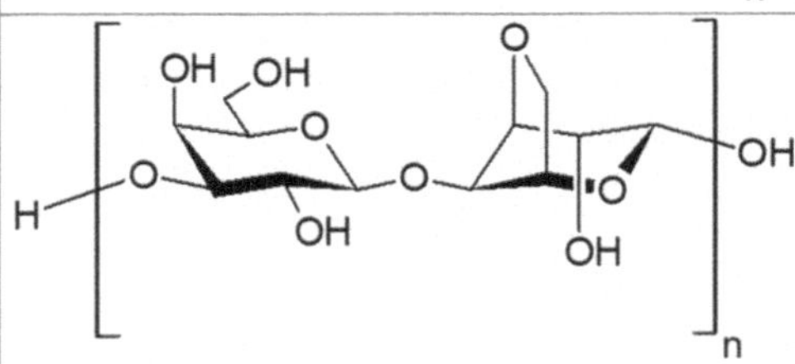
Alginate	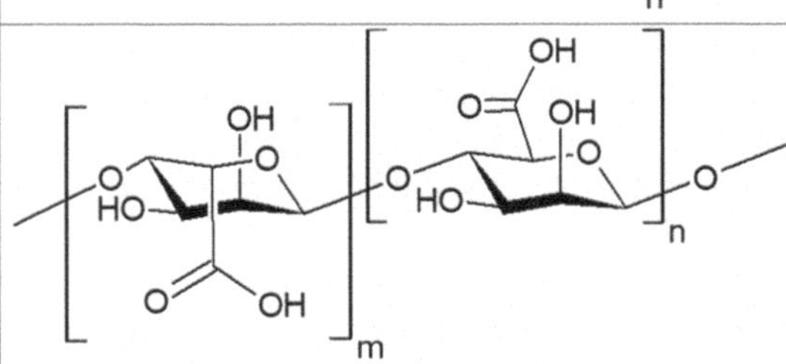

5.3.2.2 Alginate

Brown seaweed is usually the source of alginate, a naturally occurring anionic polymer. Because of its moderate gelation when divalent cations like Ca^{2+} are added, low toxicity, biocompatibility, and relative affordability, it has been thoroughly researched and used in a wide range of biomedical applications.

Structurally, alginate is an anionic polysaccharide obtained from brown algae, consisting of a copolymer of β-D-mannuronate (M) and α-L-guluronate (G) blocks. While M-segments exhibit linear and flexible conformations, G-blocks induce steric hindrance and form ionic bridges that determine the mechanical properties of the entire polymer structure [29]. Alginate is a non-toxic, non-inflammatory polysaccharide that closely resembles the polysaccharides of the extracellular matrix but shows limited cell interaction and do not undergo in vivo degradation.

Alginate hydrogels have been widely researched for wound dressings, controlled release systems, and in tissue engineering applications due to their biocompatibility, availability, and controllable gelation rates. Studies have shown that alginate is more effective when used in combination with other biopolymers.

5.3.2.3 Chitosan

Chitosan is a cationic polysaccharide derived from the partial deacetylation of chitin, a glucose derivative found in the exoskeletons of invertebrates. It has a linear structure composed of *N*-acetyl-D-glucosamine and β-(1–4)-linked D-glucosamine, making it the second most abundant natural biopolymer after cellulose. The molecular weight and degree of deacetylation of chitosan directly influence its physical and mechanical properties.

Chitosan hydrogels are produced by acidifying chitosan solutions and then neutralizing them with basic agents like glycerophosphate. Because chitosan is water-insoluble, various chemical modifications are employed to make it soluble. Modified chitosan hydrogels exhibit mucoadhesive properties and easily form gels at lower pH levels.

Chitosan is widely studied as a hydrogel due to its biocompatibility, biodegradability, non-toxicity, bioactivity, and controllable degradability by adjusting the degree of deacetylation [30]. Its disadvantages include weak mechanical properties, which can be modified by crosslinking with chemical agents or gelatin and by adding some inorganic materials [31].

Chitosan's antibacterial properties and its stimulatory effects on leukocytes make it a promising natural polymer for wound healing applications. Additionally, its controllable degradability and short degradation time under physiological conditions make chitosan an excellent candidate for producing hydrogels with micro-/nano-dimensions for the delivery of drugs, proteins, enzymes, and other bioactive agents [32]. While chitosan supports bone formation, its low mechanical properties necessitate combining it with other materials like alginate, calcium phosphate, or HA for orthopedic applications [14, 33].

5.3.2.4 Dextran

Dextran is a homopolysaccharide produced by bacteria, consisting of a linear 1,6-glycosidic bond with 1,3 branching. Even at concentrations above 20% by weight, dextran forms low-viscosity solutions in water. It can be modified by adding aldehyde, (meth)acrylate, thiol, phenol, maleimide, or vinyl sulfone groups. Despite its high cost and limited availability, dextran finds applications in tissue engineering scaffolds, as an antithrombolytic agent and a bioadhesive. Injectable dextran-based hydrogels have also been developed as site-specific, traceable, and chemotherapeutic delivery devices [14].

5.3.2.5 Cellulose

Cellulose, the most abundant natural biopolymer, is synthesized by plants and certain specific bacteria. As a hydrophobic biopolymer, cellulose can be chemically modified through esterification or etherification of its hydroxyl groups. Common forms of cellulose include methylcellulose (MC), hydroxypropyl MC (HPMC), ethyl cellulose, hydroxyethyl cellulose (HEC), and sodium carboxymethyl cellulose (NaCMC). While pure cellulose is water-insoluble, modifying the hydroxyl groups with more hydrophilic units, such as methyl or hydroxypropyl groups, renders it water-soluble. Cellulose has been successfully applied in burn and wound healing, cardiovascular, neural, cartilage, and bone tissue regeneration studies [34, 35]. It is regarded as an attractive candidate for a wide range of applications in the biomaterials field.

5.3.3 Decellularized Tissues

Decellularizing tissues and organs aims to provide an optimal environment for cells while preserving the micro- and macrostructures close to that of natural tissue. The preparation of the extracellular matrix during decellularization is influenced by factors such as the source of the tissue, types of crosslinkers used, and sterilization methods. Therefore, multiple factors must be considered and adjusted at each stage [36].

Specific tissues require customized approaches, considering factors such as thickness, shape, micro- and macro-architecture, and cell density. Ensuring the complete removal of residues from decellularized organs or tissues is critical in preventing immune reactions.

Collagen and elastin, present in the extracellular matrix, are essential for maintaining physiological elasticity. Preserving these polymers is crucial during decellularization processes for soft tissues. The extracellular matrix promotes cell proliferation, migration, differentiation, maturation, and induction. Proteoglycans within the extracellular matrix also play roles in biological activities such as cell growth, differentiation, angiogenesis, and anti-angiogenesis, making decellularized materials advantageous as tissue scaffolds.

Hydrogels derived from decellularized tissues are created by reducing dry material and enzymatically digesting it with collagenase and/or pepsin in acidic environments over varying durations, depending on the type of tissue or organ. This process yields water-soluble, low-molecular-weight biopolymers. The pre-gel is diluted before undergoing the gelation phase at approximately 37 °C, resulting in a hydrogel concentration tailored to the desired application. There are various applications of decellularized tissues in human body for treatment purposes, and one example is human iliac artery [37].

5.4 Conclusion

Polymers and hydrogels derived from natural sources require extensive research and optimization for clinical use, particularly regarding mechanical properties, standardization, and cell interactions. Natural materials derived from animal sources carry the risk of immune responses. Bioinks are key components of 3D bioprinting processes. The chemical structures and combinations of cell containing hydrogels directly affect the feasibility of 3D printing and bioprinting as well as the usefulness of the biomaterial. Natural polymers are particularly advantageous as bioinks due to their excellent compatibility with the cells. However, further advancements are needed to prepare them for clinical applications.

References

1. Jain, S., Jain, A., Jain, R., Chauhan, N.S.: Potential of natural polymeric materials in pharmaceutics. Pharmacol. Res. Nat. Prod. **2**, 100014 (2024). https://doi.org/10.1016/j.prenap.2024.100014
2. Hasirci, V., Hasirci, N.: Soft tissue applications (Ch:15). In: Fundamentals of Biomaterials, 2nd edn, pp. 231–249. Springer (2024)
3. Malikmammadov, E., Hasirci, N.: Dual and multi-stimuli responsive polymers for biomedical applications (Chapter 9). In: Aguilar, M.R., Román, J.S. (eds.) Smart Polymers and Their Applications, 2nd edn, pp. 255–278. Elsevier Woodhead Pub (2018)
4. Hasirci, V., Yilgor, P., Tanir, T.E., Eke, G., Hasirci, N.: Polymer fundamentals: polymer synthesis. In: Ducheyne, P., Healy, K.E., Hutmacher, D.W., Grainger, D.W., Kirkpatrick, C.J. (eds.) Comprehensive Biomaterials, 2nd edn, pp. 478–506. Elsevier (2017)
5. Garcia-Garcia, A., Muñana-González, S., Lanceros-Mendez, S., Ruiz-Rubio, L., Alvarez, L.P., Vilas-Vilela, J.L.: Biodegradable natural hydrogels for tissue engineering, controlled release, and soil remediation. Polymers. **16**(18), 2599 (2024). https://doi.org/10.3390/polym16182599
6. Tamay, D.G., Hasirci, N.: Bioinks—materials used in printing cells in designed 3D forms. J. Biomater. Sci. Polym. Ed. **32**(89), 1072–1106 (2021). https://doi.org/10.1080/09205063.2021.1892470
7. Satchanska, G., Davidova, S., Petrov, P.D.: Natural and synthetic polymers for biomedical and environmental applications. Polymers. **16**(8), 1159 (2024). https://doi.org/10.3390/polym16081159
8. Singh, S.K., Dhyani, A., Juyal, D.: Hydrogel: preparation, characterization and applications. Pharm. Innov. **6**(6, Part A), 25 (2017)
9. Usta, U., Asmatulu, R.: Hydrogels in various biomedical applications. In: Polymer Science: Research Advances, Practical Applications and Educational Aspects, pp. 248–257 (2015)

10. Khoeini, R., Nosrati, H., Akbarzadeh, A., Eftekhari, A., Kavetskyy, T., Khalilov, R., et al.: Natural and synthetic bioinks for 3D bioprinting. Adv. NanoBiomed. Res. **1**(8), 2000097 (2021)
11. Varaprasad, K., Raghavendra, G.M., Jayaramudu, T., Yallapu, M.M., Sadiku, R.: A mini review on hydrogels classification and recent developments in miscellaneous applications. Mater. Sci. Eng. C. **79**, 958–971 (2017)
12. Liu, X., Zheng, C., Luo, X., Wang, X., Jiang, H.: Recent advances of collagen-based biomaterials: multi-hierarchical structure, modification and biomedical applications. Mater. Sci. Eng. C. **99**, 1509–1522 (2019)
13. Achilli, M., Lagueux, J., Mantovani, D.: On the effects of UV-C and pH on the mechanical behavior, molecular conformation and cell viability of collagen-based scaffold for vascular tissue engineering. Macromol. Biosci. **10**, 307–316 (2010)
14. Gyles, D.A., Castro, L.D., Silva Jr., J.O.C., Ribeiro-Costa, R.M.: A review of the designs and prominent biomedical advances of natural and synthetic hydrogel formulations. Eur. Polym. J. **88**, 373–392 (2017)
15. Zorlutuna, P., Hasirci, N., Hasirci, V.: Nanopatterned collagen tubes for vascular tissue engineering. J. Tissue Eng. Regen. Med. **2**, 373–377 (2008)
16. Bektas, C.K., Kimiz, I., Urkmez, A.S., Hasirci, V., Hasirci, N.: A bilayer scaffold prepared from collagen and carboxymethyl cellulose for skin tissue engineering applications. J. Biomater. Sci. Polym. Ed. **29**(14), 1764–1784 (2018). https://doi.org/10.1080/09205063.2018.1498718
17. Halili, A.N., Hasirci, N., Hasirci, V.: A mechanically functional collagen-based construct designed as a meniscus substitute. J. Biomater. Tissue Eng. **3**(2), 173–184 (2013)
18. Ulubayram, K., Çakar, N., Korkusuz, P., Ertan, C., Hasirci, N.: EGF containing gelatin based wound dressings. Biomaterials. **22**(11), 1345–1356 (2001). https://doi.org/10.1016/s0142-9612(00)00287-8
19. Bilgic, H., Demiriz, M., Ozler, M., Ide, T., Dogan, N., Gumus, S., Kiziltay, A., Endogan, T., Hasirci, V., Hasirci, N.: Effects of EGF containing gelatin sponges as wound healing scaffolds. J. Biomater. Tissue Eng. **3**(2), 205–211 (2013)
20. Zhang, X., Qin, M., Jia, J., Ahmed, A., Zhao, L., Lan, W., Wei, Y., Liang, Z., Ma, X., Shi, Y., Huang, D.: A natural gelatin/casein hydrogel with on-demand adhesion via chitosan solution for wound healing. Int. J. Biol. Macromol. **290**, 139112 (2025). https://doi.org/10.1016/j.ijbiomac.2024.139112
21. Khansari, M.M., Sorokina, L.V., Mukherjee, P., Mukhtar, F., Shirdar, M.R., Shahidi, M., Shokuhfar, T.: Classification of hydrogels based on their source: a review and application in stem cell regulation. JOM. **69**, 1340–1347 (2017)
22. Malikmammadov, E., Tanir, T.E., Kiziltay, A., Hasirci, N.: Preparation and characterization of poly(ε-caprolactone) scaffolds modified with cell-loaded fibrin gel. Int. J. Biol. Macromol. IJBIOMAC. **125**, 683–689 (2019). https://doi.org/10.1016/j.ijbiomac.2018.12.036
23. Han, F., Liu, S., Liu, X., Pei, Y., Bai, S., Zhao, H., Lu, Q., Ma, F., Kaplan, D.L., Zhu, H.: Woven silk fabric-reinforced silk nanofibrous scaffolds for regenerating load-bearing soft tissues. Acta Biomater. **10**(2), 921–930 (2014)
24. Xue, X.-Y., Han, L., Cai, H.-Q., Zhang, K., Dong, S.-L., Wang, Y.-L., Yang, S.-S., Sun, Z.-C., Liu, R.-P., Wei, Y.: 3D printed conductive hydrogel based on silk fibroin and tetramer-grafted-polyethylenimine micelle for body-motion monitoring. Colloids Surf. A Physicochem. Eng. Aspects. **705**(Part 2), 135671 (2025). https://doi.org/10.1016/j.colsurfa.2024.135671
25. Mithieux, S.M., Rasko, J.E., Weiss, A.S.: Synthetic elastin hydrogels derived from massive elastic assemblies of self-organized human protein monomers. Biomaterials. **25**(20), 4921–4927 (2004)
26. Ahmed, E.M.: Hydrogel: preparation, characterization, and applications: a review. J. Adv. Res. **6**, 105–121 (2015)
27. Li, Y., et al.: Magnetic hydrogels and their potential biomedical applications. Adv. Funct. Mater. **23**, 660–672 (2013)
28. Burdick, J.A., Prestwich, G.D.: Hyaluronic acid hydrogels for biomedical applications. Adv. Mater. **23**(12), H41–H56 (2011)

29. Varghese, S.A., Rangappa, S.M., Siengchin, S., Parameswaranpillai, J.: Natural polymers and the hydrogels prepared from them. In: Hydrogels Based on Natural Polymers, pp. 17–47. Elsevier (2020)
30. Huang, Y., Onyeri, S., Siewe, M., Moshfeghian, A., Madihally, S.V.: In vitro characterization of chitosan–gelatin scaffolds for tissue engineering. Biomaterials. **26**(36), 7616–7627 (2005)
31. Ucar, S., Ermis, M., Hasirci, N.: Modified chitosan scaffolds: proliferative, cytotoxic, apoptotic and necrotic effects on Saos-2 cells and antimicrobial effect on E. coli. J. Bioact. Compat. Polym. **31**(3), 304–319 (2016). https://doi.org/10.1177/0883911515627471
32. Bernkop-Schnürch, A., Dünnhaupt, S.: Chitosan-based drug delivery systems. Eur. J. Pharm. Biopharm. **81**(3), 463–469 (2012)
33. Kara, F., Aksoy, E.A., Yuksekdag, Z., Aksoy, S., Hasirci, N.: Enhancement of antibacterial properties of polyurethanes by chitosan and heparin immobilization. Appl. Surf. Sci. **357**, 1692–1702 (2015). https://doi.org/10.1016/j.apsusc.2015.09.227
34. Kopeček, J.: Swell gels. Nature. **417**(6887), 389–391 (2002)
35. Kawecki, M., Łabuś, W., Klama-Baryla, A., Kitala, D., Kraut, M., Glik, J., et al.: A review of decellurization methods caused by an urgent need for quality control of cell-free extracellular matrix' scaffolds and their role in regenerative medicine. J. Biomed. Mater. Res. B Appl. Biomater. **106**(2), 909–923 (2018)
36. Catoira, M.C., Fusaro, L., Di Francesco, D., Ramella, M., Boccafoschi, F.: Overview of natural hydrogels for regenerative medicine applications. J. Mater. Sci. Mater. Med. **30**, 1–10 (2019)
37. Abad-Contreras, D.E., Martínez-Ortiz, A.K., Martínez-López, V., Laparra-Escareño, H., Martínez-García, F.D., Pérez-Calixto, D., Vazquez-Victorio, G., Sepúlveda-Robles, O., Rosas-Vargas, H., Piña-Barba, C., Rodríguez-López, L.A., Giraldo-Gomez, D.M., Hinojosa, C.A.: Decellularization of human iliac artery: a vascular scaffold for peripheral repairs with human mesenchymal cells. Tissue Cell. **93**, 102686 (2025). https://doi.org/10.1016/j.tice.2024.102686

Chapter 6
Synthetic Hydrogels and Bioinks

Tuğba Endoğan Tanır

Abstract Hydrogels are made from hydrophilic homopolymers or copolymers. Due to their high water retention capacity, they possess a soft and flexible tissue-like quality, resulting in low friction with surrounding tissues. They also exhibit low adhesion properties with mucosal surfaces and tissues, allowing them to be easily removed from a wound when applied. Because of their water absorption capability, they are used for the excretion of excessive amounts of body fluids. In the preparation of hydrogels, synthetic polymers are preferred in some cases due to the difficulties associated with the chemical modification of natural polymers and the potential of natural polymers for inducing an immunological response in the human system. Synthetic hydrogels are used in various applications, including tissue engineering scaffolds, drug delivery systems, wound dressings, lubricants, contact lenses, etc. Synthetic hydrogels can be designed and synthesized to demonstrate a diverse range of features and perform specific functions based on their chemical structure, synthesis methods, and types of crosslinking. In this chapter synthesis, characteristics, and biomedical uses of synthetic hydrogels are covered.

Keywords Poly(acrylic acid) · Poly(ethylene oxide) · Poly(vinylpyrrolidone) · Poly(vinyl alcohol) · Polymer-peptide hydrogels · PEO · PVP · PVA

6.1 Synthesis of Synthetic Hydrogels

Hydrogels are synthesized through the crosslinking of one or more monomers or polymer chains, and the chemical characteristics of the used monomers and polymers determine the properties of the resulting hydrogel. Since water absorption is the most important property of hydrogels, they are prepared from specific

T. E. Tanır (✉)
Central Laboratory, Middle East Technical University, Ankara, Türkiye
e-mail: endogan@metu.edu.tr

N. Hasirci, V. Hasirci (eds.), *Hydrogels and Bioinks in Tissue Engineering*, Biological and Medical Physics, Biomedical Engineering,
https://doi.org/10.1007/978-3-031-90966-5_6

monomers and polymers having functional groups such as hydroxide (–OH), carbonyl (–COOH), amine (–NHR), etc. Table 6.1 lists the most common monomers used to synthesize synthetic hydrogels.

Any method that can be used to synthesize a crosslinked polymer can also be applied to prepare hydrogels. Free-radical polymerization is generally used for the synthesis of synthetic hydrogels. The hydrogel synthesis takes place according to the principle of mixing monomers (or polymers), initiators, and crosslinkers in specific ratios and conditions (such as pH and temperature) to form network-structured polymer chains. Examples of crosslinkers used in synthetic hydrogel synthesis include ethylene glycol dimethacrylate, *N,N′*-methylene bisacrylamide, and divinylbenzene. Some of the crosslinkers used in synthetic polymeric hydrogel synthesis are listed in Table 6.2. In the preparation of hydrogels, the structure and composition of the chemicals, the type and amount of crosslinking agent, the

Table 6.1 The monomers utilized to synthesize synthetic hydrogels

Monomer	Chemical structure	Examples
Acrylic acids	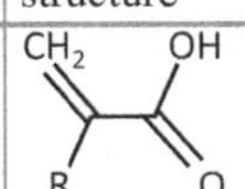	Methacrylic acid, acrylic acid
Acrylamides	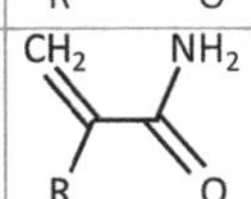	Methacrylamide, acrylamide
N-substituted acrylamides	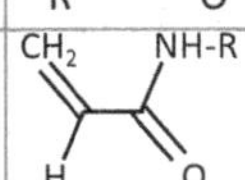	*N*-isopropylacrylamide
Acrylates	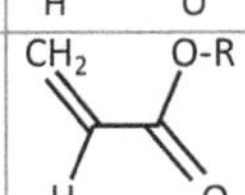	2-Hydroxyethyl acrylate
Methacrylates	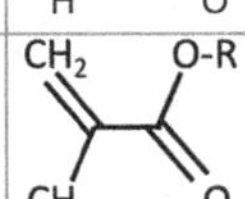	Butyl methacrylate, 2-hydroxyethyl methacrylate, 2-methoxyethyl methacrylate
Glycols	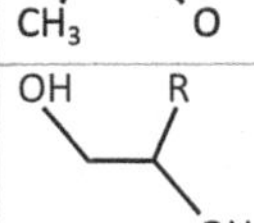	Propylene glycol, ethylene glycol
Vinyls	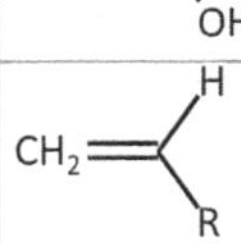	*N*-vinyl pyrrolidone, vinyl alcohol, *N*-vinylcaprolactone
Lactones	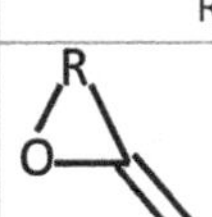	Caprolactone

Table 6.2 Examples of crosslinkers used to synthesize synthetic polymeric hydrogels

Polymer	Crosslinking agents
Poly(hydroxyethyl methacrylate)	Ethylene glycol dimethacrylate
Poly(acrylic acid)	Methylene bisacrylamide
Poly(vinyl alcohol)	Glutaraldehyde, sodium borate/boric acid
Poly(ethylene glycol)	Poly(ethylene glycol dimethacrylate)
Poly(acrylamide)	Methylene bisacrylamide
Poly(*N*-isopropylacrylamide)	Methylene bisacrylamide, poly(ethylene glycol dimethacrylate)
Poly(*N*-vinyl-2-pyrrolidone)	Methylene bisacrylamide
Poly(methacrylic acid)	Methylene bisacrylamide
Polyurethane	Polyols

crosslinking method, and even small changes in synthesis conditions (such as temperature and time) can lead to hydrogels with different properties.

Hydrogels are categorized according to their physical characteristics, ionic charge, source, response to stimuli, crosslinking, and preparation method [1]. Natural polymers, synthetic polymers, or a mixture of the two can be used to produce hydrogels. Depending on the ionic charge, hydrogels can be cationic, anionic, neutral, or amphoteric, with their charge determined by the polymer used. Hydrogels are classified into four types based on their preparation method:

(a) **Homopolymeric hydrogels:** These structures are formed by crosslinking of hydrophilic monomer or polymer chains (e.g., poly(acrylic acid), poly(vinyl alcohol)).
(b) **Copolymeric hydrogels:** These structures are formed by crosslinking of two different monomers or polymer chains. One of the monomers or polymers must be hydrophilic (e.g., poly(hydroxyethyl methacrylate-co-methyl methacrylate)).
(c) **Multipolymer hydrogels:** Structures formed by the reaction of three or more comonomers or polymers (e.g., poly(ethylene glycol-co-caprolactone-co-ethylene glycol), poly(sodium acrylate-co-*N*-isopropylacrylamide-co-acrylamide)).
(d) **Interpenetrating polymer network (IPN) hydrogels:** These are structures produced by swelling and crosslinking the prepared crosslinked network structure in a monomer or polymer solution (e.g., β-chitosan-modified crosslinked chitosan/PNIPAM/PEGDA hydrogels). Semi-IPN hydrogels are formed if the second structure is only diffused into the first network but not crosslinked. Schematic representation of semi-IPN hydrogels is given in Fig. 6.1.

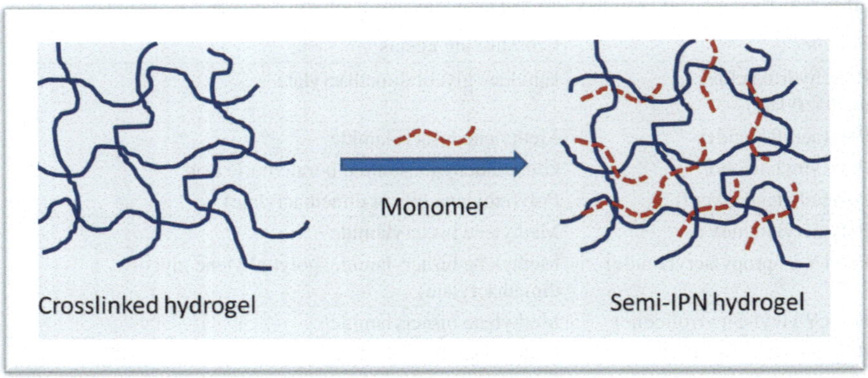

Fig. 6.1 Semi-IPN hydrogel

6.2 Crosslinking Methods of Hydrogels

Crosslinkers are used in hydrogel synthesis to prevent the dissolution of polymer chains. Crosslinking of polymer chains affects important features of the synthesized hydrogel, like elasticity, viscosity, solubility, mechanical properties, toughness, and thermal properties [2]. The selection of a crosslinking technique depends on the kind of polymer being used and crosslinking can be either chemical or physical. Crosslinking can be achieved through various methods like UV and gamma irradiation, bulk polymerization, free radical polymerization, solution casting, simple mixing, and interpenetrating network formation.

Physically crosslinked hydrogels are those that do not contain covalent bonds and do not require a crosslinking agent for their synthesis. Hydrogels prepared by methods such as ionic interactions, heating/cooling of a polymer solution, freeze-thaw cycling, complex coacervation, and hydrogen bonding are physically crosslinked [3]. In physical crosslinking, reversible and temporary (transient) bonding points form due to changes in the polymer's conformation. This is because the bonds formed during physical crosslinking arise from physical interactions or entanglement of the polymer chains. Examples of physically crosslinked hydrogels are polymethacrylic acid-polyethylene glycol and polyethylene glycol-poly(γ-benzyl L-glutamate) [4].

Hydrogels crosslinked by covalent bonds are known as chemically crosslinked hydrogels. The functional groups of the polymer, such as –COOH, $-NH_2$, and –OH, react with the crosslinker to produce chemical crosslinking. During chemical crosslinking, changes occur in the polymer's configuration, resulting in permanent bonding points. Examples of chemically crosslinked hydrogels include acrylic acid or acrylamide crosslinked with methylene bisacrylamide. Chemically crosslinked hydrogels can also be prepared using high-energy radiation and enzymes. In the radiation crosslinking method, the biocompatibility of the polymer is preserved as no chemical additives are needed [5]. In this method, the polymer is exposed to high-energy sources such as X-rays, gamma rays, or electron beams, generating free radicals. Polyethylene glycol, polyacrylic acid, and polyvinyl alcohol can be given as examples of hydrogels crosslinked by this method [6]. There are studies indicating that radiation-crosslinked hydrogels show improved strength and elongation properties [7].

6.3 Properties of Synthetic Hydrogels

The parameters under which hydrogels are synthesized, including the concentration of the crosslinker, the chemistry of the monomers forming the network structure, and the ratio of the monomers, have a significant impact on their structural properties. These factors allow the hydrogels to be synthesized with properties suitable for their intended purpose.

The crosslinks between polymer chains of hydrogels affect their physical characteristics, like glass transition temperature (Tg), melting point, durability, viscosity, solubility, and elasticity [2]. For instance, the glass transition temperature of the polymer increases since the crosslinks restrict the rotational motion between the polymer chains. The most important parameter affecting the swelling of hydrogels is the crosslinking density. Hydrogels having high crosslinking density possess a dense network structure; therefore, they exhibit less swelling. If the crosslinking density is low, the hydrogel structure is loose, and they swell more. Increasing the degree of crosslinking restricts the mobility of the polymer chains, thus reducing the gel's swelling capacity.

The factor that enables swelling in polymer chains is the presence of hydrophilic groups. The more hydrophilic groups a hydrogel has, the higher its water absorption capacity. However, excessive water content reduces the mechanical strength of the hydrogel. By varying the degree of crosslinking, a hydrogel's mechanical properties can be adjusted for the intended use. As the degree of crosslinking increases, the hydrogel becomes mechanically stronger. However, higher crosslinking reduces the percentage of elongation, making the hydrogels more brittle. It has been noted that IPN (interpenetrating network) hydrogels produced by copolymerization using hydrophobic monomers exhibit improved mechanical properties [8].

The hydrogel's degradation is critical for biomedical applications. For example, the controlled degradation of the hydrogel should support cell secretion and matrix remodeling in tissue engineering applications. The hydrogel should preserve its stability until the new tissue forms. The hydrogel should degrade and be removed from the body when it is no longer required. In this way, the hydrogel can be removed from the body without requiring surgery. Moreover, the controlled degradation of the hydrogel ensures the controlled delivery of the loaded agents at the desired dose. Enzymatic hydrolysis, hydrolysis of ester bonds, or photolytic cleavage are the three main processes that lead to the degradation of a hydrogel. Therefore, the main chain, crosslinks, and side groups of the polymer should be designed to degrade through these pathways [9]. Degradation can occur either as bulk or surface erosion

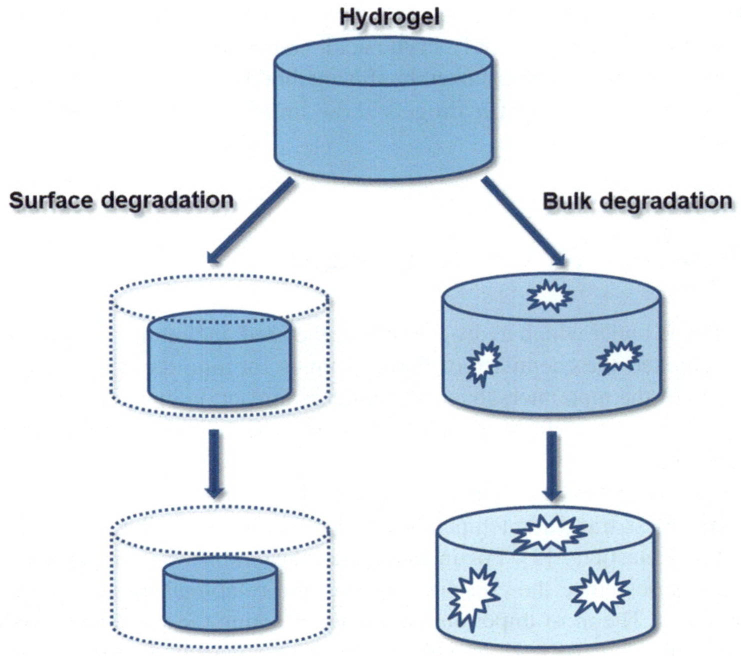

Fig. 6.2 Schematic representation of degradation of a hydrogel

(Fig. 6.2). Surface degradation begins at the hydrogel's surface, whereas bulk degradation occurs all over the hydrogel.

Some hydrogels undergo volumetric changes depending on changes in environmental conditions like temperature, pH, solvent, magnetic field, electric field, ionic strength, and biological molecules. These hydrogels are known as stimuli-responsive hydrogels or smart hydrogels. The properties of these hydrogels will be discussed in Chap. 8.

Water has a significant impact on the physicochemical characteristics of hydrogels. The hydrophilic groups such as carboxyl (–COOH), hydroxyl (–OH), amide (–$CONH_2$), and sulfonic (–SO_3H) in the hydrogel structure assist in the absorption of water [10]. Particular interactions between polymer chains and water molecules, as well as temperature, determine how much water a hydrogel absorbs overall. The swelling of a hydrogel takes place in three stages: the first stage involves the primary binding of water molecules to hydrophilic groups; the second stage includes the secondary binding where water molecules interact with existing hydrophobic groups; and the third stage is the free water that fills voids when swelling equilibrium is reached [11].

6.4 Types of Synthetic Hydrogels

Compared to natural polymers, physical and chemical properties of synthetic polymers are easier to control, which make synthetic polymers preferable in the preparation of synthetic hydrogels. Although synthetic polymer hydrogels exhibit lower biological activity than natural hydrogels, they are suitable for imparting biological functions. Some of the synthetic polymers utilized in the synthesis of hydrogels are discussed in the sections below.

6.4.1 Poly(Acrylic Acid) and Derivatives

Poly(acrylamide) (PAAm) hydrogels are synthesized using acrylamide as a monomer and typically *N,N′*-methylenebisacrylamide (MBAAm) as a crosslinker. Since the vinyl structure of the acrylamide monomer polymerizes very easily, PAAm is a frequently preferred hydrogel. PAAm exhibits excellent adhesive properties, can retain large amounts of water, and is biocompatible. Poly(acrylic acid) (PAA) is a homopolymeric hydrogel. Commercially available PAA hydrogels are made up of 97.5% water and 2.5% PAA. They are durable and possess ideal elasticity properties.

Poly(hydroxyethyl methacrylate) (PHEMA) hydrogels were the first synthetic hydrogels developed by Wichterle and Lim. Their first biomedical application occurred in the 1960s when Wichterle and Lim proposed that crosslinked poly(hydroxyethyl methacrylate) could be used to produce soft contact lenses [12]. Their high oxygen permeability, appropriate refractive index, and strong mechanical stability are the primary factors that enable their use in contact lenses. In addition to their high water retention capacity, PHEMA hydrogels also possess flexibility, porosity, responsiveness to stimuli, softness, and similarity to natural tissue. They are resistant to degradation, not absorbed by the body, and can be sterilized with heat.

Poly(*N*-isopropylacrylamide) (PNIPAM) hydrogels are temperature-sensitive. This polymer can be used in various forms, such as simple chains in solution, microscopic and macroscopic gels, membranes, thin films, coatings, or fibers. PNIPAM has gained particular importance in applications like controlled drug delivery, enzyme immobilization, and protein separation. PNIPAM exhibits a volume change around its lower critical solution temperature (LCST). If the temperature is increased, it changes from its swollen state, that is, its structure with high water content, to its deflated state by losing its water. This transition is due to the composition of the material, which consists of hydrophilic (amide group) and hydrophobic (isopropyl group) components. Hydrogen bonds between water molecules and hydrophilic segments dominate below the LCST. In this state, water molecules bind to the polar groups of the hydrogel and organize around the hydrophobic groups. Above the LCST, the opposite occurs: hydrogen bonds weaken, hydrophobic interactions become stronger, and the molecules fold inwards, expelling water molecules from the hydrogel structure. The transition temperature, which depends on the microstructure of the hydrogel, is determined by synthesis conditions and is typically between 30 and 35 °C for PNIPAM. PNIPAM is very desirable for use in biomedical applications due to its transition temperature being quite close to body temperature. Designing the microstructure of the hydrogel and adjusting the temperature during synthesis can be tailored according to the target application. The transition temperature can also be changed by adjusting the ratio of hydrophilic/hydrophobic groups. For instance, copolymerization with hydrophilic components like acrylic acid (AAc) increases the hydrogen bonding capacity relative to hydrophobic bonds, thus raising the transition temperature. Copolymerization of PNIPAM with other monomers, or formation of polymers with interpenetrating network (IPN) structures, can yield hydrogels with enhanced flexibility, sensitivity to other stimuli, or faster responses to temperature changes.

6.4.2 Poly(Ethylene Oxide) (PEO) and Copolymers

Poly(ethylene glycol) (PEG) is an oligomer of ethylene oxide and is a synthetic polymer often employed in the preparation of synthetic hydrogels, especially in medical applications. FDA has authorized the usage of PEGs in biomedical field because they are biocompatible, non-immunogenic, and resistant to absorption by protein molecules. PEG is widely used in wound healing, tissue engineering, bone prostheses, and many other medical fields. Examples of derivatives of this synthetic polymer include polyethylene glycol methacrylate (PEGMA), polyethylene glycol diacrylate (PEGDA), and polyethylene glycol dimethacrylate (PEGDMA). PEG-based hydrogels are used as drug delivery systems because of their biocompatibility, non-toxicity to surrounding tissues, and water solubility.

6.4.3 Poly(Vinylpyrrolidone) (PVP)

PVP, synthesized by free radical polymerization, is a synthetic hydrogel component commonly employed in biomedical applications due to its water solubility, low toxicity, and film-forming properties. The PVP monomer, *N*-vinylpyrrolidone, is copolymerized with acrylic acid, methacrylates, and other vinyl monomers to be used as contact lenses, controlled drug release, and drug immobilization systems. Its hydrophilic and hydrophobic functional groups make it soluble in both water and various organic solvents, such as methanol, ethanol, chloroform, and propanol. This property enables its use as a carrier for certain hydrophilic and hydrophobic drugs in the pharmaceutical industry.

6.4.4 Poly(Vinyl Alcohol) (PVA)

Poly(vinyl alcohol) (PVA) is a biocompatible and water-soluble polymer having hydroxyl functional groups that can be crosslinked to create hydrogel structures. PVA-based hydrogels exhibit good mechanical properties and can maintain a long-term moist environment due to their water-retention properties. PVA hydrogels are utilized in the production of contact lenses, wound dressings, and drug delivery systems. It has been shown that PVA hydrogels prepared by freeze-thaw cycles exhibit better mechanical properties than those prepared using UV radiation or crosslinkers. Moreover, compared to PEG, the functional groups of PVA structures are easier to access, allowing for the introduction of desired properties.

6.4.5 Polymer-Peptide Hydrogels

These hydrogels consist of short-chain biological molecules formed by amino acid monomers linked by peptide (amide) bonds. Additionally, they can be combined with synthetic polymers. Hydrogels made of polymer-peptide conjugate (PPC) are physically bonded structures having high water absorption capacity. Like proteins, secondary structures like α-helices and β-sheets, as well as tertiary structures, may form as a result of physical bonding. PPC hydrogels offer the benefit of integrating the greatest features of synthetic polymers and peptides [13].

6.5 Biomedical Applications of Synthetic Hydrogels

Hydrogels are preferred by researchers for biomedical applications because of their ease of use, low immunogenicity, biodegradability, and biocompatibility. By controlling hydrogel's physicochemical features and their crosslinking reactions, they can be processed as solid, semi-solid, or liquid [14].

Thanks to their structural properties, hydrogels are suitable for many biomedical applications, such as sensors, contact lenses, artificial organs, drug delivery systems, antimicrobial applications, wound dressings, and cancer treatments. The application areas of some synthetic hydrogels are given in Table 6.3. It has been reported that PMMA-based gel electrolytes are the most preferred coating materials for electrochromic windows due to their high transparency and excellent gelation and solvent retention properties [15]. Agarose and poly(acrylamide) gels are materials used in electrophoresis processes to separate, identify, and purify DNA or RNA fragments. PHEMA was the first potential synthetic hydrogel used for biomedical applications. Shortly after its discovery, it was used in the production of contact lenses by Professor Lim and Wichterle in 1955. It has also been used in many drug delivery systems [16]. The LCST value of PNIPAM in water is approximately 32 °C

Table 6.3 Application areas of synthetic hydrogels

Synthetic hydrogel	Application areas
PAA and derivatives	Diapers, ion exchange resins and adhesives, drug delivery systems, thickener, dispersing, suspending, and emulsifying agents used in pharmaceuticals, cosmetics, and paints
PHEMA	Contact lens, artificial skin production, drug delivery systems, bone marrow, and spinal cord cell regeneration scaffolds
PEO and derivatives	Cell culture, biosensor, drug carrier
PVP	Synthetic plasma, drug delivery systems
PVA	Tissue engineering, drug delivery systems, optical films
Polypeptide	Injectable polymeric systems, cell culture, spinal cord treatment

and can be adjusted to body temperature using copolymerization methods [17]. Therefore, PNIPAM and its copolymers are highly suitable for tissue engineering applications, as its solution consisting of cells and polymer can be easily prepared at room temperature and injected into the desired region. PEG, being non-toxic to the body, is an FDA-authorized polymer for human use. It is appropriate for biological applications and can be injected into the body without side effects. The applications of various synthetic polymeric hydrogels are listed in Table 6.3.

6.5.1 *Tissue Engineering Applications*

Hydrogels are being investigated extensively in the tissue engineering field. Hydrogels designed as scaffold structures for tissue engineering can be tailored to contain pores large enough to support cell survival and proliferation, to release growth factors or drugs, and to degrade as needed. The most important advantages of hydrogel scaffolds are their ability to protect cells and fragile structures like peptides, proteins, and DNA in aqueous environments; easy transport of nutrients into cells and cellular metabolites outward; easy modification with ligands that enhance cell adhesion; in vivo injectability as a liquid that gels at body temperature; and biocompatibility. However, an important drawback of hydrogels is that they have low mechanical properties and there are some difficulties in sterilization, processing, and application.

6.5.2 *Wound Dressings*

In the form of gauze, gels, hydrogels, and hydrocolloids, wound dressings are mostly composed of polymeric materials. The most promising method for wound healing among these is the use of hydrogels. Hydrogels are ideal materials when used as a wound dressing material because at the wound site they can preserve a moist environment, remove exudates from the wound area, prevent infection, and provide a favorable environment for tissue regeneration [18]. In order to prevent complications and scarring, hydrogel-based wound care products are commercially available. There are various products commercially available in the market as wound dressings prepared as hydrogel films, hydrogel sheets, amorphous gels, and hydrogel-impregnated gauze [19]. Suprasorb® G is one of commercially available synthetic hydrogel wound dressings in the form of film which is made up of acrylic polymers, polyethylene, and phenoxyethanol with 70% water. It can be used for dry wounds, pressure ulcers, as well as first- and second-degree burns. Simpurity™ hydrogel wound dressing produced by Safe n' Simple is in the form of sheets. It contains polyethylene oxide, polyvinyl alcohol, acrylate, polyurethane, and purified water. This hydrogel is mainly used for dry scabs, skin burns, and dry wounds. DermaGauze™, hydrogel impregnated gauze dressing, is the product of DermaRite industries. Acrylate polymer is the main component of this hydrogel and it is used for acute or chronic wounds. AquaDerm™ hydrogel sheet dressing is also a product of DermaRite industries. Apart from pressure ulcers it can be used for minor burns and radiation tissue damage.

6.5.3 Contact Lenses

Contact lenses are in direct contact with the eye for purposes such as vision correction, drug delivery in ophthalmic diseases, and cosmetic. Biocompatibility, wettability, oxygen permeability, refractive index, and transmittance are crucial necessities for hydrogel used as contact lenses. Contact lenses are available in two main types: soft and rigid. Polymethyl methacrylate (PMMA) is used to make rigid contact lenses. In 1960, Wichterle and Lim developed soft contact lenses using PHEMA hydrogel [12]. Later, studies have been conducted to develop hydrogels from hydrophilic polymers such as PVA and PAN [20]. Soft contact lenses are characterized by their oxygen permeability, ability to transfer water, and high water content [21]. Oxygen permeability is critical to ensure the effective functioning of the cornea. The cornea utilizes atmospheric oxygen to preserve its structure, function, and clarity [22]. In 1971, Bausch and Lomb introduced the first soft contact lenses made of PHEMA to the market. The disposable SofLens Daily, iconnect New Biotrue®, PureVision2 HD, and Ultra Contact Lens are commercial contact lenses manufactured by Bausch and Lomb. Other companies like Ciba Vision, CooperVision, and Johnson & Johnson also produce a variety of soft contact lenses to be used for different eye diseases. For example, AcuvueOasys® contact lenses produced by Johnson & Johnson are used for presbyopia while Clariti® 1 day silicone contact lenses with ultra-high oxygen permeability are used for short-sighted and long-sighted eye problems.

6.5.4 Drug Delivery Systems

Hydrogels, because of their high-water retention capacity, are extensively used as drug delivery systems. The significant water content helps to trap hydrophilic drugs, while their porous structure allows for controlled drug release [23, 24]. Hydrogels need to be optimized before they can be employed as a drug delivery system. Drugs can be loaded into carrier hydrogels by two ways. In the first approach, the drug and the polymer are mixed, then a polymerization reaction takes place in the presence of a crosslinking agent. In the second approach, in order to load drug, the prepared hydrogel is placed in a solution carrying the drug [25]. The most significant disadvantages of using hydrogels as drug delivery systems are their low mechanical strength and the complications of hydrophobic drug loading [26]. For a controlled drug release study, a poloxamer hydrogel loaded with tetramethylpyrazine (TMP), which has antioxidant and anti-inflammatory effects, was developed. Poloxamer is compatible with drugs that are hydrophilic or hydrophobic. Xia et al. found that the drug release rate of TMP decreased due to micelle encapsulation, while increasing the temperature enhanced the release rate [27]. Another study highlighted the use of a paclitaxel-loaded, temperature-sensitive, in-situ forming hydrogel for localized prevention of tumor recurrence at the tumor resection site. Paclitaxel is a chemotherapy drug used to treat various cancers [28]. Cervidil® is a vaginal hydrogel containing prostaglandin E2, which has a significant function in cervical maturation. After placement in the vagina, dinoprostone is released at the desired dose (0.3 mg/h). SUPPRELIN® LA, containing a hydrogel reservoir with the drug histrelin acetate, is a subcutaneous implant used for the treatment of precocious puberty. Drug release occurs through controlled diffusion over 12 months. Examples of hydrogels prepared using synthetic polymers in recent years for the purposes of biomaterial applications are summarized in Table 6.4.

6.6 Conclusion

Hydrogels can be prepared using natural polymers, synthetic polymers, or combinations. Synthetic polymers have great potential for the production of hydrogels since their physical and chemical properties are more controllable than natural polymers. Compared to natural hydrogels, synthetic polymer hydrogels exhibit less biological activity. However, synthetic polymeric hydrogels can be modified to obtain the desired functionality. In this chapter, monomers, polymers, and crosslinking agents used in the synthesis of synthetic polymeric hydrogels, preparation methods, properties, and biomedical applications of them have been discussed. Synthetic polymeric hydrogels differ in chemical and physical properties because of their diverse chemical structures, synthesis techniques, water content, and crosslinking methods. A minor change in any of the conditions like chemical composition, synthesis method, water content, crosslinking agent, and crosslinking method can lead to

Table 6.4 Examples of synthetic hydrogels prepared recently

Application	Hydrogel	Explanation	Reference
Gastric cancer	PDLLA-PEG-PDLLA	For intraoperative adjuvant combination chemotherapy of gastric cancer, a 5-fluorouracil and cis-platinum co-delivery system was designed	[29]
Cartilage	PAAm	It was shown that prepared hydrogel efficiently lubricates both native and degraded cartilage explants	[30]
Tissue engineering	Poly(NIPAAm-co-(*N*-acryloxysuccinimide)-co-(polylactide/hydroxy methacrylate)-co-(oligo (ethylene glycol) (PNPHO)	By crosslinking PNPHO with a synthetic peptide, an injectable thermoresponsive hydrogel was developed	[31]
Wound dressing	Copolymerized *N,N′*-bis(acryloyl) cystamine, acrylic acid, allyl polyethylene glycol 500 and PCL	The porous hydrogel carrying bFGF promoted angiogenesis	[32]
Cell culture	Polyisocyanide	As a coculture system, polyisocyanide hydrogels were employed, that mimic the mechanical properties of the extracellular matrix	[33]
Tissue engineering	Methacrylic acid-polyethylene glycol	Islets injected into degradable hydrogels reversed diabetes in mice	[34]
Tissue engineering	PLGA-PEG-PLGA, synthetic silicate nanoplatelets and gold nanorods	An injectable conductive nanocomposite hydrogel was injected into the myocardium of MI rats	[35]
Nerve tissue engineering	GelMA-PHEMA	Schwann cells were seeded on a nerve guide hydrogel	[36]
Bone regeneration	PHEMA	Composite biomaterials of PHEMA/titanium-doped bioactive glasses were prepared by applying the intermediate water concept	[37]
Hemostasis	PAA/PAAm/quaternate chitosan	A self-gelling hemostasis powder was produced to improve the hemostatic effect	[38]
Cell encapsulation	Methylcellulose/PVA	A temperature-sensitive hydrogel appropriate for cell encapsulation was produced	[39]
Wound dressing	PU/BSP	Hydrogels containing *Bletilla striata* polysaccharide/PU can rapidly absorb tissue exudates	[40]

(continued)

Table 6.4 (continued)

Application	Hydrogel	Explanation	Reference
Drug delivery	PU/curcumin	Curcumin, a hydrophobic drug, incorporated PU hydrogels were synthesized by in-situ copolymerization	[41]
Drug delivery	PVA/PAA	Bilayer nanofibers loaded with doxorubicin and clarithromycin for osteosarcoma treatment were produced by electrospinning	[42]
Drug delivery	PAA/cellulose nanocrystals	Cisplatin carrying a mucoadhesive hydrogel was synthesized for the treatment of colorectal cancer	[43]
Cell culture	PNIPAM	Human embryonic kidney cells' responses to hydrogel surfaces made of PNIPAM and copolymers were examined	[44]
Injectable drug delivery systems	Poly(2-alkyl-2-oxazoline) (PAOx)	In order to induce in situ gelation, cyclooctyne and azide functional side chains were introduced to PAOx copolymers	[45]
Spinal cord regeneration	Lysine containing peptide conjugated hydrogels	For the treatment of spinal cord injury, injectable lysine-based hydrogel was prepared	[46]
Cell culture	PEG	Biodegradable microporous PEG hydrogel was synthesized to efficiently differentiate 3D human bone cell networks	[19]
Soft tissue	Poly(HEMA-co-MMA)	By changing the ratios of HEMA and MMA, hydrogels having improved mechanical strength and appropriate fibroblast cell adhesion and proliferation behavior was synthesized	[47]
Musculoskeletal tissue	PLA-cell laden GelMA	Hybrid culture system was prepared by filling 3D printed PLA scaffold with cell laden GelMA	[48]
Cell culture	Ethylene glycol grafted isocyanopeptide-based polymers (PIC)	Physicochemical properties of PIC-peptide conjugates were studied	[49]
Regenerative medicine	PEG-cyclic RGD	Short peptide containing PEG hydrogel assisted differentiation of hiPSCs to definitive endoderm	[50]

obtaining hydrogels with completely different properties. It is possible to design new hydrogels from synthetic hydrogels with specific properties tailored to their intended application areas. There is much research about synthetic hydrogels intended to be used in biomedical applications. However, only a few of the hydrogels developed to be used as contact lenses, wound dressings, drug delivery systems and cosmetic products have proceeded to the level of commercialization.

References

1. Ullah, F., Bisyrul, M., Javed, F., Akil, H.: Classification, processing and application of hydrogels: a review. Mater. Sci. Eng. C. **57**, 414–433 (2015)
2. Maitra, J., Shukla, V.: Crosslinking in hydrogels—a review. Am. J. Polym. Sci. **4**, 25–31 (2014)
3. Hasirci, V., Hasirci, N.: Polymers as biomaterials. In: Fundamental of Biomaterials, 2nd edn, pp. 83–101. Springer (2024). https://doi.org/10.1007/978-3-031-54046-2_6
4. Zafar, S., Hanif, M.F., Azeem, M., Mahmood, K., Gondal, S.A.: Role of crosslinkers for synthesizing biocompatible, biodegradable and mechanically strong hydrogels with desired release profile. Polym. Bull. **79**, 9199–9219 (2022)
5. Ovais, M., Nadhman, A., Khalil, A.T., Raza, A., Khuda, F., Sohail, M.F., Zakiullah, I.N.U., Sarwar, H.S., Shahnaz, G., Ahmad, I., Saravanan, M., Shinwari, Z.K.: Biosynthesized colloidal silver and gold nanoparticles as emerging leishmanicidal agents: an insight. Nanomedicine. **12**, 2807–2819 (2017)
6. Hennink, W.E., van Nostrum, C.F.: Novel crosslinking methods to design hydrogels. Adv. Drug Deliv. Rev. **64**(1), 223–236 (2012)
7. Wach, R.A., Mitomo, H., Yoshii, F., Kume, T.: Hydrogel of radiation-induced crosslinked hydroxypropylcellulose. Macromol. Mater. Eng. **287**, 285–295 (2002)
8. Alam, A., Zhang, Y., Kuan, H.C., Lee, S.H., Ma, J.: Polymer composite hydrogels containing carbon nanomaterials—morphology and mechanical and functional performance. Prog. Polym. Sci. **77**, 1–18 (2018)
9. Kharkar, P.M., Kiick, K.L., Kloxin, A.M.: Designing degradable hydrogels for orthogonal control of cell microenvironments. Chem. Soc. Rev. **42**, 7335–7372 (2013)
10. Blackman, L.D., Gunatillake, P.A., Cass, P., Locock, K.E.S.: An introduction to zwitterionic polymer behavior and applications in solution and at surfaces. Chem. Soc. Rev. **48**, 757–770 (2019)
11. Gibas, I., Janik, H.Z.: Review: synthetic polymer hydrogels for biomedical applications. Ch&ChT. **4**, 297–304 (2010)
12. Wichterle, O., Lim, D.: Hydrophilic gels for biological use. Nature. **185**, 117–118 (1960)
13. Klok, H.A.: Peptide/protein−synthetic polymer conjugates: quo vadis. Macromolecules. **42**(21), 7990–8000 (2009)
14. Varaprasad, K., Raghavendra, G.M., Jayaramudu, T., Yallapu, M.M., Sadiku, R.: A mini review on hydrogels classification and recent developments in miscellaneous applications. Mater. Sci. Eng. C Mater. Biol. Appl. **79**, 958–971 (2017)
15. Deepa, M., Agnihotry, S.A., Gupta, D., Chandra, R.: Ion-pairing effects and ion-solvent-polymer interactions in $LiN(CF_3SO_2)_2$-PC-PMMA electrolytes: a FTIR study. Electrochim. Acta. **49**, 373–383 (2004)
16. Lu, S.X., Anseth, K.S.J.: Photopolymerization of multilaminatedpoly(HEMA) hydrogels for controlled release. J. Control. Release. **57**, 291–300 (1999)
17. Heskins, M., Guillet, J.E.: Solution properties of poly(N-isopropylacrylamide). J. Macromol. Sci. A Chem. **2**, 1441–1455 (1968)

18. Gupta, B., Agarwal, R., Alam, M.S.: Hydrogels for wound healing applications. In: Biomedical Hydrogels. Woodhead Publishing (2011)
19. Zauchner, D., Müller, M.Z., Horrer, M., Bissig, L., Zhao, F., Fisch, P., Lee, S.S., Zenobi-Wong, M., Müller, R., Qin, X.H.: Synthetic biodegradable microporous hydrogels for in vitro 3D culture of functional human bone cell networks. Nat. Commun. **15**, 5027 (2024)
20. Lai, Y.C., Wilson, A.C., Zantos, S.G.: Contact Lenses. Kirk-Othmer Encyclopedia of Chemical Technology, 5th edn. Wiley Online Library (2014)
21. Muncan, J., Mileusnic, I., Sakota Rosic, J., Vasic-Milovanovic, A., Matija, L.: Water properties of soft contact lenses: a comparative near-infrared study of two hydrogel materials. Int. J. Polym. Sci. **2016**, 1–8 (2016)
22. Nicolson, P.C., Vogt, J.: Soft contact lens polymers: an evolution. Biomaterials. **22**(24), 3273–3283 (2001)
23. Li, J., Mooney, D.J.: Designing hydrogels for controlled drug delivery. Nat. Rev. Mater. **1**(12), 16071 (2016)
24. Narayanaswamy, R., Torchilin, V.P.: Hydrogels and their applications in targeted drug delivery. Molecules. **24**(3), 603 (2019)
25. Mishra, B., Upadhyay, M., Reddy, A.S., Vasant, B., Muthu, M.: Hydrogels: an introduction to a controlled drug delivery device, synthesis and application in drug delivery and tissue engineering. Austin J. Biomed. Eng. **4**(4), 1037–1041 (2017)
26. Hoare, T.R., Kohane, D.S.: Hydrogels in drug delivery: progress and challenges. Polymer. **49**(8), 1993–2007 (2008)
27. Xia, H., Jin, H., Cheng, Y., Cheng, Z., Xu, Y.: The controlled release and anti-inflammatory activity of a tetramethylpyrazine-loaded thermosensitive poloxamer hydrogel. Pharm. Res. **36**(4), 52 (2019)
28. Shive, M., Bichara, A., Berrada, M., Ruel-garie, E.: A thermosensitive chitosanbased hydrogel for the local delivery of paclitaxel. Eur. J. Pharm. Biopharm. **57**, 53–63 (2004)
29. Chen, W., Shi, K., Liu, J., Yang, P., Han, R., Pan, M., Yuan, L., Fang, C., Yu, Y., Qian, Z.: Sustained co-delivery of 5-fluorouracil and cis-platinum via biodegradable thermo-sensitive hydrogel for intraoperative synergistic combination chemotherapy of gastric cancer. Bioact. Mater. **23**, 1–15 (2023)
30. Vishwanath, K., McClure, S.R., Bonassar, L.J.: Polyacrylamide hydrogel lubricates cartilage after biochemical degradation and mechanical injury. J. Orthop. Res. **41**(1), 63–71 (2023)
31. Calder, D., Fathi, A., Oveissi, F., Maleknia, S., Abrams, T., Wang, Y., Maitz, J., Tsai, K.H.Y., Maitz, P., Chrzanowski, W., Canoy, I., Menon, V.A., Lee, K., Ahern, B.J., Lean, N.E., Silva, D.M., Young, P.M., Traini, D., Ong, H.X., Mahmoud, R.S., Montazerian, H., Khademhosseini, A., Dehghani, F.: Thermoresponsive and injectable hydrogel for tissue agnostic regeneration. Adv. Healthc. Mater. **11**(23), 2201714 (2022)
32. Wu, Q., Gu, H., Cui, H.: Porous synthetic hydrogel carrying basic fibroblast growth factor with controllable and rapid degradation rate to promote wound healing. Cell. Polym. **41**(6), 231–242 (2022)
33. Zhang, Z., Chen, W., Tiemessen, D.M., Oosterwijk, E., Kouwer, P.H.: A temperature-based easy-separable (tempeasy) 3D hydrogel coculture system. Adv. Healthc. Mater. **11**(10), 2102389 (2022)
34. Kinney, S.M., Ortaleza, K., Vlahos, A.E., Sefton, M.V.: Degradable methacrylic acid-based synthetic hydrogel for subcutaneous islet transplantation. Biomaterials. **281**, 121342 (2022)
35. Zhu, K., Jiang, D., Wang, K., Zheng, D., Zhu, Z., Shao, F., Qian, R., Lan, X., Qin, C.: Conductive nanocomposite hydrogel and mesenchymal stem cells for the treatment of myocardial infarction and non-invasive monitoring via PET/CT. J. Nanobiotechnol. **20**, 211 (2022)
36. Usal, T.D., Yucel, D., Hasirci, V.: A novel GelMA-pHEMA hydrogel nerve guide for the treatment of peripheral nerve damages. Int. J. Biol. Macromol. **121**, 699–706 (2019)
37. Mabrouk, M., Beherei, H.H., Shiomoto, S., Tanaka, Y., Osama, L., Tanaka, M.: Effect of titanium-doped bioactive glass on poly(2-hydroxyethyl methacrylate) hydrogel compos-

ites: bioactivity, intermediate water, cell proliferation, and adhesion force. Ceram. Int. **49**, 13469–13481 (2023)
38. Tan, Z., Li, X., Yu, C., Yao, M., Zhao, Z., Guo, B., Liang, L., Wei, Y., Yao, F., Zhang, H., Li, J.: A self-gelling powder based on polyacrylic acid/polyacrylamide/quaternate chitosan for rapid hemostasis. Int. J. Biol. Macromol. **232**, 123449 (2023)
39. Huang, Z., Xiao, X., Jiang, X., Yang, S., Niu, C., Yang, Y., Yang, L., Li, C., Feng, L.: Preparation and evaluation of a temperature-responsive methylcellulose/polyvinyl alcohol hydrogel for stem cell encapsulation. Polym. Test. **119**, 107936 (2023)
40. Chen, T., Xu, G., Bao, J., Huang, Y., Yang, W., Hao, W.: One-pot preparation of hydrogel wound dressings from Bletilla Striata polysaccharide and polyurethane with dual network structure. Eur. Polym. J. **181**, 111648 (2022)
41. Feng, Y., Xiao, K., He, Y., Du, B., Hong, J., Yin, H., Lu, D., Luo, F., Li, Z., Li, J., Tan, H., Fu, Q.: Tough and biodegradable polyurethane-curcumin composited hydrogel with antioxidant, antibacterial and antitumor properties. Mater. Sci. Eng. C. **121**, 111820 (2021)
42. Chee, B.S., de Lima, G.G., de Lima, T.A.M., Seba, V., Lemarquis, C., Pereira, B.L., Bandeira, M., Cao, Z., Nugent, M.: Effect of thermal annealing on a bilayer polyvinyl alcohol/polyacrylic acid electrospun hydrogel nanofibres loaded with doxorubicin and clarithromycin for a synergism effect against osteosarcoma cells. Mater. Today Chem. **22**, 100549 (2021)
43. Vakili, M.R., Mohammed-Saeid, W., Aljasser, A., Hopwood-Raja, J., Ahvazi, B., Hrynets, Y., Betti, M., Lavasanifar, A.: Development of mucoadhesive hydrogels based on polyacrylic acid grafted cellulose nanocrystals for local cisplatin delivery. Carbohydr. Polym. **255**, 117332 (2021)
44. Capella, V., Liaudat, A.C., Broglia, M.F., Barbero, C.A., Bosch, P., Rodríguez, N., Rivarola, C.R.: Biointerfacialanalysis of HEK293 cells in contact with hydrogels based on poly-N-isopropylacrylamide and copolymers. Adv. Mater. Interfaces. **9**(34), 2201162 (2022)
45. Park, J.R., Bolle, E.C.L., Santos Cavalcanti, A.D., Podevyn, A., Van Guyse, J.F.R., Forget, A., Hoogenboom, R., Dargaville, T.R.: Injectable biocompatible poly(2-oxazoline) hydrogels by strain promoted alkyne-azide cycloaddition. Biointerphases. **16**(1), 011001 (2021)
46. Zhang, L., Yao, K., Wei, J., Li, G., Lin, Y., Zhou, Y.L., Yang, Y.: Convenient in situ synthesis of injectable lysine-contained peptide functionalized hydrogels for spinal cord regeneration. Appl. Mater. Today. **27**, 101506 (2022)
47. Kim, J.R., Cho, Y.S., Park, J.H., Kim, T.H.: Poly(HEMA-co-MMA) hydrogel scaffold for tissue engineering with controllable morphology and mechanical properties through self-assembly. Polymers. **16**(21), 3014 (2024)
48. Simon, E., Lennard, S.K., Andreas, E., Dominik, L., Wolfgang, E., Michael, S., Carsten, R., Tilo, D., Michal, J., Jacob, S.: Additively manufactured 3D clamp-culture system for the investigation of material-cell interactions in multi-material hybrid scaffolds for musculoskeletal tissue defects. J. Biomed. Mater. Res. B Appl. Biomater. **112**(11), e35494 (2024)
49. Westra van Holthe, N.J., Wang, Z., Lauko, J., Gilbert, E.P., Gangasandra, V., Rowan, A.E.: Controlling ligand density and viscoelasticity in synthetic biomimetic polyisocyanide hydrogels for studying cell behaviours: the key to truly biomimetic hydrogels. Mater. Adv. **5**(23), 9458–9470 (2024)
50. Adriana, M.R., Ana, M.B., Elijah, M.N., Morgan, Z., Michael, H., Andres, G.: Synthetic hydrogel substrate for human induced pluripotent stem cell definitive endoderm differentiation. Biomaterials. **315**, 122920 (2025)

Chapter 7
Hydrogels as Extracellular Matrix Mimics

Özge Erdemli and Ayşen Tezcaner

Abstract In tissue engineering approaches, the use of tissue scaffolds that mimic the properties of tissue-specific extracellular matrices by providing structural support and facilitating cellular processes has gained importance for producing three-dimensional (3D) functional tissues and organs to repair or replace damaged tissues. For cell adhesion, proliferation, migration, differentiation, and tissue formation, cells must receive the necessary biochemical and/or mechanical signals from their environment. Therefore, tissue scaffolds that can provide a suitable 3D environment and the necessary water content in terms of biochemical, biomechanical, and topographic aspects are needed. In the formation of tissue scaffolds, the synthetic or natural biomaterials that have porosity, wettability, biodegradability, mechanical, and surface properties suitable for mimicking natural extracellular matrix are preferred. In this way, the interactions between cells and biomaterials enable the tissue regeneration process to be directed correctly. Up to today, no hydrogel system has been developed that mimic the native tissue closely enough. However, advances in tissue engineering and biomaterials will surely enable the development of novel hydrogel approaches that mimic the tissue structurally, mechanically, and also in terms of chemical composition and dynamic properties.

Keywords Extracellular matrix · Hydrogels · Collagen · Glycoproteins · Proteoglycans · Glycosaminoglycans · ECM · GAGs

---

Ö. Erdemli
Department of Molecular Biology and Genetics, Baskent University, Ankara, Türkiye

A. Tezcaner (✉)
Department of Engineering Sciences, Middle East Technical University, Ankara, Türkiye

BIOMATEN Center of Excellence in Biomaterials and Tissue Engineering, Middle East Technical University, Ankara, Türkiye
e-mail: tezcaner@metu.edu.tr

N. Hasirci, V. Hasirci (eds.), *Hydrogels and Bioinks in Tissue Engineering*, Biological and Medical Physics, Biomedical Engineering,
https://doi.org/10.1007/978-3-031-90966-5_7

7.1 Extracellular Matrix

All tissues have an extracellular matrix (ECM) that contains cells as well as non-cellular components and well-organized fiber protein networks. The ECM provides mechanical support to cells and plays a role in regulating various cellular processes. The ECM has a 3D porous structure containing fibrous protein networks such as collagen, fibronectin and elastin, embedded in a viscoelastic gel with a high-water content composed of proteoglycans and polysaccharides (Fig. 7.1). Although the types, amounts, and spatial distributions of the components contained in the ECM vary depending on the tissue type, fibrous proteins such as collagen, elastin, fibronectin, and laminin, as well as glycoproteins, proteoglycans, and glycosaminoglycans (GAGs), are the main components of the ECM [1].

Collagen is the major protein component that constitutes more than 30% of the ECM. Collagens, which have a characteristic triple helical morphology composed of homo- or heterotrimeric α-chains, provide tissue flexibility and stability while also functioning as supportive tissue structures. Collagen types I and III are the most common types of collagens found in most tissues. Additionally, some collagen types, such as collagen types VIII and VI, are also prevalent in tissues. In connective tissues, collagen IX and II are abundant, while collagen types XIII and XVII are frequently seen in epithelial tissues.

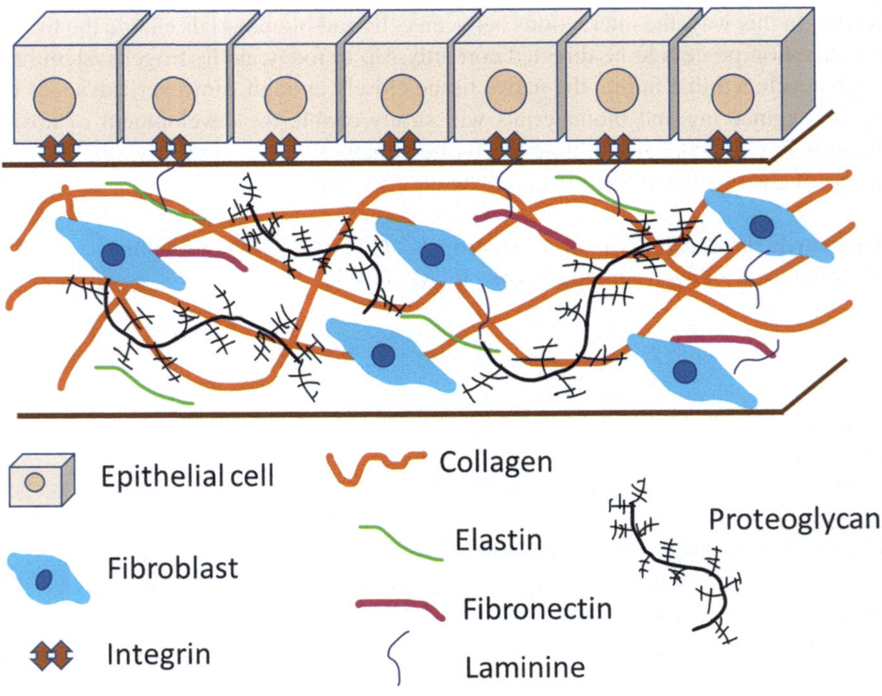

Fig. 7.1 Extracellular matrix structure

Fibronectin, the second most abundant protein in the ECM, is a high molecular weight dimeric glycoprotein and provides an attachment site for many cells. Fibronectin binds to cells primarily via integrin receptors. Another family of cell surface receptors that bind to fibronectin are syndecans. Fibronectin binds to cell surface receptors and provides communication between the intracellular and extracellular environment. Laminins, large heterotrimeric cross-shaped glycoproteins, are molecules that facilitate cell adhesion and are found in the basement membrane of almost every tissue. They also play a critical role in cell differentiation, migration, proliferation, and angiogenesis. Another important ECM protein, elastin, is the main component of elastic fibers that form a randomly oriented and interconnected fiber system found in elastic tissues and organs such as blood vessels, lungs, and skin. Elastin also has the ability to participate in cell signaling, playing a role in cell adhesion and migration.

Proteoglycans, one of the most important structural and functional biomacromolecules found in the ECM, consist of a protein core to which one or more negatively charged GAG chains of the same or different types are covalently attached. Proteoglycans, which have both structural and biological roles, also serve to provide mechanical resistance against compressive forces in tissues, facilitate tissue hydration, and trap growth factors in the ECM. Because of their ability to interact with other ECM molecules and cells, proteoglycans are important molecules for the organization of the ECM. The GAG chains attached to proteoglycans generally contain 20–100 disaccharide repeat units. Six major types of GAGs have been identified in mammalian cells: chondroitin sulfate, dermatan sulfate, keratan sulfate, heparan sulfate, heparin, and hyaluronic acid. Except for hyaluronic acid, GAGs are anionic polysaccharides containing sulfate groups and they are attached to core proteins. Hyaluronic acid is not bound to a protein core and is not esterified with sulfate groups. Hyaluronic acid contributes to maintenance of the structural integrity of the tissue and retention of water in the tissue structure.

The ECM, the non-cellular component of tissues, is a hierarchically organized three-dimensional structure with different chemical compositions, different structural arrangements, and a high proportion of water, depending on the type and function of the tissue in which it is located. While the ECM has long been considered a structure that merely provides physical support for cells, studies have shown that in addition to its physical properties, the spatial organization and biochemical complexity of its components are also crucial in determining cell behaviors [2, 3]. The ECM has a dynamic structure due to the secretion of extracellular components such as proteins or soluble factors by cells at various stages of tissue formation and development processes or during wound healing, and its biochemical composition and physical properties constantly change as this structure undergoes remodeling [1, 3, 4]. Changes in the physical properties of the ECM, such as morphology, topography, water content, and biomechanical properties, often affect processes such as cell-cell adhesion, proliferation, and migration. Changes in the biochemical composition of the ECM mostly affect processes that determine cell fate, such as survival, proliferation, or differentiation [3]. Moreover, physical and biochemical properties are also interconnected and regulate cell-ECM interactions, including molecular recognition, mechanotransduction, electrical sensing, and soluble molecule transfer.

Due to their high-water content and similar physical and mechanical properties, hydrogels are widely used to mimic the natural ECM [1–4]. Hydrogels mimicking ECM properties can be prepared from natural biopolymers, synthetic polymers, or their hybrids. Synthetic hydrogels are 3D polymer networks formed by covalent or ionic crosslinking of hydrophilic homopolymers or copolymers.

Synthetic polymers such as poly(vinyl alcohol) (PVA), poly(ethylene glycol) (PEG), poly(ethylene oxide) (PEO), poly(2-hydroxyethyl methacrylate) (PHEMA), poly(acrylic acid) (PAA) and poly(acrylamide) (PAAm), etc. are widely used in the preparation of hydrogels by many different methods ranging from monomer polymerization to crosslinking [5]. Although synthetic hydrogels have advantages over natural polymers in terms of controllable chemical, physical, mechanical, and biodegradation properties, they have lower biological activity than natural hydrogels. To overcome this limitation, immobilization of natural ECM proteins or cell adhesion peptides derived from these proteins on the surface of synthetic hydrogels or incorporation of growth factors and other bioactive molecules into the structure of synthetic hydrogels are frequently used methods [6].

Natural hydrogels prepared from peptide, protein, polysaccharide, and nucleic acid-based biopolymers derived from renewable natural resources such as plants, animals, humans and microorganisms are one of the most promising biomaterials for regenerative medicine and tissue engineering applications due to their biocompatibility, degradability, and natural bioactivity [7]. Natural hydrogels containing cell-binding motifs that provide bioactive cues for cellular processes, or natural hydrogels that may exhibit structural features similar to the natural ECM, can regulate cell adhesion, proliferation, migration, differentiation, remodeling, and tissue formation processes [7]. Despite all these advantages, poor mechanical properties and differences in chemical composition from isolation to isolation due to their natural origin are the main limitations encountered in the use of natural hydrogels.

Although frequently used for tissue engineering applications, synthetic polymers and biopolymers derived from plants and microorganisms lack the biochemical properties and 3D structure of the native mammalian ECM. For this reason, studies with hydrogel-based tissue scaffolds containing natural ECM components, prepared by isolation of mammalian ECM components or development of decellularization techniques, in which the bioactive properties of the 3D structure and components are preserved, have been common in recent years [8, 9].

7.2 Mimicking the Extracellular Matrix

The structural, biochemical, and mechanical properties of the native cell microenvironment in the target tissue should be considered in the development of scaffolds that can mimic the properties of the ECM (Fig. 7.2). The biochemical properties of the native ECM can be mimicked by incorporating one or more components of the

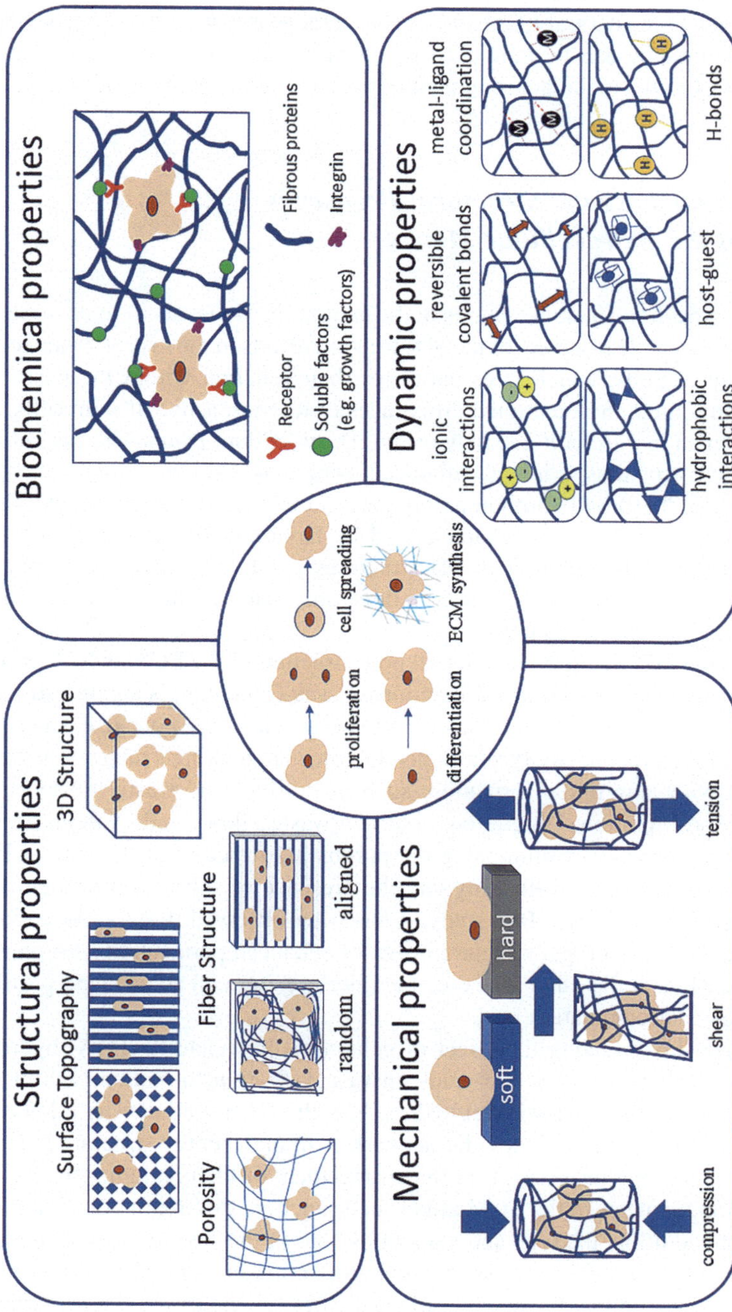

Fig. 7.2 Basic approaches used to mimic the extracellular matrix

native ECM into the scaffold structure, by changing the types or ratios of these components, and/or by adding various growth factors to the medium. The mechanical properties of the natural ECM and its structural properties such as shape, porosity, and topography can be mimicked by changing the parameters of the surface modifications or the manufacturing methods used to produce the scaffold.

7.2.1 Mimicking the Structural Properties of the Extracellular Matrix

One of the most fundamental features of the natural ECM that needs to be mimicked is its 3D structure. The neglect of the 3D structure results in the loss of characteristic morphology of cells, which is essential for their function within the tissue. For example, structural and functional differences of rat mesenchymal stem cells were examined in an in vitro culture study using 2D cell culture plates with flat surfaces and 3D microstructures (Nichoid) produced using a hybrid organic-inorganic photopolymer (SZ2080) by two-photon polymerization [10]. The morphology of cells cultured on flat surfaces was characterized by large and spread nuclei and cytoplasm, whereas cells cultured in 3D structures exhibited smaller and spherical nuclei, as well as spatial organization in three dimensions, which is crucial for the maintenance of stem cell function.

Besides the 3D structure, the internal pore structure of the ECM, which plays an important role in cell survival and functional tissue formation, is another structural feature to be mimicked. Pore architectural features, including pore size, shape, distribution, and interconnectivity, play a pivotal role in regulating a range of processes associated with new tissue formation, such as nutrient and waste transport, cell adhesion, proliferation, migration, nutrient/waste flow and vascularization. Hydrogels are capable of mimicking the porous 3D structure of the natural ECM due to their high water swellability and the presence of submicron or nano-sized crosslinked network pores. In a study, it has been observed that the formation of macropores in the hydrogel structure provides enhanced permeability for nutrient/waste flow while at the same time creating interfaces for cell attachment, proliferation, and ECM accumulation [11].

In a study investigating the effect of gelatin methacrylate (GelMA) hydrogels with different densities of micro-voids on vascularization formation both in vitro and in vivo, it has been shown that hydrogels with higher void density allowed the formation of interconnected vascular networks within the entire structure [12].

Tissue anisotropy resulting from the arrangement of fibers ranging from 10 nm to 300 μm in diameter in the ECM affects cell function, differentiation, and migration [13]. In addition, the type and amount of fiber in the structure plays an important role in determining the mechanical properties of tissues and ECM. For example, it has been observed that the quantity of elastic fibers increases in tissues subjected to tensile forces, whereas the quantity of collagen fibers increases in harder and more durable tissues [13]. Changes in the amount, structure, size, and arrangement of ECM fibers are also observed in aging and disease conditions.

In addition to determining the mechanical properties of the tissue, nanofiber structure also influences growth factor signaling and thus cellular behavior. The binding of growth factors, such as vascular endothelial growth factor (VEGF), bone morphogenic protein (BMP), transforming growth factor-β (TGF-β), and epidermal growth factor (EGF), to collagen and fibronectin fibers in the ECM facilitates interactions between integrin receptors on the cell surface [13]. Various in vitro and theoretical modeling studies have demonstrated that the non-linear stress-strain behavior of the 3D fiber network within the ECM influences the sensing of mechanical signals by cells and facilitates cell-cell interactions through tissue-specific fiber position and arrangement [13]. To mimic the natural ECM structure, tissue scaffolds with fiber structures have been produced from natural or synthetic polymers using various techniques such as electrospinning, self-assembly, and 3D printing. Composite tissue scaffolds formed by combining fibers produced by different methods with hydrogels have mechanical and degradation properties that support the cell behaviors necessary for healing as shown by various studies [14, 15].

Fiber proteins with micro- and nanoscale properties in the native ECM provide topographic stimuli to regulate cellular behavioral processes. Another strategy is the production of micro- and nano-topographies on the surface of tissue scaffolds that have been prepared to mimic the natural ECM. Hydrogels patterned by various methods such as mold casting, photolithographic patterning, electrospinning, nanoimprint lithography, multiphoton patterning, e-beam lithographic patterning, and 3D printing have been shown to mimic the ECM, enhance cell adhesion, and regulate cell morphology, migration, and orientation [16].

7.2.2 Mimicking the Biochemical Properties of the Extracellular Matrix

Collagen, the most abundant protein in mammalian tissues, is the major basic component of the ECM, and more than 28 different types of collagens, varying in tissue-specificity, have been shown to play a structural role in cell adhesion [3]. In addition to structural support, collagen also has a role at the biomolecular level in cell adhesion, survival, differentiation, and different cell signaling mechanisms [3, 17]. Because of these properties, collagen is frequently used in tissue engineering and 3D cell culture studies [17–19]. Glycoproteins such as fibronectin, laminin, and vitronectin with multiple binding sites that can interact with collagen and other proteoglycans are involved in cell adhesion. Through these proteins, the ECM provides a surface to which cells can adhere via surface receptors such as integrin, selectin, CD44, and syndecan. ECM proteins such as collagen, fibronectin, and laminin have motifs that increase the adhesion of cells to the ECM through integrin binding.

The RGD peptide motif (Arg-Gly-Asp), which is found in ECM proteins and promotes cell adhesion by interacting with integrins, is one of the most widely investigated recognition motifs. Some other receptors, including discoidin domain receptors, dystroglycan, syndecanes, CD44, and lectin, play a role in regulating cell-ECM interactions by interacting with various ECM proteins or polysaccharides [3]. The proteins in the fiber structure and the proteins that provide adhesion are embedded in a highly water-containing viscoelastic gel structure formed by proteoglycans and GAGs. Heparan and chondroitin sulfate, the major proteoglycans in the ECM, play a role in determining the stiffness of the hydrogel-like structure of the ECM and thus in controlling cell fate [3]. In addition, many ECM components such as collagen, fibronectin, heparin, and heparan sulfate glycosaminoglycans also have binding sites for growth factors and thus cellular activities are regulated by the release of growth factors into the environment by the ECM. Growth factors generally include the transforming growth factor beta (TGF-β), epidermal growth factor (EGF), fibroblast growth factor (FGF), and platelet-derived growth factor (PDGF) families. The proteolytic degradation of the ECM by enzymes such as matrix metalloproteinases (MMPs), plasmin and elastase secreted by cells is also a key feature of tissue repair and remodeling processes.

The most basic biological functions of the ECM, which provides a suitable microenvironment for cell-cell communication, cell growth, and tissue remodeling, are cell adhesion, proteolytic degradation, and growth factor binding, as mentioned above. The most common strategies used to mimic the properties of the ECM that enable cell adhesion are the production of hydrogels containing protein and/or polysaccharide components involved in cell adhesion or the modification of hydrogel surfaces with recognition motifs [3]. Hydrogels composed of components of the ECM, such as collagen [18–20], hyaluronic acid [21, 22], and fibrin [20, 23], or a mixture of ECM proteins, such as Matrigel™, are frequently studied because they mimic the ECM and support cell adhesion and differentiation. In addition, various synthetic or natural hydrogels that provide various advantages in terms of physicochemical and mechanical properties but are inadequate in terms of cell adhesion can be provided with functionality and effectiveness for biological processes through bioconjugation. In the bioconjugation approach, ECM components or peptide motifs that play a role in directing cellular processes are added to structures that are insufficient for cell attachment by covalent or non-covalent binding or physical entrapment [6]. The hydrogels obtained from bioconjugation not only retain the advantages of physicochemical and mechanical properties provided by the polymers in their structure, but also gain the functional properties necessary for cellular processes provided by the ECM components. The hydrogels obtained from bioconjugation both retain the physicochemical and mechanical property advantages provided by the polymers in their structure and gain the functional properties necessary for cellular processes to be provided by ECM components.

Various growth factors can be loaded into hydrogels to provide the biochemical signals necessary to promote tissue formation and improve vascularization. Thus, the ECM's ability to bind and release growth factors is mimicked, allowing growth factors to be delivered to the target site in a controlled manner, in a specific amount and for a specific duration. Growth factors loading into hydrogels can be achieved by physical loading and by chemical or non-chemical binding.

The ability of the ECM to be degraded is crucial for tissue development. Therefore, the degradation properties of hydrogels to be used as tissue scaffolds must be compatible with the natural ECM accumulation and cell migration produced by the cells adhering to them. Therefore, synthetic polymers and natural polymers with biodegradable properties are preferred in the production of hydrogel tissue scaffolds. The biodegradability of hydrogels can be engineered using different production methods. Although degradation profiles are predictable in synthetic hydrogels that degrade predominantly hydrolytically, hydrolytic degradation does not represent the dynamic processes occurring in the native ECM. The predominant cellular process is proteolytic degradation and the hydrolysis of proteins into peptides and amino acids. The proteolytic degradation property of the ECM can be mimicked by modifying synthetic hydrogels with various proteolytic enzyme-sensitive peptides derived from extracellular proteins [24]. In a study in which human mesenchymal stem cells were loaded into poly(ethylene glycol) (PEG) hydrogels crosslinked with matrix metalloproteinase (MMP)-responsive peptide (KCGPQG↓ IWGQCK), it was shown that cell migration and remodeling were promoted by initiation of hydrogel degradation in response to MMPs secreted by the cells [25].

7.2.3 Mimicking the Mechanical Properties of the Extracellular Matrix

The ECM determines the fate of cells by directing cell growth and differentiation by transmitting mechanical stimuli from the environment to the cells. Cells sense the stiffness of the ECM through a number of force-sensing proteins present in the ECM and also through a process known as “cellular mechanosensing” that occurs through focal adhesion. Detected mechanical signals cause conformational changes in the structural motifs of the cell and are mediated through a series of biochemical pathways known as mechanotransduction. Changes in gene and protein expression during mechanotransduction initiate various cellular processes, including cell growth, differentiation, and death. In tissues with different physiological and functional roles, the types, quantitative ratios, and spatial arrangements of the ECM components cause the ECM to have different stiffness values [26]. Collagen fibers, which constitute a large part of the structure of the ECM, provide tensile strength to the ECM. Proteins with elastic properties, such as elastin and proteoglycans, are responsible for the elasticity of the ECM. In the structure of hard tissues such as bone, the amount and variety of fiber proteins is much higher than elastic proteins. On the other hand, blood vessels must have more elastic properties in order to withstand the tensile forces exerted by blood flow and therefore contain more elastin in their structure. The hierarchical architecture of the ECM is another parameter that determines the stiffness of the structure. For example, the hierarchical anisotropic arrangement of collagen fibers in the tendon structure, formed by the undulating pattern of collagen fibers within the fascicles, is of great importance for the unique mechanical properties of the tendon. In contrast, the dura mater, the thickest and outermost layer of the cerebral membranes surrounding the brain tissue and spinal cord, has very low stiffness due to the irregular arrangement of collagen fibers in its structure.

Mimicking the structural and biochemical properties as well as the mechanical properties of the ECM is very important for various cellular properties such as cell adhesion, spreading, proliferation, differentiation, and apoptosis. Many studies in the literature on mesenchymal stem cells have shown that these cells differentiate into different cell types such as osteocytes, chondrocytes, myocytes, or neurons on surfaces with different hardness [27]. Therefore, the physical and chemical properties of hydrogels mimicking the ECM need to be adjusted to match the viscoelastic properties of the native ECM [28]. It is possible to adjust the viscoelastic properties of hydrogels as desired by changing the production parameters such as molecular weights, component ratios, environmental conditions, and crosslinking methods of the polymers used in hydrogel production or by adding various nanomaterials to the hydrogel structures [26–28]. Hydrogel stiffness can be modified using physical and chemical crosslinking methods. Hydrogels physically crosslinked using ionic bonds, hydrogen bonds, electrostatic interactions are of great interest in tissue engineering applications due to their biocompatibility and the mild conditions under which the crosslinking process takes place. However, due to various factors affecting crosslinking, including temperature, ion concentration, and pH, physical crosslinking cannot be fully controlled, thus causing difficulties in achieving the desired hydrogel hardness [26]. Chemical crosslinking, which involves covalent bonds that are stronger than ionic bonds, allows hydrogel hardness to be controlled and makes hydrogels more mechanically durable. There are studies that mimic the ECM with hybrid structures having adjustable stiffness and bioactive properties obtained by adding inorganic bioactive filler materials such as calcium phosphate, bioceramics, clay, carbon nanotubes, graphene oxide, superparamagnetic iron oxide nanoparticles (SPIONs) to the hydrogel structure [29, 30]. There are also studies in the literature in which various fiber structures were added to hydrogel scaffolds to mimic the fiber-based structure of the ECM [14, 15]. Fiber arrangement, fiber aspect ratio, material properties, and fiber-hydrogel interactions are used to control the mechanical properties of fiber-hydrogel composite structures to mimic the mechanical environment adequate for cell attachment, migration, differentiation, and proliferation [15].

7.2.4 *Mimicking the Dynamic Properties of the Extracellular Matrix*

The structural, biochemical, and mechanical cues of the ECM during physiological processes and pathological events are highly dynamic and significantly influence cellular processes such as cell adhesion, spreading, proliferation, differentiation, and apoptosis. Although they provide advantages in terms of stability and mechanical properties, covalently crosslinked hydrogels are deficient in terms of the dynamic properties required for cell behavior and remodeling of the ECM. Therefore, to promote tissue formation, hydrogel scaffolds made of polymers that can provide dynamic changes and are biodegradable are preferred. Such hydrogels can be produced by various methods and can mimic the ECM in terms of structural, biochemical, and mechanical properties mentioned in the previous sections. However, the physical support provided to the cells over time is limited because the changes caused by the degradation of the hydrogel structures lead to a decrease in mechanical properties. In recent years, hydrogels with reversible bindings and the ability to adapt their properties according to environmental changes have been developed to mimic the dynamic changes of the natural ECM [31]. Dynamic hydrogels formed by supramolecular chemistry that includes physical interactions such as reversible covalent bonds, electrostatic interactions, H-bonds, host-guest, and metal-ligand coordination provide dynamic cues through shear thinning and self-healing as well as tunable elastic and viscoelastic properties [31, 32].

7.3 Strategies Used in the Production of Extracellular Matrix-Like Hydrogels

Hydrogel-based tissue scaffolds, which can mimic ECM properties, have been prepared from many different materials using various fabrication methods such as in situ gelation, electrospinning, 3D printing, microfabrication, and microfluidic technologies depending on the properties of the target tissue and the type of application to be used. Depending on the characteristics of the target tissue and the type of application to be used, the final properties of the end product can be adjusted to meet the needs of the application by changing various chemical or physical surface properties.

7.3.1 *In Situ Gelation*

In situ gelling hydrogels, which are formed by crosslinking of the hydrogel solution spontaneously or with external stimuli after the addition of cells and/or bioactive components to the precursor hydrogel solution, are used in drug delivery systems and tissue scaffold applications [33, 34]. Such hydrogels solidify at the application site after injection, providing a temporary structure on which cells can adhere and form a functional new tissue. Such 3D hydrogels have the advantages of being easily loaded with cells and being applied to the living system with minimal invasiveness [33]. In addition, hydrogels that can gel in place can easily take the shape of the injected area and adhere to surrounding tissues. In a study, it was shown that graphene oxide-gelatin hydrogels, which can be gelled in situ by enzyme-mediated crosslinking after injection, have the mechanical properties required to mimic the ECM, can adhere to the surrounding tissue, and support the proliferation of human dermal fibroblasts (hDFBs) [35]. The first strategy used to prepare in situ gelling hydrogels was the polymerization of small molecules using initiators and chemical cross-linkers. Chemical crosslinking can be achieved by various reactions such as enzymes, high-energy radiation, photoactivation, or click reactions. Crosslinking of polymer chains with external environmental stimuli such as temperature, pH, enzymes, and redox reactions is also used to prepare in situ gelling hydrogels. It is possible to prepare self-healing and in situ gelling hydrogels with desired mechanical properties thanks to the covalent bonds involved in Schiff base reactions, physical bonds such as hydrogen bonds, and crosslinking formed by reversible bonds [36]. It has been shown that the oxidized sodium alginate-chitosan fiber hydrogels obtained by mixing chitosan microfibers prepared by electrostatic self-organization method with oxidized sodium alginate solution are very similar to the natural ECM in terms of hierarchical arrangement [37]. Moreover, the reversibility of the Schiff base reaction and its sensitivity to hydrogen ions (H^+) provided pH-responsive and dynamic self-healing properties to the oxidized sodium alginate-chitosan fiber hydrogels.

7.3.2 Crosslinking

The method used in crosslinking hydrogels that can mimic the properties of the ECM should be compatible with biological factors such as cells and growth factors. In addition, the preparation of such hydrogels must take place under mild conditions. Finally, the degradation products must be non-toxic to cells and tissues. In physically crosslinked hydrogels, non-covalent forces (secondary interactions) such as hydrogen bonding, ionic forces, van der Waals interactions, polyelectrolyte complexation, stereocomplexation, and hydrophobic forces are used to transition from the solution phase to the gel phase. The most important advantages of physical crosslinking are that it enables the preparation of hydrogels at room temperature without requiring functionalization of the precursor polymers and without causing cytotoxicity. Because they are crosslinked by weak secondary interactions, such hydrogels show reversible responses to environmental changes [5]. Physically crosslinked hydrogels may also enable the development of stimuli-responsive systems with room-temperature self-healing and injectable properties [38]. However, since they are crosslinked by secondary interactions that are not very strong, these hydrogels are mechanically weak and brittle against external stimuli [5].

The presence of ionic groups in polysaccharides such as alginate, chitosan, and cellulose derivatives enables the production of hydrogels by physically crosslinking through interactions of opposite electrical charges. For example, if alginate, a natural polymer, reacts with divalent cations such as calcium (Ca^{2+}), barium (Ba^{2+}), and magnesium (Mg^{2+}), the negatively charged guluronate blocks in the alginate chains interact with these ions and form crosslinks. Chitosan, a natural polycationic polymer, can easily form polyelectrolyte complexes through electrostatic interactions with natural anionic polymers such as pectin, chondroitin sulfate, and alginate, or with some synthetic polymers such as polylactic acid, polyacrylic acid, and polyphosphoric acid [38]. Amide, urea, carboxylic acid, pyrrole, carbazole, and hydroxyl groups can provide cross-linking by forming hydrogen bonds among themselves or by interacting with electron-donating groups such as pyridine and imidazole groups. However, instead of a single hydrogen bond, multiple and multivalent hydrogen bonds must be established in the structure for hydrogel formation. Hydrophobic interactions are used for crosslinking water-soluble polymers with hydrophobic end groups, side chains, or monomers. Hydrophobic interactions enable gelation in temperature-induced transition polymers based on methylcellulose and Pluronic, which are soluble at low temperature and self-assemble at higher temperature. Methods using thermal induction or ultrasonic treatments based on low critical solution temperature (LCST) or upper critical solution temperature (UCST) induce hydrophobic interactions and enable the production of physical crosslinked hydrogels.

In chemically crosslinked hydrogels, the transition from the solution phase to the gel phase is achieved by using various crosslinkers such as small binding molecules, enzymes, or radiations, various chemical reactions such as oxidation-reduction reactions, radiation-initiated reactions, free radical polymerization, Michael reaction, Schiff base reaction, and click reaction [5]. Chemically crosslinked hydrogels have better stability than physically crosslinked hydrogels due to the presence of strong covalent bonding between the polymer chains and do not react reversibly to environmental changes. Additionally, as a result of crosslinking involving strong covalent bonds, there are significant improvements in the thermal, mechanical, chemical, and surface properties of hydrogels. Radiation-initiated crosslinking reactions using UV, near-infrared radiation, and in some cases visible light occur at unsaturated vinyl, acrylate, methacrylate, or allylic groups of precursor polymers.

7.3.3 Using Extracellular Matrix Components

Various ECM components such as collagen, elastin, fibrin, and hyaluronic acid are frequently used to construct hydrogel-based tissue scaffolds. In addition, as mentioned before, by adding ECM components to the structure, the biochemical and physical properties of the natural ECM are better mimicked, and synthetic or natural hydrogels that are inadequate in terms of cell adhesion can be enabled to gain functionality and effectiveness for biological processes. Although the ECM is frequently used as a biomaterial because it provides structural support, attachment sites, and biological and mechanical cues to cells due to its tissue-specific microstructural properties and components, it has some disadvantages due to its properties, such as low reproducibility in isolation, limited control in structural and mechanical properties, and easy degradation [39].

To create ECM-based hydrogel scaffolds, methods such as isolation of ECM components from relevant tissues or decellularization of relevant organs or tissues are used [8, 9]. There are two basic steps in the preparation of extracellular matrices obtained as a result of isolation: (1) homogenization of the ECM and/or dissolution of its basic components with enzymes, acids, or other solvents for certain periods of time; (2) neutralization of the obtained soluble ECM by controlling temperature, ionic strength, and/or pH to ensure that it becomes a homogeneous gel [8]. Although components obtained from natural ECM by isolation methods are frequently used in the production of hydrogel-based tissue scaffolds, isolated components cannot provide the three-dimensional structure and regional release of growth factors in natural tissue [9].

In recent years, with the help of recombinant technology, ECM proteins such as collagen, laminin, and elastin are produced in bacteria and eukaryotic cell lines under in vitro conditions by making specific modifications [9]. In the final stage of this process, which is more reproducible and less expensive than isolation methods from natural tissues, proteins obtained by decellularization methods largely retain their structure and bioactivity.

Another method to prepare hydrogel-based tissue scaffolds mimicking the properties of the ECM is decellularization. In this method, cells in the tissue are removed using physical, chemical, biological approaches, and their combination to obtain decellularized ECM. Although different methods are used, the main purpose of tissue decellularization is to almost completely remove the nuclear content and cellular components from the medium while ensuring minimum damage to the components and structural properties of the ECM [40]. Therefore, it is necessary to choose a decellularization method that is appropriate for the cellular and other component content, density, and thickness of the target tissue. Physical methods used in decellularization include disruption of the cell membrane and lysis of cells by freezing-thawing, osmotic pressure, mechanical agitation, and high hydrostatic pressure [39–42]. In physical approaches, the structure of the ECM is less damaged compared to other approaches, but the remains of the disintegrated cells do not leave the environment [41]. Additional chemical or biological methods are required to completely remove cellular fragments.

Chemical methods applied in decellularization use detergents, acids, bases, and hypertonic and hypotonic solutions to disrupt cellular attachments and remove cellular components [39–42]. Ionic detergents such as sodium dodecyl sulfate (SDS), sodium deoxycholate (SDC), sodium lauryl sulfate, and Triton X-200 disrupt lipid-lipid, lipid-protein, deoxyribonucleic acid (DNA)-protein, and protein-protein interactions, resulting in decellularization [42]. However, disruption of protein-protein interactions by ionic detergents may also cause damage to the ECM structure and bioactive components. Non-ionic detergents such as Triton X-100 provide decellularization by dissolving the cell membrane and disrupting lipid-lipid, lipid-protein, and DNA-protein interactions. Although non-ionic detergents, which do not damage protein-protein interactions, provide milder conditions than ionic detergents, they are often insufficient to remove nuclear components such as DNA [42]. Zwitterionic detergents, which exhibit the properties of both non-ionic and ionic detergents, are preferred for the decellularization of thin tissues with a thickness of less than 1.5 mm, such as urinary bladder and lung tissue. Acids such as acetic acid, peracetic acid, hydrochloric acid, and sulfuric acid, which catalyze the hydrolytic degradation of biomolecules, cytoplasmic components, and the degradation of nucleic acids, and bases such as ammonium hydroxide, calcium hydroxide, sodium sulfide, and sodium hydroxide are also used in the decellularization of tissues. However, acids and bases that cause harsh decellularization conditions can cause damage to the ECM structure and bioactive components. Hypertonic and hypotonic salt solutions decellularize tissues by causing cell lysis, dehydration, and death as a result of osmotic effects. Although such solutions facilitate the removal of chemical and cellular wastes from the ECM, they may cause tissue disintegration or swelling [42]. However, they cannot completely remove cellular debris unless combined with other decellularization methods.

Enzymes such as nuclease, trypsin, dispase, lipase, and phospholipidase provide decellularization of tissues with high specificity as they can specifically remove cells and unwanted ECM components. However, in long-term procedures, the mechanical stability of the ECM may decrease, and various ECM components may be damaged by enzymatic treatments [43].

The ECM derived from tissue by decellularization methods can form a tissue scaffold that preserves most of the main protein components and the 3D structure of the native tissue that can support cellular processes. In a previous study, it was observed that a hydrogel obtained from decellularized human dental pulp promoted the adhesion, migration, and proliferation of dental pulp stem cells and facilitated the differentiation of cells into odontoblast-like, neural-like, and angiogenic cells during 14 days of in vitro cell culture [44]. The efficacy of hydrogels produced from ECM obtained from decellularized porcine adipose tissue loaded with adipose stromal cells on full-layer excisional wounds in mice in vivo was investigated, and it was shown that these hydrogels had positive effects on fibroblast activation and VEGF secretion and could significantly provide re-epithelialization, angiogenesis, and skin appendage regeneration at the wound site [45].

ECM components produced in cell lines with recombinant technology approaches or obtained by isolation or decellularization from native tissues exhibit improved host response and integration compared to synthetic materials. However, the inadequacy of the mechanical strength of these natural materials may be insufficient to provide the long-term structural support required for cellular activities. To overcome such limitations, synthetic polymers in the form of organized networks and fibers can be combined with natural extracellular components. It has been shown that a biohybrid scaffold consisting of electrospun biodegradable poly(ester urethane)urea microfibers dispersed in an ECM hydrogel obtained from decellularized porcine skin tissue has more favorable mechanical properties than the ECM hydrogel alone, with high flexibility and soft tissue-like behavior [46]. In full thickness abdominal wall injuries applied in vivo in rats, biohybrid scaffolds facilitated cell migration and provided tissue formation with mechanical properties similar to native abdominal wall tissue.

7.3.4 Bioprinting

Three-dimensional (3D) bioprinting is a computer-controlled method that involves layer-by-layer spatial deposition or modeling of biomaterials to create 3D tissue scaffolds with well-defined microstructures. In recent years, the combination of hydrogel and bioprinting methods has been used to prepare tissue scaffolds that can mimic the properties of the ECM. Thus, the production of 3D tissue scaffolds with biologically and anatomically correct, more controlled pore sizes, inter-pore junctions, appropriate mechanical properties, biodegradability, and functional tissue formation supporting properties is provided.

Cell-laden hydrogels, called bioinks, play a very important role in determining cell-biomaterial interactions and subsequent functional tissue formation. The purpose of using bioink during bioprinting is to support cellular processes such as adhesion, proliferation, and differentiation by mimicking the ECM. Achieving the appropriate balance between cell density and bioink concentration is very important for high cell viability and structural stability. Briefly, an ideal bioink should have biocompatibility, suitable printability, components that support cellular processes, appropriate biodegradation, mechanical integrity, and stability properties. The subject of 3D bioprinting, its techniques, and bioink properties are detailed in Chaps. 3 and 9.

Decellularized ECM-based bioinks are used because they contain growth factors, ECM protein and polysaccharide components, and have suitable 3D bioprinting topologies [41]. In one study, it was reported that decellularized skeletal muscle ECM-based bioink provided a myogenic environment, resulting in high viability and contractile function of myoblast cells [47]. In addition, compared to collagen bioink, this bioink was found to increase the formation and maturation of myotubes and respond better to electrical stimulation.

Decellularized ECM-based bioinks can be modified before or after 3D printing process. The mechanical properties of decellularized ECM-based bioinks, where the 3D layouts and biological activities of the natural components are largely preserved, are generally poor. Therefore, the mechanical properties of the final product can be improved by adding several chemical and biological crosslinking agents during or after 3D printing process. For example, bioink was produced as a result of vitamin B2-induced UVA crosslinking and thermal gelation of the ECM obtained by decellularization of porcine heart tissue with sodium dodecyl sulfate (SDS) and Triton-X100 [48]. In this study, it was stated that the bioink with improved mechanical properties after the crosslinking process provided high cell viability, active proliferation of cardiac progenitor cells, and cardiomyogenic differentiation. In another study, a bioink that was produced by mixing the ECM formed by decellularization of liver tissue with gelatin methacryloyl (GelMA) and cross-linked with light was produced [49]. The tissue scaffold, created through digital light processing-based 3D bioprinting, was obtained with high precision and exhibited very good durability, largely due to the rapid crosslinking of GelMA incorporated into the structure. This decellularized ECM-based system was used as an in vitro cancer model by providing a liver-specific microenvironment to hepatocellular cancer cells loaded during 3D bioprinting.

In order to improve the mechanical properties of hydrogel-based bioinks, it is possible to use frameworks printed with high mechanical strength biomaterials such as thermoplastic polymers, apart from crosslinking methods. In one study, poly(caprolactone (PCL) frameworks were used to improve the printability of decellularized ECM bioinks derived from adipose, cartilage, and heart tissues and the mechanical integrity of the printed structures, and 3D tissue scaffolds were prepared [50]. Human adipose stem cells and human mesenchymal stromal cells interacted with the prepared tissue scaffolds, and high cell viability and functionality was reported in in vitro conditions.

7.3.5 Micro/Nano-patterning Techniques

Despite their various advantages, some problems due to their surface properties, such as unexpected bacterial adhesion and undesirable protein adsorption, limit hydrogel applications. To solve such problems and improve the surface properties of hydrogels, chemical, biochemical, and topographic surface modifications are used. ECM components or peptide motifs that play a role in directing cellular processes can be added to synthetic or natural hydrogel structures that are inadequate for cell adhesion by various chemical and biochemical surface modification methods. In surface topographical modification, a surface roughening method or different micro/nano-patterning methods are used [16]. Therefore, methods such as negative mold casting, photolithographic patterning, nanoprinting, 3D printing, ion-induced nano-patterning, and swelling-induced patterning are preferred for patterning hydrogels containing crosslinkers [16]. In a study investigating the neuronal differentiation of human mesenchymal stem cells (hMSCs) and mouse embryonic stem cells (mESCs) on microgrooved polyacrylamide hydrogels prepared by the pouring method, hMSCs cells maximally elongated and expressed neuronal markers in hydrogels with 10/15 μm grooves, whereas mESCs cells were unable to detect topographical features [51]. These results suggest that different stiffness and topographic cues cause different neuronal differentiation behaviors in cells. In another study, human corneal endothelial cell interactions were investigated with UV cross-linked gelatin methacrylate (GelMA) hydrogels with micron and submicron pillar patterns on their surfaces using nanoimprint lithography [52]. It has been shown that hydrogel groups with patterns on the surface to mimic ECM surface topography have higher cell density and more homogeneous cell sizes compared to the unpatterned group. The pattern structure allowed for the formation of transplantable human corneal endothelial cell monolayers with very high functional properties.

7.4 Conclusion

Developments in the field of biomaterials show that the structural, biochemical, and mechanical properties of the natural ECM can be mimicked by hydrogels. To date, no hydrogel has been developed that can completely mimic the natural ECM in terms of its inherent properties, but studies on hydrogels that can mimic several properties in combination have become more prominent in recent years. Advances in tissue engineering and biomaterials will enable the development of novel hydrogel approaches that mimic the chemical, structural, mechanical, and dynamic properties of the natural ECM with high accuracy.

References

1. Theocharis, A.D., Skandalis, S.S., Gialeli, C., Karamanos, N.K.: Extracellular matrix structure. Adv. Drug Deliv. Rev. **97**, 4–27 (2016)
2. Akhmanova, M., Osidak, E., Domogatsky, S., Rodin, S., Domogatskaya, A.: Physical, spatial, and molecular aspects of extracellular matrix of in vivo niches and artificial scaffolds relevant to stem cells research. Stem Cells Int. **2015**, 167025 (2015)
3. Nicolas, J., Magli, S., Rabbachin, L., Sampaolesi, S., Nicotra, F., Russo, L.: 3D extracellular matrix mimics: fundamental concepts and role of materials chemistry to influence stem cell fate. Biomacromolecules. **21**, 1968–1994 (2020)
4. Doyle, A.D., Nazari, S.S., Yamada, K.M.: Cell–extracellular matrix dynamics. Phys. Biol. **19**, 021002 (2022)
5. Bashir, S., Hina, M., Iqbal, J., Rajpar, A.H., Mujtaba, M.A., et al.: Fundamental concepts of hydrogels: synthesis, properties, and their applications. Polymers. **12**, 2702 (2020)
6. Unal, A.Z., West, J.L.: Synthetic ECM: bioactive synthetic hydrogels for 3D tissue engineering. Bioconjug. Chem. **31**, 2253–2271 (2020)
7. Gomez-Florit, M., Pardo, A., Domingues, R.M.A., Graça, A.L., et al.: Natural-based hydrogels for tissue engineering applications. Molecules (Basel, Switzerland). **25**, 5858 (2020)
8. Saldin, L.T., Cramer, M.C., Velankar, S.S., White, L.J., Badylak, S.F.: Extracellular matrix hydrogels from decellularized tissues: structure and function. Acta Biomater. **49**, 1–15 (2017)
9. Xing, H., Lee, H., Luo, L., Kyriakides, T.R.: Extracellular matrix-derived biomaterials in engineering cell function. Biotechnol. Adv. **42**, 107421 (2020)
10. Remuzzi, A., Bonandrini, B., Tironi, M., Longaretti, L., Figliuzzi, M., et al.: Effect of the 3D artificial nichoid on the morphology and mechanobiological response of mesenchymal stem cells cultured in vitro. Cells. **9**(8), 2020 (1873)
11. De France, K.J., Xu, F., Hoare, T.: Structured macroporous hydrogels: progress, challenges, and opportunities. Adv. Healthc. Mater. **7**, 1700927 (2018)
12. Fu, J., Fan, C., Lai, W.S., Wang, D.: Enhancing vascularization of a gelatin-based microcavitary hydrogel by increasing the density of the micro-cavities. Biomed. Mater. **11**(5), 055012 (2016)

13. Prince, E., Kumacheva, E.: Design and applications of man-made biomimetic fibrillar hydrogels. Nat. Rev. Mater. **4**(2), 99–115 (2019)
14. Xu, S., Deng, L., Zhang, J., Yin, L., Dong, A.: Composites of electrospun-fibers and hydrogels: a potential solution to current challenges in biological and biomedical field. J. Biomed. Mater. Res. B Appl. Biomater. **104**, 640–656 (2016)
15. Pei, B., Wang, W., Fan, Y., Wang, X., Watari, F., Li, X.: Fiber-reinforced scaffolds in soft tissue engineering. Regen. Biomater. **4**(4), 257–268 (2017)
16. Cui, L., Yao, Y., Yim, E.K.F.: The effects of surface topography modification on hydrogel properties. APL Bioeng. **5**, 031509 (2021)
17. Sapudom, J., Pompe, T.: Biomimetic tumor microenvironments based on collagen matrices. Biomater. Sci. **6**, 2009–2024 (2018)
18. Dinescu, S., Kaya, M.A., Chitoiu, L., Ignat, S., Kaya, D.A., Costache, M.: Collagen-based hydrogels and their applications for tissue engineering and regenerative medicine. In: Mondal, I.H. (ed.) Cellulose-Based Superabsorbent Hydrogels, pp. 1–21. Springer (2018)
19. Marques, C.F., Diogo, G.S., Pina, S., Oliveira, J.M., Silva, T.H., Reis, R.L.: Collagen-based bioinks for hard tissue engineering applications: a comprehensive review. J. Mater. Sci. Mater. Med. **30**(3), 32 (2019)
20. Schneider-Barthold, C., Baganz, S., Wilhelmi, M., Scheper, T., Pepelanova, I.: Hydrogels based on collagen and fibrin—frontiers and applications. BioNanoMaterials. **17**(1–2), 3–12 (2016)
21. Amorim, S., Reis, C.A., Reis, R.L., Pires, R.A.: Extracellular matrix mimics using hyaluronan-based biomaterials. Trends Biotechnol. **39**, 90 (2021)
22. An, S., Choi, S., Min, S., Cho, S.-W.: Hyaluronic acid-based biomimetic hydrogels for tissue engineering and medical applications. Biotechnol. Bioprocess Eng. **26**(4), 503–516 (2021)
23. Yu, Z., Li, H., Xia, P., Kong, W., Chang, Y., et al.: Application of fibrin-based hydrogels for nerve protection and regeneration after spinal cord injury. J. Biol. Eng. **14**, 22 (2020)
24. Zhu, J., Hu, J., Marchant, R.E.: 9—Biomimetic hydrogels as scaffolds for tissue-engineering applications. In: Ruys, A.J. (ed.) Biomimetic Biomaterials, pp. 238–275. Woodhead Publishing (2013)
25. Schultz, K.M., Kyburz, K.A., Anseth, K.S.: Measuring dynamic cell–material interactions and remodeling during 3D human mesenchymal stem cell migration in hydrogels. Proc. Natl. Acad. Sci. U. S. A. **112**(29), E3757–E3764 (2015)
26. Yi, B., Xu, Q., Liu, W.: An overview of substrate stiffness guided cellular response and its applications in tissue regeneration. Bioact. Mater. **15**, 82–102 (2022)
27. Luo, T., Tan, B., Zhu, L., Wang, Y., Liao, J.: A review on the design of hydrogels with different stiffness and their effects on tissue repair. Front. Bioeng. Biotechnol. **10**, 817391 (2022)
28. Narasimhan, B.N., Horrocks, M.S., Malmström, J.: Hydrogels with tunable physical cues and their emerging roles in studies of cellular mechanotransduction. Adv. NanoBiomed. Res. **1**(10), 2100059 (2021)
29. Utech, S., Boccaccini, A.R.: A review of hydrogel-based composites for biomedical applications: enhancement of hydrogel properties by addition of rigid inorganic fillers. J. Mater. Sci. **51**, 271–310 (2016)
30. Heid, S., Boccaccini, A.R.: Advancing bioinks for 3D bioprinting using reactive fillers: a review. Acta Biomater. **113**, 1–22 (2020)
31. Tong, X., Yang, F.: Recent progress in developing injectable matrices for enhancing cell delivery and tissue regeneration. Adv. Healthc. Mater. **7**, 1701065 (2018)
32. Morgan, F.L.C., Fernández-Pérez, J., Moroni, L., Baker, M.B.: Tuning hydrogels by mixing dynamic cross-linkers: enabling cell-instructive hydrogels and advanced bioinks. Adv. Healthc. Mater. **11**, 2101576 (2022)
33. Yang, J.-A., Yeom, J., Hwang, B.W., Hoffman, A.S., Hahn, S.K.: In situ-forming injectable hydrogels for regenerative medicine. Top. Issue Biomater. **39**(12), 1973–1986 (2014)
34. Sontyana, A.G., Mathew, A.P., Cho, K.-H., Uthaman, S., Park, I.-K.: Biopolymeric in situ hydrogels for tissue engineering and bioimaging applications. Tissue Eng. Regen. Med. **15**, 575–590 (2018)

35. Ryu, S.B., Park, K.M., Park, K.D.: In situ graphene oxide-gelatin hydrogels with enhanced mechanical property for tissue adhesive and regeneration. Biochem. Biophys. Res. Commun. **592**, 24–30 (2022)
36. Tu, Y., Chen, N., Li, C., Liu, H., Zhu, R., et al.: Advances in injectable self-healing biomedical hydrogels. Acta Biomater. **90**, 1–20 (2019)
37. Li, K., Zhu, Y., Zhang, Q., Shi, X., Liang, F., Han, D.: A self-healing hierarchical fiber hydrogel that mimics ecm structure. Materials. **13**(22), 5277 (2020)
38. Hu, W., Wang, Z., Xiao, Y., Zhang, S., Wang, J.: Advances in crosslinking strategies of biomedical hydrogels. Biomater. Sci. **7**(3), 843–855 (2019)
39. Krishtul, S., Baruch, L., Machluf, M.: Processed tissue–derived extracellular matrices: tailored platforms empowering diverse therapeutic applications. Adv. Funct. Mater. **30**(18), 1900386 (2020)
40. Golebiowska, A.A., Intravaia, J.T., Sathe, V.M., Kumbar, S.G., Nukavarapu, S.P.: Decellularized extracellular matrix biomaterials for regenerative therapies: advances, challenges and clinical prospects. Bioact. Mater. **32**, 98–123 (2024)
41. Kabirian, F., Mozafari, M.: Decellularized ECM-derived bioinks: prospects for the future. Decellularization and recellularization strategies for translational medicine. Methods. **171**, 108–118 (2020)
42. Zhang, X., Chen, X., Hong, H., Hu, R., Liu, J., Liu, C.: Decellularized extracellular matrix scaffolds: recent trends and emerging strategies in tissue engineering. Bioact. Mater. **10**, 15–31 (2022)
43. Milan, P.B., Masoumi, F., Biazar, E., Jalise, S.Z., Mehrabi, A.: Exploiting the potential of decellularized extracellular matrix (ECM) in tissue engineering: a review study. Macromol. Biosci. **25**, 2400322 (2024)
44. Li, J., Rao, Z., Zhao, Y., Xu, Y., Chen, L., Shen, Z., et al.: A decellularized matrix hydrogel derived from human dental pulp promotes dental pulp stem cell proliferation, migration, and induced multidirectional differentiation *in vitro*. J. Endod. **46**(10), 1438–1447 (2020)
45. Tan, Q.-W., Tang, S.-L., Zhang, Y., Yang, J.-Q., Wang, Z.-L., et al.: Hydrogel from acellular porcine adipose tissue accelerates wound healing by inducing intradermal adipocyte regeneration. J. Invest. Dermatol. **139**(2), 455–463 (2019)
46. Hong, Y., Huber, A., Takanari, K., Amoroso, N.J., et al.: Mechanical properties and in vivo behavior of a biodegradable synthetic polymer microfiber-extracellular matrix hydrogel biohybrid scaffold. Biomaterials. **32**, 3387–3394 (2011)
47. Choi, Y.-J., Kim, T.G., Jeong, J., Yi, H.-G., Park, J.W., et al.: 3d cell printing of functional skeletal muscle constructs using skeletal muscle-derived bioink. Adv. Healthc. Mater. **5**, 2636–2645 (2016)
48. Jang, J., Kim, T.G., Kim, B.S., Kim, S.-W., Kwon, S.-M., Cho, D.-W.: Tailoring mechanical properties of decellularized extracellular matrix bioink by vitamin B2-induced photocrosslinking. Acta Biomater. **33**, 88–95 (2016)
49. Ma, X., Yu, C., Wang, P., Xu, W., Wan, X., Lai, C.S.E., et al.: Rapid 3D bioprinting of decellularized extracellular matrix with regionally varied mechanical properties and biomimetic microarchitecture. Biomaterials. **185**, 310–321 (2018)
50. Pati, F., Jang, J., Ha, D.-H., Won Kim, S., Rhie, J.-W., et al.: Printing three-dimensional tissue analogues with decellularized extracellular matrix bioink. Nat. Commun. **5**, 3935 (2014)
51. Sthanam, L.K., Saxena, N., Mistari, V.K., Roy, T., Jadhav, S.R., Sen, S.: Initial priming on soft substrates enhances subsequent topography-induced neuronal differentiation in ESCs but not in MSCs. ACS Biomater. Sci. Eng. **5**, 180–192 (2019)
52. Rizwan, M., Peh, G.S.L., Ang, H.-P., Lwin, N.C., Adnan, K., et al.: Sequentially-crosslinked bioactive hydrogels as nano-patterned substrates with customizable stiffness and degradation for corneal tissue engineering applications. Biomaterials. **120**, 139–154 (2017)

Chapter 8
Smart Hydrogels and Applications

Eda Ayse Aksoy

Abstract Three-dimensional networks of polymers called hydrogels have the ability to absorb vast volumes of water or biological fluid. Since their introduction as soft contact lenses in the 1960s, their use in the medical field has significantly expanded. Many hydrogels are widely used in both laboratory studies and clinical applications. Smart hydrogels, also known as environmentally responsive hydrogels, are materials capable of reversible changes in response to single or multiple external stimuli. Smart hydrogels are the theme of significant and competitive research in various scientific fields, including biomaterials, tissue engineering, biotechnology, and pharmaceuticals. Due to their chemical design diversity, they are utilized as next-generation materials in smart and multi-drug carriers, soft biomimetic structures, biorecognition and bioseparation systems, biosensors, nanoscale devices, nanorobots, tissue engineering scaffolds, injectable systems, and as 3D and 4D printing bioinks.

Keywords pH-responsive · Glucose-responsive · Redox-responsive · Temperature-responsive · Light-responsive · Electrically responsive · Self-healing · Multiple stimuli

E. A. Aksoy (✉)
Department of Basic Pharmaceutical Sciences, Faculty of Pharmacy, Hacettepe University, Ankara, Türkiye

Department of Polymer Science and Technology, Graduate School of Science and Engineering, Hacettepe University, Ankara, Türkiye
e-mail: edaayse@hacettepe.edu.tr

N. Hasirci, V. Hasirci (eds.), *Hydrogels and Bioinks in Tissue Engineering*, Biological and Medical Physics, Biomedical Engineering,
https://doi.org/10.1007/978-3-031-90966-5_8

8.1 Properties of Smart Hydrogels

In earlier years, hydrogel research focused on developing structurally stable hydrogels through chemical crosslinking techniques for use after surgical applications. These types of traditional hydrogels were generally utilized in implantable applications. However, the catalysts and crosslinkers used for permanent crosslinks in conventional hydrogel systems could cause protein denaturation, damage to cells, and ultimately reduce biocompatibility. Furthermore, traditional hydrogels lack the ability to produce smart responses to external environmental stimuli and exhibit weak mechanical properties. For these reasons, the development of smart hydrogels that can respond to external factors from their environment has become highly attractive.

Smart hydrogels are made to react to external changes (stimuli) in the environment by changing their electrical and magnetic characteristics, reversible volume phase transitions (VPT), or sol-gel phase transitions. The creation of smart hydrogels that react to different stimuli is made possible by the addition of macromolecular segments and stimuli-sensitive functional groups to the hydrogel network. The development of hydrogels that break down in response to suitable environmental stimuli when in contact with or implanted in the body, the slow response times of smart hydrogels to stimuli and the hysteresis that goes along with them, and the development of extracellular matrix (ECM) mimetics to enhance biocompatibility and prevent immune system reactions are all major challenges for smart hydrogel designs. Smart hydrogels that can respond to one or more features of the surrounding tissue environment, such as pH, temperature, or enzymes, at the intended in vivo application site have created a breakthrough in developing new therapeutic options for diseases like cancer. Due to their ability to deliver, target, and release therapeutics in a spatially and temporally controlled manner within in vivo environments, smart hydrogels have garnered significant attention in the pharmaceutical field. This approach largely prevents therapeutics from entering systemic circulation, thereby minimizing potential side effects.

The diversity of bonds and functional groups that can be introduced into the chemical structures of smart and responsive hydrogels enhances their induction and response sensitivities. This increases their success in lateral flow-based diagnostic kits, lab-on-chip systems, and biosensors. In tissue engineering, smart hydrogels offer a wide range of applications in injectable forms or as 3D and 4D printed tissue scaffold structures for the regeneration of bones, cartilage, vertebrae, muscles, nerves, corneas, skin, and cardiac tissues.

8.2 Development of Smart Hydrogels

"Smart" or "stimuli-responsive hydrogels" exhibit swelling and shrinking behaviors in aqueous environments due to external stimuli from their surroundings. The research studies on hydrogels started in 1894. The term "smart" was introduced to the scientific community in 1948 by Kuhn and his colleagues. The term "smart" gained importance due to their ability to undergo structural and functional changes in response to specific triggers (stimuli). The first paper reporting on a smart polymer was on poly(acrylic acid) (PAA), which showed structural changes depending on the pH of the environment [1].

Smart hydrogel systems can be classified by a wide range of criteria, including natural or synthetic origin of polymer, biodegradability, and crosslinking mechanisms. Chemically crosslinked hydrogel networks can be obtained by covalent bonding of polymer chains or polymerizing water-soluble monomers (or mixtures). Physically crosslinked structures, which are the focus of this section, are designed using non-covalent supramolecular forces to create fast and reversible networks. In studies focusing on drug delivery systems, the type of external stimuli is of primary importance. The responses of smart hydrogel systems can be induced by different types of physical or chemical stimuli. Physical stimuli include pressure, light, temperature, and magnetic or electric fields. Chemical or biochemical stimuli involve ionic strength, pH, ions, and specific molecules [2]. Smart/stimuli-responsive hydrogels undergo volumetric and structural transitions in response to these stimuli.

Smart hydrogels capable of responding to factors like temperature, pH, and light are used in biomaterials, pharmaceutical materials, biosensors, drug delivery and targeting systems, hydrogeological systems, soil water retention, and wastewater treatment systems. The most analyzed smart systems are pH- and temperature-sensitive hydrogels since both stimuli are repeatable and easily controlled. These hydrogels respond to changes in the pH or temperature by undergoing volumetric or permeability changes. An example of a temperature-sensitive one is the one that gels at body temperature. The harmful effects of the crosslinking chemicals often employed in injectable hydrogels, which were found to adversely affect proteins and cells, made smart hydrogels that spontaneously gelled after injection more popular in the 2000s. Temperature-sensitive smart hydrogels that gel under physiological conditions have provided increasing advantages in surgical implantations, as they reduce risks such as infection and scarring [3].

8.3 Factors Affecting Smart Hydrogels and Classification

Smart hydrogels react to a range of chemical (pH, glucose, and ionic strength), biological (enzymes, antigens, and antibodies), and physical (temperature, light, electromagnetic fields, pressure, and ultrasonic radiation) stimuli. The classification of smart hydrogels based on these stimuli is presented in Table 8.1. In addition to these types of smart hydrogels, dual- or multi-responsive hydrogels, self-healing hydrogels, and shape-memory hydrogels are also synthesized as "next-generation smart hydrogels."

The classification of external stimulus types for smart hydrogels and the reversible changes observed in the hydrogel network in response to stimuli are schematically summarized in Fig. 8.1.

Table 8.1 Types of external stimuli for smart hydrogels

Physical	Chemical	Biological
Temperature	pH	Enzyme
Light	Ionic strength	Antigen/antibody
Electrical, magnetic field	Redox	Reactive oxygen species (ROS)
Pressure	Water	Glucose
Radiation (UV, laser vb.)	Solvent type	
Sound wave		
Shear, stress		

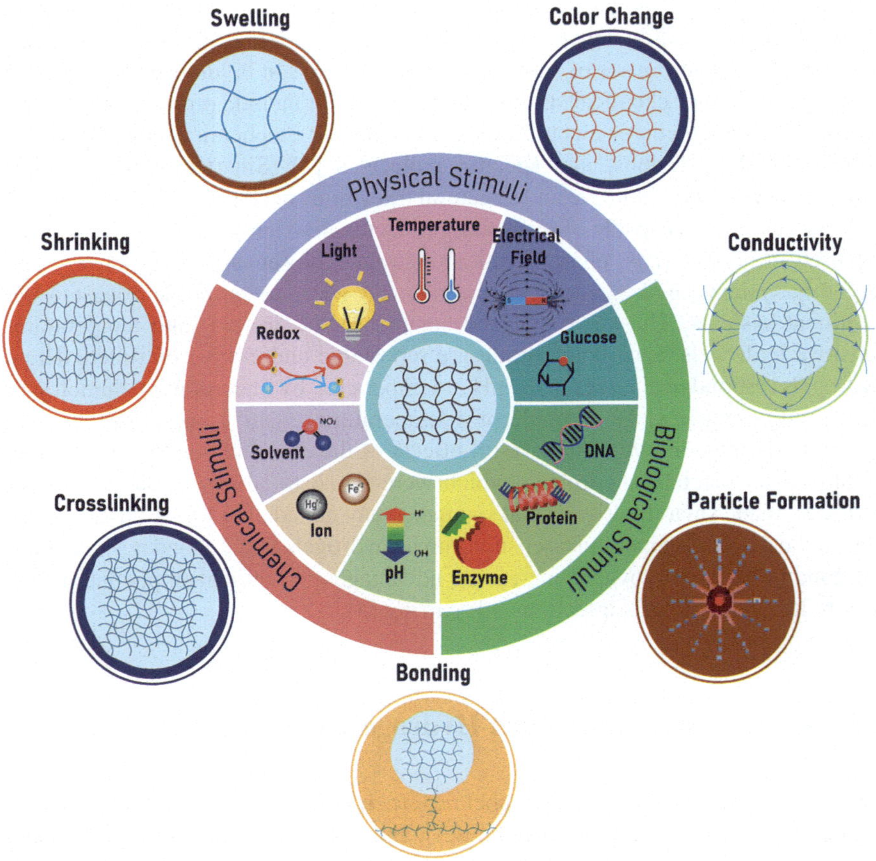

Fig. 8.1 Schematic representation of different smart/stimuli-responsive hydrogels

8.3.1 pH-Responsive Hydrogels

The interactions between polymer chains and between solvent molecules are directly impacted by changes in the environment's pH or acidity. A polymer chain may have ionizable groups that, in reaction to pH variations, donate or receive protons, much like macromolecules like proteins and nucleic acids. This may cause swelling, shrinkage, structural alterations, or changes in solubility. Acidic (such as carboxylic and sulfonic acids) or basic (such as ammonium salts) groups are structurally present in pH-responsive polymers, and they react to pH variations in their surroundings by gaining or losing protons. Polymers with several ionizable groups are called polyelectrolytes. For example, poly(acrylic acid) (PAA) is an anionic polyelectrolyte with a pKa of 4.25. At pH values above 4.25, PAA loses protons and ionizes (deprotonates), significantly increasing electrostatic repulsion forces between polymer chains, allowing water molecules to penetrate into the network, and causing hydrogel swelling. Conversely, in acidic environments with pH below 4.25, the acidic protons of PAA reduce charge density and cause polymer chains to contract (Fig. 8.2).

A cationic polyelectrolyte like poly(*N*,*N*-diethylaminoethyl methacrylate) ionizes in more acidic environments than its pKa value. The entry of water between polymer chains due to electrostatic repulsion increases the hydrodynamic volume of the polymer, causing it to swell (Fig. 8.2). Figure 8.2 illustrates the conformational states and ionization reactions of pH-responsive poly(acrylic acid) and poly(*N*,*N*-diethylaminoethyl methacrylate) under varying pH conditions [1, 3].

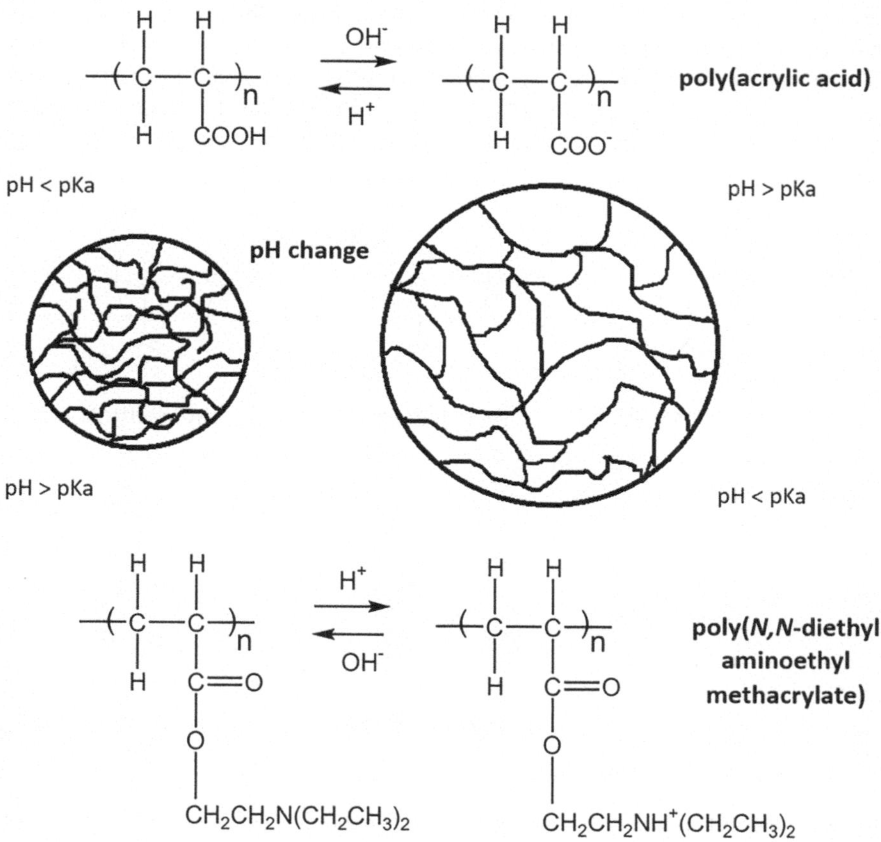

Fig. 8.2 Conformational transitions and ionization reactions of pH-responsive poly(acrylic acid) and poly(*N*,*N*-diethylaminoethyl methacrylate) under pH changes

Protonation/deprotonation reactions of weak Brønsted acids and bases primarily occur in the pH range of 4–8, making them suitable for use in medical devices within the body. Both acidic and basic units are present in amphiphilic hydrogels, which show biphasic transitions in acidic and basic environments as opposed to neutral ones. Transitions between shrinking and swelling occur within a narrow range near the pKa value.

Carboxylic functional groups are commonly found in pH-responsive polyacids, which have pKa values between 5.0 and 6.0. Carboxylic groups deprotonate at higher pH levels, which causes swelling, increased hydrodynamic volume, and electrostatic repulsion between molecular chains. They continue to exist as non-ionized macromolecules at lower pH values.

Figure 8.3 displays the most prevalent pH-responsive polyacids and polybases. With pKa values between 3.0 and 11.0, polyacids containing sulfonamide groups enable hydrogen ionization from amidic nitrogen. Amino groups in polybases generate polyelectrolytes by accepting protons at low pH and releasing them under basic circumstances. Examples of weak polybases include poly(*N*,*N*-dimethylaminoethyl methacrylate), poly(*N*,*N*-diethylaminoethyl methacrylate), poly(L-lysine), and aromatic bases like 4-vinylpyridine, 2-vinylpyridine, and poly(vinyl imidazole). Natural polymers such as alginate, albumin, chitosan, pectin, and gelatin also exhibit pH-responsive activity [4].

As summarized in Table 8.2, significant pH differences exist in various regions of the healthy human body. Additionally, certain diseases can also change pH levels in the body. These pH-sensitive hydrogels' ability to respond to dynamic

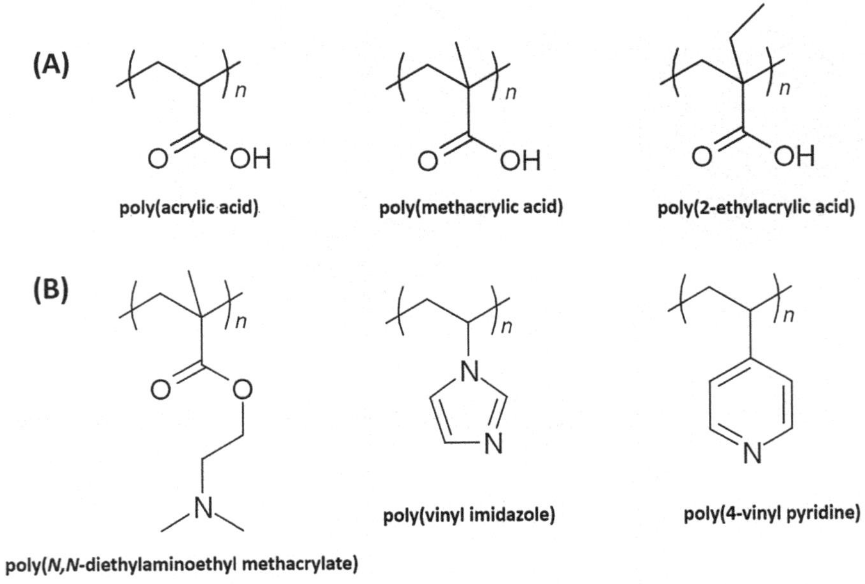

Fig. 8.3 Chemical structures of some significant pH-responsive polymers: (**a**) polyacids, (**b**) polybases

Table 8.2 pH values in different regions of the human body

Healthy tissue/organ/cell	pH values
Stomach	1.0–3.0
Colon	7.0–7.5
Vagina	4.0–5.0
Blood	7.4–7.5

environments is particularly important for biomedical and pharmaceutical applications, especially in drug targeting systems. pH-responsive hydrogels are widely used in cancer drug targeting. The pH in tumor tissues (pH = 5.5–6.0) is more acidic compared to normal tissues (pH = 7.4). This creates a significant advantage for targeting tumors with pH-responsive controlled drug delivery systems [5, 6].

8.3.2 Glucose-Responsive Hydrogels

Systems that react to elevated glucose levels are known as glucose-responsive hydrogels, and they are frequently employed in the management of diabetes. These smart hydrogels can be used as mesoporous nanoparticles, vesicles, micelles, microgels, and nanogels. Glucose oxidase and concanavalin A are the most basic examples of glucose-responsive substances. Additionally, phenylboronic acid and its derivatives react selectively with glucose, making them glucose-responsive substances.

Concanavalin A is a naturally occurring glucose-binding lectin that has been isolated from specific plants, whereas glucose oxidase is an enzyme that catalyzes glucose-related processes. These data allow for the classification of glucose-responsive hydrogels into three primary groups: phenylboronic acid hydrogels, concanavalin A hydrogels, and glucose oxidase hydrogels. The most widely used method for creating automated devices that respond to glucose and release insulin is to encapsulate the enzymes glucose oxidase and catalase in nanoparticles and embed them in a pH-sensitive hydrogel that has been saturated with insulin. Under hyperglycemic conditions, the enzymatic conversion of glucose to gluconic acid triggers the swelling of the hydrogel, expansion of the network pores, and subsequent release of insulin from the hydrogel matrix. After insulin release, blood glucose levels decrease, and as a result, the increase in pH halts further insulin release. This glucose-responsive hydrogel system operates as a self-activated valve, synchronizing insulin release in both normal glycemic and hyperglycemic conditions [7].

8.3.3 Ionic Strength and Redox-Responsive Hydrogels

Ionic balance and ion sensitivity are critical for bodily functions, making it essential to understand pharmacological systems and design polymeric networks accordingly. The primary determinant of an ionic hydrogel system's degree of swelling is the level of ionization inside the polymeric network. The ionic strength, pH, and ionization level of the surrounding medium are some of the variables that affect the dynamics of equilibrium and swelling behavior. When the ionic strength of the fluid around a pH-sensitive hydrogel interacts with the pH, the hydrogel swells. For example, ocular drug delivery systems exist where the presence of ions like Na^+, K^+, Mg^{2+}, and Ca^{2+} in tears triggers the hydrogel.

Preparing hydrogels for various biomedical applications using ionic crosslinking techniques is a straightforward approach. Biomedical applications make extensive use of ionically crosslinked chitosan hydrogels. The injection of in situ gel-forming biopolymers makes the development of therapeutic implants appealing. Chenite et al. reported that cationic polysaccharide solutions gelled with polyol-phosphate salts containing anionic groups could form pH-sensitive thermogels [8]. Synergistic forces, such as hydrogen bonding, electrostatic contacts, and hydrophobic interactions, were produced by the combination of polyol-phosphate salts and cationic polysaccharide chitosan that are appropriate for thermogel production. Chitosan glycerol-phosphate solutions exhibited a rheological transition into a solid gel structure under physiological pH ($\approx$7.15) and temperature (37 °C) [8].

In the literature, in situ chitosan hydrogels have also been prepared under physiological conditions using ionic/covalent co-crosslinking. For this, a chitosan solution was first neutralized with the ionic crosslinking agent glycerol phosphate disodium salt complex, and then the chemical crosslinking agent genipin was added. The resulting hydrogel retained the thermosensitive character typical of ionic crosslinking and exhibited improved mechanical properties and chemical stability due to the presence of strong covalent bonds [9].

Redox potential, a biological factor influenced by many pharmacological conditions, also changes in diseases like inflammation and cancer. Redox-responsive hydrogels can be obtained by integrating organometallic structures or thiol-disulfide exchange principles into the hydrogel network. Wegner et al. developed 4-arm histidine-modified PEG hydrogels that are cobalt-cross-linked and redox-responsive (4A-PEG-His). The oxidation of cobalt ions from Co^{2+} to Co^{3+} caused these hydrogels to change from a self-healing viscoelastic fluid to a stable elastic solid [10]. For the delivery of drug-loaded liposomes in thiol-rich environments, Liang and Kiick created redox-responsive PEG-based polymeric hydrogels [11]. For the controlled and sequential release of chemotherapeutics doxorubicin and cytochrome-c, this study created glutathione-sensitive hybrid hydrogels based on a reversible Michael-type addition reaction between arylthiol-functionalized 4-arm PEG and maleimide-functionalized liposomes [11].

To identify reactive oxygen species (ROS) and protect cells from oxidation brought on by ROS, redox-responsive hydrogels are also being created. ROS are highly reactive substances released during cellular metabolism and are abundant in the human body. Despite their destructive oxidative effects on cellular fluids, nucleic acids, and proteins, ROS significantly contribute to cell signaling. Hu et al. prepared an antibacterial ROS-sensitive hydrogel based on polyacrylic acid containing silver nanoparticles and polyglutamic acid containing Fe^{2+}/Fe^{3+}. They reported that while the hydrogel maintained stability under normal conditions, it rapidly degraded when exposed to ROS and inhibited pathogenic *P. aeruginosa* bacteria [12].

8.3.4 Temperature-Responsive Hydrogels

One of the key stimuli for smart polymer systems is temperature. Its natural variability and ease of monitoring in vitro and in vivo create biomedical advantages. Thermosensitive hydrogels are smart polymeric systems that respond to specific temperatures with changes in volume (swelling/shrinking) or permeability. These hydrogels have hydrophobic and hydrophilic groups within their polymer chains, and the precise balance between these hydrophobic and hydrophilic components is what gives them their sensitivity.

Temperature changes alter the balance and interaction of these groups with water, resulting in volume-phase transitions. These transitions lead to sol-gel phase changes as temperature affects the solubility of the crosslinked network. The critical temperature (Tc) defines this transition point. The sol phase is a fluid liquid, while the gel phase is a non-flowing, solid-like state that maintains structural integrity. At the critical temperature, the solvency, conformational forms, and solubility capacity of the polymer chains exhibit abrupt changes. Consequently, smart hydrogels swell or shrink in response to the critical temperature. In the design of temperature-responsive hydrogels, at least one component of the polymer system must exhibit temperature-dependent solubility in the solvent (typically water). To display a distinct change in swelling degree with temperature, the components must become insoluble below or above a specific temperature. This volume change is reversible depending on the polymer chain's composition and the hydrogel's ionization degree. At the critical temperature, the polymer undergoes a phase change. The lowest temperature at which the polymer dissolves is called the lower critical solution temperature (LCST), while the highest temperature is called the upper critical solution temperature (UCST).

Smart hydrogels exhibit phase separation or gelation at their LCST or UCST. Below the LCST, the polymer is soluble; above the LCST, it begins to shrink, becoming hydrophobic and insoluble, resulting in gel formation. Conversely, hydrogels with UCST dissolve upon heating. More specifically, near the critical temperature, the polymer in solution undergoes a phase transition from a soluble to an insoluble state (e.g., precipitation or micelle formation). LCST can be adjusted by altering the hydrophilic-to-hydrophobic ratio of the hydrogel [13]. The LCST values of some polymers are summarized in Table 8.3.

Among temperature-sensitive hydrogels, poly(*N*-isopropylacrylamide) (PNIPAAm) is probably the most researched. PNIPAAm homopolymers and copolymers are utilized in drug delivery systems and tissue engineering research. PNIPAAm exhibits a sharp phase transition at an LCST of approximately 32 °C. Both a hydrophilic amide group and a hydrophobic isopropyl group are present in its structure. Below the LCST, hydrogen bonds form between the polar amide

Table 8.3 LCST values of selected polymers

Polymer	LCST (°C)
Poly(*N*-isopropylacrylamide) (PNIPAAm)	32–34
Poly(*N*-isopropylmethacrylamide)	38–42
Poly (*N*-ethyl-*N*-methylacrylamide)	56
Poly (*N*,*N*-diethylacrylamide) (PDEAAm)	32
Poly (*N*,*N*-dimethyl aminoethyl methacrylate) (PDMAEMA)	50
Poly(vinylcaprolactam)	32–40
Poly(methylvinylether) (PVME)	34–38
Poly(vinylisobutyroamide)	35
Poly(ethylene oxide) (PEO)	96
Hydroxypropylcellulose (HPC)	44

groups and water molecules, resulting in an expanded network conformation of PNIPAAm chains. Above the LCST, the hydrogen bonds within the network break, causing PNIPAAm hydrogels to rapidly shrink by expelling water. This leads to a transition from a fluid-like state to a gel-like solid.

A temperature-sensitive polymer's LCST is mostly related to hydrophilic and hydrophobic units in its polymer chain. Increased hydrophilicity raises the LCST, while the presence of more hydrophobic chemical groups lowers it. Copolymers of PNIPAAm with more hydrophilic monomers increase the polymer's hydrophilicity, resulting in stronger polymer-water interactions and a higher LCST. Conversely, adding more hydrophobic monomers results in a decrease of LCST compared to PNIPAAm. It is possible to tune the LCST for specific applications according to the desired requirements. The ratio of hydrophilicity to hydrophobicity in the polymer chain can be changed in a variety of ways. One common and successful method is designing the molecular structure of the monomer. *N*-substituted polyacrylamides are often preferred, as the hydrophilic/hydrophobic balance is primarily governed by the *N*-groups. For example, in PNIPAAm, the hydrophilic amide group is complemented by a hydrophobic isopropyl group. Poly(*N*-ethyl-*N*-methylacrylamide) has a higher LCST of about 56 °C because the bulky isopropyl group is substituted with linear alkyl groups [3].

Temperature sensitivity can also be observed in natural polymers. Cellulose is an insoluble natural polysaccharide. By supplementing its hydroxyl groups with additional hydrophobic groups, such as methyl or hydroxypropyl, cellulose can become water soluble. Methylcellulose, with an LCST of 60–80 °C, is a typical example of a thermally reversible gel. To adjust the LCST of methylcellulose closer to human body temperature, grafting can be employed. Liu et al. studied the effects of NIPAAm-co-methylcellulose grafting ratios and observed that lower percentages of methylcellulose reduced the LCST [14].

Chitosan, a polysaccharide derived from the alkaline hydrolysis of chitin, has reactive amine (–NH_2) and hydroxyl (–OH) groups along its chain, allowing crosslinking with various agents. The first thermosensitive chitosan-based systems were reported in the literature by crosslinking chitosan with β-glycerophosphate [6]. These structures remain in a liquid state at room temperature but undergo gelation upon reaching physiological temperatures (37 °C). The gelation temperature is unaffected by the chitosan's molecular weight. The degree of deacetylation, the concentration of chitosan, and the concentration of β-glycerophosphate all have a direct impact on the gelation temperature and should be adjusted for specific uses [15].

Drug delivery systems frequently use temperature-sensitive hydrogels. Certain polymer-based hydrogels with critical transition temperatures within the body temperature range are particularly attractive for biomedical applications. Studies have also been conducted on thermosensitive cell culture surfaces for tissue engineering and cell culture applications. Hatakeyama et al. prepared temperature-sensitive tissue culture plate (TCP) surfaces by electron beam-induced graft copolymerization of 2-carboxyisopropylacrylamide (CIPAAm), a carboxylated derivative of *N*-isopropylacrylamide (IPAAm). Insulin was immobilized on TCP surfaces functionalized with thermosensitive hydrogels, and it was demonstrated that insulin immobilization supported cell proliferation in serum-supplemented media. Detachment of bovine carotid artery endothelial cells as a monolayer from the insulin-immobilized surfaces was achieved by lowering the culture temperature from 37 to 20 °C after 24 h of culture [16].

8.3.5 Light/Photo-Responsive Hydrogels

Light-responsive hydrogels are a type of smart hydrogel that responds to light stimuli by containing photosensitive functional units in their polymeric structure. Their physicochemical properties change upon exposure to light. The light stimulus can originate from near-infrared (NIR) or ultraviolet (UV) light sources. Commonly used light-sensitive molecules respond well to near-UV light and blue light, offering minimal tissue penetration. However, the activation of biomaterials modified with light-sensitive groups is generally limited to areas just beneath the skin in vivo. Therefore, approaches effective in vivo applications require low-energy, long-wavelength light capable of significant tissue penetration [17]. As a result, while light-responsive hydrogels produce successful results in the laboratory, their applicability rarely extends beyond in vitro cell culture. Fundamental limitations, such as tissue opacity and light penetration, greatly restrict their use in in vivo tissue applications. Nonetheless, they find applications in cornea and contact lens treatments.

Designing materials with degradation and reformation qualities in response to a suitable stimulus is made possible by reversible covalent chemistry. Because the intensity and duration of light may be readily adjusted, light-sensitive modules are appealing for this function. Diels-Alder cycloaddition products, disulfides, boronic esters, light-sensitive cyclic groups, and Schiff bases can all refunctionalize materials with reversible covalent bonds. To precisely control the time and light-guided processes, light-responsive reversible covalent bonds are essential. The intended reactions and reaction rates are guaranteed by carefully choosing the light's wavelength and intensity. The reversible creation or photolytic cleavage of polymer networks is made possible by the reversible dimerization reactions that functional groups such as coumarin rings, anthracene derivatives, and pyrimidine bases have with light. Coumarin derivatives are employed as crosslinking groups in polymeric materials because of their capacity to carry out reversible cycloaddition/cycloreversion processes and the modest conditions needed for dimerization. When coumarin derivatives interact with long-wave UV light (365 nm), they create cyclobutane-linked dimers through [2 + 2] cycloaddition, which can be reversed by cycloreversion using short-wave UV light (254 nm), as shown by Kabb et al. [16]. Thus, coumarin-crosslinked polymers are regarded as superior options for light-responsive reversible covalent gels. In their investigation, they created light-responsive copolymers by radical copolymerizing *N,N*-dimethylacrylamide (DMA) with coumarin-containing acrylic (7-(2-acryloyloxyethoxy)-4-methylcoumarin) or acrylamido (7-(2-acrylamidoethoxy)-4-methylcoumarin) monomers. Short-wave UV light caused the resultant linear copolymers to become water-soluble, while long-wave UV light caused them to gel [18].

When exposed to optical stimuli, photochromic components induce photoreactions such as photoisomerization, cleavage, and/or dimerization. Hydrogels containing light-sensitive units like o-nitrobenzyl and azobenzene in their network structures are optically responsive. Another example of light-responsive hydrogels is composite hydrogels that incorporate nanostructures capable of absorbing near-infrared (NIR) light, such as nanoshells, nanorods, and carbon nanotubes [19].

UV irradiation can trigger photopolymerization and photodegradation in polymeric systems. Photopolymerization utilizes the photon energy of light to prepare polymers or hydrogels. Photopolymerization can cause in situ gelation even though it is a step in hydrogel production. Benzoin derivatives, benzoketals, acetophenones, and hydroxyalkylphenones are examples of photoinitiators that are exposed to visible or ultraviolet light in order to produce free radicals, which facilitate photopolymerization. Using photosensitive compounds like cinnamic acid, coumarin, or benzyl diethyldithiocarbamate to functionalize hydrophilic polymers is an additional strategy. These molecules undergo intermolecular photodimerization when exposed to ultraviolet light [1].

The utilization of photosensitive crosslinkers, such as azobenzene and o-nitrobenzyl derivatives, is the fundamental idea behind photodegradation in light-sensitive hydrogels. When exposed to light, these molecules undergo photoisomerization or photooxidation, which alters the structure of the hydrogel matrix and causes the release of medications or biomolecules. When exposed to visible light (400–500 nm) or UV radiation at long wavelengths (365 nm), the o-nitrobenzyl ether group undergoes irreversible photoisomerization. However, various degradation rates are caused by changes in this photosensitive crosslinking agent's absorption and quantum yield at different wavelengths. Therefore, choosing the right wavelength is essential for managing medication release from these sensitive systems. It is important to note that there are certain drawbacks to using UV light as a physical stimulus for smart hydrogel drug delivery. Certain bodily areas are poorly penetrated by UV light, which can lead to carcinogenesis, particularly after extended exposure [1].

8.3.6 Electrically/Magnetically Responsive Hydrogels

Electromagnetic-sensitive hydrogels are smart hydrogel systems that exhibit physical changes, such as shrinking, swelling, or bending, in response to small electrical or magnetic field changes. These systems are typically ionizable polyelectrolyte hydrogels. An electric field is one of the most common external stimuli used to modulate drug release from smart hydrogels. Electroconductive hydrogels are formed by combining polyelectrolytes or a three-dimensional crosslinked polymer network with a conductive polymer. When an electric field is applied, electrochemical oxidation/reduction reactions may occur, leading to structural changes in the hydrogel and subsequent polymer network erosion, which releases the embedded drug [19]. Various polymers are used to develop electroactive and magnetically sensitive hydrogels. Examples include conductive polymers like polypyrrole, poly(3,4-ethylenedioxythiophene) (PEDOT), polyaniline (PANI), and synthetic or natural polymers such as poly(vinyl alcohol) (PVA), sulfonated polystyrene, acrylates/vinyl sulfonates, alginate, hyaluronic acid, and chitosan. To impart electroconductive properties, additives like gold nanoparticles, graphene, and carbon nanotubes are often incorporated into hydrogels [19].

To synthesize magnetically responsive hydrogels, magnetic nanoparticles are usually embedded into crosslinked polymer networks. Co-precipitation or encapsulating, grafting, and mixing procedures are the three primary methods used. Within the hydrogel framework, the magnetic component functions as a composite, and thermal, mechanical, and acoustic balances are immediately upset upon magnetic stimulation. By manipulating the magnetic field to arrange the magnetic nanoparticles inside the hydrogels, magnetic field-responsive hydrogel polymers' magnetothermal capacity and hyperthermic behavior can be regulated [20].

Magnetic iron oxide nanoparticles are added to polymer matrices to improve magnetic-sensitive hydrogels. By using gelatin and magnetic nanoparticles (10–250 nm in diameter), Hu et al. created a mechanically strong and flexible "ferrogel." This ferrogel demonstrated pulsatile drug release triggered by high-frequency magnetic fields [21]. Moreover, highly controlled and repeatable drug release was achieved using high-frequency cyclic magnetic stimulation. A critical parameter for such ferrogel systems is the ratio of polymer to magnetic iron oxide nanoparticles. Drug release under magnetic fields may be reduced by an overly high ratio, which lowers the saturation magnetization value.

These hydrogels are often studied in combination with thermoresponsive hydrogels. The use of magnetic iron oxide nanoparticles within responsive polymers relies on their capacity to convert electromagnetic energy into heat under an external high-frequency field. Drug release can be kinetically and thermally controlled with the appropriate sensitivity by applying microstructural and molecular changes, such as shrinking the pore size of the thermoresponsive polymeric matrix.

8.3.7 Antigen/Antibody-Responsive Hydrogels

Antibodies attach to particular antigens via Van der Waals forces, hydrogen bonds, and hydrophobic and electrostatic interactions. When particular antigens or antibodies are detected, the hydrogel network undergoes structural alterations, which is the fundamental idea behind antigen/antibody-responsive hydrogels. The analyte concentration directly correlates with the effectiveness of antigen/antibody-responsive hydrogels. The hydrogel matrix's chemical functional groups and the porous structure's surface area are important components for increasing the sensitivity of these hydrogels.

These hydrogels are prepared by either physically trapping antigens or antibodies within the hydrogel or chemically binding them to the network structure. Lim et al. developed a poly(acrylic acid)-based hydrogel capable of detecting Hepatitis B core antigen (HBcAg) [20]. The hydrogel's swelling property was triggered by binding HBcAg antibodies (anti-HBc) to the antigen. This hydrogel demonstrated high sensitivity to HBcAg, capable of detecting concentrations up to 4 mg/mL, as validated through quartz crystal microgravimetry. Its reversible swelling and shrinking behavior also suggested potential applications as a biosensor [22].

8.3.8 Self-Assembled Hydrogels via Supramolecular Chemistry

Supramolecular chemistry is based on non-covalent intermolecular interactions. Researchers frequently concentrate on physical interactions (such as electrostatics and hydrogen bonding) when developing hydrogels. These interactions are relevant to the creation of smart hydrogels due to the dynamic nature of supramolecular chemistry. Supramolecular chemistry allows for variability in interactions, rearrangement of components, and tunability of connections within a supramolecular structure.

Dynamic covalent bonds can be utilized to build stimuli-responsive systems in addition to non-covalent ones. Supramolecular hydrogels can respond to environmental stimuli by experiencing reversible gel-sol transitions and exhibiting certain capabilities, including optoelectronic characteristics, bioactivity, self-healing ability, and shape-memory capability. Consequently, they find use in gene transfection, drug delivery systems, bioimaging, protein carrier systems, and disease detection and therapy [23]. Due to their 3D crosslinked architectures, superior biocompatibility, biodegradability, and intelligent reactions to physiological stimuli, fluorescent bioimaging probes based on supramolecular nanohydrogel exhibit significant promise in the diagnosis of cancer [23]. In the literature, fluorescent supramolecular hydrogels modified with the fluorophore molecule 4-nitro-2,1,3-benzoxadiazole have been produced and demonstrated efficacy in imaging HeLa cells for cancer detection [24]. It has been demonstrated that self-assembling peptide hydrogels, which are extremely sensitive to pH variations in their surroundings, are promising instruments for pH-controlled drug delivery in acidic tumor tissues. Amphiphilic peptide derivatives have been used to create supramolecular hydrogels that can release anti-inflammatory, anti-HIV, and anti-cancer medications. Exogenous nucleic acids can be delivered to particular tumor cells or tissues using cationic supramolecular hydrogels. Ca^{2+} ion-tunable supramolecular hydrogels have been created, drawing inspiration from biomineralization [23]. For example, the synthetic compound [Boc-L-DOPA(OBn)$_2$-OH] (S)-*N*-tert-butoxycarbonyl-3,4-bis-(benzyloxy)-phenylalanine] facilitates hydrogelation by means of π-π stacking and electrostatic interactions. Alkaline circumstances produce hydrogelation based on Boc-L-DOPA(OBn)$_2$-OH. The concurrent production of $CaCO_3$ crystals within the fibrillar network was noted when hydrogels were produced in the presence of both Na_2CO_3 and $CaCl_2$. This behavior provided a model for biomineralization studies due to its potential to form calcified regions and supply Ca^{2+} ions [25].

8.4 Self-Healing Hydrogels

Hydrogels capable of self-healing after physical deformations or responding to environmental stimuli and, in some cases, exhibiting shape memory significantly expand the range of smart hydrogel applications. In recent years, self-healing hydrogels have been developed for a wide variety of applications, from soft robotics to tissue engineering. When tissue is damaged, it reacts through certain mechanisms to repair itself and restore its functions. The design of self-healing hydrogels can draw inspiration from these biological phenomena. By integrating self-healing functionality with a three-dimensional hydrogel structure, novel, smart, tissue-like hydrogels can be created.

Self-healing hydrogels are categorized into two groups: autonomic self-healing hydrogels and non-autonomic self-healing hydrogels. The first group can repair damage without requiring any external stimulus. The second group, however, requires stimuli such as light, magnetic fields, or heat to initiate the repair process [26]. Self-healing capability refers to a material's ability to recover from physical damage. Self-healing polymers are created using both physical and chemical methods. These include secondary interactions (such as hydrogen bonds or electrostatic interactions), composites, diffusion and flow effects, dynamic chemistry, covalent bond reformation, recombination, and heterogeneous systems [26]. Polar functional groups that direct hydrogen bonding between two distinct hydrogel structures or between the shattered fragments of a hydrogel can be used to create self-healing hydrogels. The functional groups must be accessible across the interface, and the hydrogel network must have malleable characteristics.

To achieve effective and durable repair, it is critical that the chains possess sufficient length and flexibility. However, chains should not be excessively long, as steric hindrance could obstruct the convergence of functional groups, and hydrophobic collapse might occur. Therefore, the type of chains, their lengths, and the balance between hydrophilic and hydrophobic groups are crucial.

For wound healing, Deng et al. created injectable antibacterial hydrogels that are extremely self-healing [25]. Dialdehyde-modified bacterial cellulose and hydroxypropyltrimethyl ammonium chloride chitosan, also known as quaternized chitosan, were used in the creation of these hydrogels. Dynamic Schiff base reactions between the amine groups of the quaternized chitosan and the aldehyde groups of the modified bacterial cellulose allowed for the self-healing characteristics. Because of its injectability, dynamic Schiff base interactions, and potent antibacterial activity against Gram-positive and Gram-negative bacteria, including *S. aureus* and *E. coli*, the hydrogel demonstrated quick self-healing capability as a functional wound dressing [27].

Gelatin, a natural biopolymer, has been functionalized with dopamine and acrylic acid to create a self-healing bioadhesive with excellent adhesion strength and mechanical properties in wet conditions. The hydrogel, prepared by grafting dopamine onto gelatin-co-poly(acrylic acid) chains, exhibited reversible mechanical properties, high elongation, self-healing capabilities, and adhesion in wet conditions. The adhesion strength of the synthesized hydrogel to tissue surfaces was measured as 6.5 kPa underwater [28].

A rapidly self-healing poly(acrylic acid) derivative hydrogel has also been developed for use as artificial joint cartilage [27]. For this, *N*-((2-((2-aminoethyl)amino)ethyl)carbamothioyl)acrylamide (AECAm), a special carbamothioyl monomer, was initially created from acrylic acid monomers. Following that, it was copolymerized with acrylic acid using radical polymerization generated by gamma radiation at a dose of 30 kGy. Because of the electrostatic interactions and intermolecular hydrogen bonding between PAAc and PAECAm, the resultant p(AAc-co-AECAm) hydrogel showed quick self-healing activity on its own in 3 s without the requirement for any external stimuli [29].

8.5 Shape Memory Hydrogels

Shape memory hydrogels are a highly popular type of smart hydrogel. Their creation is motivated by the shape memory and elastic deformation effects of a sample that is momentarily restored by physical crosslinks or reversible covalent bonds, producing a non-permanent, reversible shape. Shape reversibility is possible when the right stimulus is present. A transient shape produced by stretching, compressing, or folding can be fixed by shape memory hydrogels. Reversing the planned deformation that is applied when the hydrogels are activated allows for control over the direction of movement.

When gold nanoparticles are loaded into PNIPAm-based hydrogels, the particles' surface plasmon resonance absorption and photothermal-induced indirect heating enable shape modification [30]. Another work created pH-induced hydrogels of DNA and polyacrylamide. These hydrogels exhibited pH-triggered swelling and shrinking characteristics through a self-assembly mechanism. At pH 8, cytosine-rich amino acid sequences randomly coil to produce a "semi-liquid" state, whereas at low pH, they self-assemble into a patterned structure [31].

8.6 Shear- and Stress-Responsive Hydrogels

Depending on the polymer network, these hydrogels exhibit either shear-thinning or shear-thickening behavior in response to shear forces, resulting in physical or structural changes. As shear rates increase, shear-thinning hydrogels show a reduction in viscosity. The capacity of shear-thinning hydrogels to flow under shear is their main benefit for biomedical applications, and this makes them ideal for 3D bioprinting. Extrusion-based bioprinting makes extensive use of polymers like gelatin and alginate since extrusion reduces their viscosity [32]. However, the viscosity of shear-thickening hydrogels rises linearly with the shear rate. These hydrogels' colloidal nature, which provides little room for liquid, causes them to behave in a strain-thickening manner. Bone tissue engineering based on hydroxyapatite systems is the primary biomedical application for shear-thickening hydrogels, which are mainly ceramic-based systems [33].

8.7 Dual/Multiple Stimuli-Responsive Hydrogels

Various techniques are employed in the synthesis of dual/multiple stimuli-responsive hydrogels. These hydrogels can be synthesized using methods such as atom transfer radical polymerization (ATRP), reversible addition-fragmentation chain transfer (RAFT), or click chemistry approaches like thiol-ene chemistry [34]. Copolymerization and the formation of interpenetrating networks (IPNs) are also utilized in the production of dual/multiple stimuli-responsive hydrogels. Semi-IPNs or full IPNs with polymer chains sensitive to different stimuli can be prepared. Additionally, the copolymerization of monomers responsive to distinct stimuli allows for the synthesis of dual/multiple stimuli-responsive hydrogels. Stimuli-sensitive polymer chains can also be grafted with monomers responsive to other stimuli via graft copolymerization.

In the literature, semi-IPN hydrogels of poly(dimethylaminoethyl methacrylate)/carboxymethyl chitosan, sensitive to dual stimuli (pH and temperature), have been produced using physical crosslinkers and have been investigated for drug loading and release with theophylline [35]. Graft-type PNIPAM-g-carboxymethyl chitosan thermo- and pH-sensitive hydrogels have been synthesized and used for cisplatin release [36]. Poly(ethylene glycol)-block-poly(L-glutamate) copolymers prepared via ring-opening polymerization have been synthesized as polypeptide hydrogels with simultaneous thermal and pH sensitivity [37].

Dual/multiple stimuli-responsive hydrogels that are mechanically robust have garnered a lot of attention lately. Robust, mechanically strong, and biocompatible pH- and temperature-sensitive nanocomposite double-network hydrogels made of poly(*N*-isopropylacrylamide)/clay laponite/carboxymethyl chitosan/genipin have been synthesized through free-radical polymerization under UV light. Instead of toxic organic molecules, clay and genipin were used as crosslinkers [38].

8.8 Smart Hydrogels in Biomedical and Tissue Engineering Applications

The application of smart hydrogel types in biomedical and tissue engineering has been ongoing for many years. Competitive advancements in polymer science, technology, and nanotechnology have significantly contributed to the development of smart hydrogel systems. A general overview of smart hydrogels and their common biomedical applications is presented in Table 8.4, while examples of commonly used smart hydrogels in tissue engineering applications are provided in Table 8.5.

Table 8.4 An overview of smart hydrogels and their common biomedical applications [39]

Hydrogel	Application
Temperature-sensitive hydrogels	Wound dressings, skin tissue engineering
Light-sensitive hydrogels	Drug delivery systems, microfluidic systems
Injectable hydrogels	Bone, cartilage, and meniscus tissue engineering, drug delivery systems, osteoarthritis treatment
pH-sensitive hydrogels	Drug-protein delivery systems, drug targeting, cancer treatment
Glucose-sensitive hydrogels	Diabetes treatment, immunoisolation devices
Biochemical-sensitive hydrogels	Smart sensors and actuators
Magnetic-sensitive hydrogels	Drug delivery systems, tissue repair, targeted MRI for disease diagnosis
Electro-sensitive hydrogels	Membrane and implant based drug delivery

Table 8.5 Tissue engineering applications of smart hydrogels

Tissue engineering type	Hydrogel type and smart features	Chemical composition and formulation	Application outcome	Reference
Cardiac	Injectable hydrogel sensitive to myocardial infarction and myocardial wall	The hydrogel contains boronic ester bonds between the carboxymethyl cellulose 3-aminobenzene boric acid group and the dihydroxy group of polyvinyl alcohol. Additionally, a formulation containing curcumin-loaded PLGA nanoparticles (anti-inflammatory drug) and rhCol III was prepared	In vitro studies with HUVEC and H9C2 cells showed that the hydrogel exhibited cytocompatibility, antioxidant, anti-apoptotic, and anti-inflammatory properties, reducing ROS, TUNEL, CD68, and TNF-α expression levels in infarct regions. It also increased cell proliferation and angiogenesis	[40]
Cardiac	Electroconductive, thermosensitive, and injectable hydrogel	Thermosensitive chitosan/poloxamers (pluronic) conjugated hydrogels include bacterial nanocellulose fibers physically bonded with gold nanoparticles for electroconductive properties	In vitro experiments with H9C2 cells showed good cell adhesion and high biocompatibility (over 90% cell viability after 72 h of incubation)	[41]
Cartilage	Liquid-sensitive differential swelling, 4D biofabrication-compatible hydrogel	Two different ink formulations: Tyramine-functionalized hyaluronic acid (HAT, high swelling) and HAT-alginate blend (AHAT, low swelling). These inks were prepared as bilayer scaffolds using 3D printing, with shape transformation triggered by immersion in a liquid as a 4D stimulus	Human bone marrow cells showed high cell viability. After 28 days in a chondrogenic environment, histology revealed clear cartilage-like matrix production	[42]
Cartilage	Self-healing, reconfigurable, supramolecular hydrogel	Hyaluronic acid (HA) chains functionalized with ureidopyrimidinone (UPy). HA molecules with UPy segments form dynamic hydrogen bonds and self-heal under shear thinning by rearranging polymer chains	Improved retention of modified HA was observed in vivo after intra-articular injection. A stable supramolecular network was reestablished by rapid UPy dimerization. Self-healing networks resisted quick dissolution in the synovial cavity	[43]

(continued)

Table 8.5 (continued)

Tissue engineering type	Hydrogel type and smart features	Chemical composition and formulation	Application outcome	Reference
Bone	Thermosensitive, injectable hydrogel	Alendronate-loaded chitosan/β--glycerophosphate hydrogels were prepared	Alendronate-loaded thermogel showed a lower inflammatory response, faster proliferation, and granulation tissue maturation compared to plain thermogel	[44]
Cornea	Dual crosslinked adhesive hydrogel	A hybrid hydrogel comprising methacrylated hyaluronic acid and acellular swine corneal stroma matrix was created by dual crosslinking utilizing phenyl-2,4,6-trimethylbenzoylphosphinate-based photocrosslinking and carboxymethyl cellulose/NHS	Experimental eyes treated with hybrid hydrogel maintained their transparency in a rabbit corneal stromal damage model. Corneal reepithelialization was promoted by prolonged retention in the stromal bed	[45]
Skeletal muscle	3D bioprinted complex hydrogel	For manufacturing an integrated muscle-tendon unit, thermoplastic polyurethane (PU) hydrogel loaded with C2C12 cells was printed for elasticity and muscle development, while PCL hydrogel loaded with NIH/3T3 cells was printed for rigidity and tendon development	The finished structure displayed stiffness on the PCL-NIH/3T3 tendon side, elastic characteristics on the PU-C2C12 muscle side, and intermediate mechanical features at the interface. On days 1 and 7, there was advanced tissue growth and high cell viability (>80%)	[46]

8.9 Conclusion

Studies on smart hydrogels for biomedical and tissue engineering applications are progressing at a competitive pace, enabling the design of smart systems with desired properties. The chemical functionalization of monomers, crosslinkers, and chains, as well as the precise tuning of intermolecular forces (in homopolymeric and copolymeric networks and their blends), enhances the sensitivity and responsiveness of smart hydrogel systems to external stimuli. Smart hydrogels can exhibit specific functionalities such as bioactivity, self-healing, supramolecular assembly, and shape memory capabilities, and they can undergo reversible gel-sol transitions in response to environmental stimuli. The application range of smart hydrogels that are sensitive to environmental triggers and capable of autonomously or stimulus-responsive self-repair after physical deformation is expected to be remarkably broad. The incorporation of cells, bioactive agents, proteins, peptides, or active pharmaceutical ingredients into smart hydrogel systems also provides versatility and creates opportunities for applications across nearly all biomedical, pharmaceutical, and tissue engineering fields. However, there remain critical challenges to address, such as ensuring that smart, reversible, and multi-stimuli-responsive properties occur within appropriate timeframes: avoiding toxicity from the chemicals used; maintaining the integrity of structures post-application or implantation; ensuring pharmacological distribution in vivo; and confirming the safety of biodegradation products.

References

1. Ferreira, N.N., Ferreira, L.M.B., Cardoso, V.M.O., Boni, F.I., Souza, A.L.R., Gremião, M.P.D.: Recent advances in smart hydrogels for biomedical applications: from self-assembly to functional approaches. Eur. Polym. J. **99**, 117–133 (2018)
2. Malikmammadov, E., Hasirci, N.: Dual and multi-stimuli responsive polymers for biomedical applications (Chapter 9). In: Aguilar, M.R., Román, J.S. (eds.) Smart Polymers and Their Applications, 2nd edn, pp. 255–278. Elsevier Woodhead Pub. (2018)
3. Kuckling, D., Doering, A., Krahl, F., Arndt, K.F.: Stimuli-responsive polymer systems. In: Matyjaszewski, K., Möller, M. (eds.) Polymer Science: A Comprehensive Reference, pp. 377–413. Elsevier (2012)
4. Sanjuan, S., Perrin, P., Pantoustier, N., Tran, Y.: Synthesis and swelling behavior of pH-responsive polybase brushes. Langmuir. **23**(10), 5769–5778 (2007)
5. Komez, A., Buyuksungur, S., Hasirci, V., Hasirci, N.: Effect of chemical structure on properties of polyurethanes: temperature responsiveness and biocompatibility. J. Bioact. Compat. Polym. **33**(5), 479–497 (2018). https://doi.org/10.1177/0883911518783233
6. Bagherifam, S., Skjeldal, F.M., Griffiths, G., Mælandsmo, G.M., Nyström, B., Hasirci, V., Hasirci, N.: pH-responsive nano carriers for doxorubicin delivery. Pharm. Res. **32**, 1249–1263 (2015). https://doi.org/10.1007/s11095-014-1530-0
7. Koetting, M.C., Peters, J.T., Steichen, S.D., Peppas, N.A.: Stimulus-responsive hydrogels: theory, modern advances, and applications. Mater. Sci. Eng. R Rep. **93**, 1–49 (2015)
8. Chenite, A., Chaput, C., Wang, D., Combes, C., Buschmann, M.D., Hoemann, C.D., et al.: Novel injectable neutral solutions of chitosan form biodegradable gels in situ. Biomaterials. **21**(21), 2155–2161 (2000)

9. Moura, M.J., Faneca, H., Lima, M.P., Gil, M.H., Figueiredo, M.M.: In situ forming chitosan hydrogels prepared via ionic/covalent co-cross-linking. Biomacromolecules. **12**(9), 3275–3284 (2011)
10. Wegner, S.V., Schenk, F.C., Witzel, S., Bialas, F., Spatz, J.P.: Cobalt cross-linked redox-responsive PEG hydrogels: from viscoelastic liquids to elastic solids. Macromolecules. **49**(11), 4229–4235 (2016)
11. Liang, Y., Kiick, K.L.: Liposome-cross-linked hybrid hydrogels for glutathione-triggered delivery of multiple cargo molecules. Biomacromolecules. **17**(2), 601–614 (2016)
12. Hu, J., Liu, Z., Yu, Q., Ma, T.: Preparation of reactive oxygen species-responsive antibacterial hydrogels for efficient anti-infection therapy. Mater. Lett. **263**, 127254 (2020)
13. Qiu, Y., Park, K.: Environment-sensitive hydrogels for drug delivery. Adv. Drug Deliv. Rev. **53**(3), 321–339 (2001)
14. Liu, W., Zhang, B., Lu, W.W., Li, X., Zhu, D., De Yao, K., Wang, Q., Zhao, C., Wang, C.: A rapid temperature-responsive sol–gel reversible poly (N-isopropylacrylamide)-g-methylcellulose copolymer hydrogel. Biomaterials. **25**(15), 3005–3012 (2004)
15. Dang, Q.F., Yan, J.Q., Lin, H., Liu, C.S., Chen, X.G., Ji, Q.X., et al.: Biological evaluation of chitosan-based in situ-forming hydrogel with low phase transition temperature. J. Appl. Polym. Sci. **132**(10) (2015)
16. Hatakeyama, H., Kikuchi, A., Yamato, M., Okano, T.: Influence of insulin immobilization to thermoresponsive culture surfaces on cell proliferation and thermally induced cell detachment. Biomaterials. **26**(25), 5167–5176 (2005)
17. Rapp, T.L., DeForest, C.A.: Visible light-responsive dynamic biomaterials: going deeper and triggering more. Adv. Healthc. Mater. **9**(7), 1901553 (2020)
18. Kabb, C.P., O'Bryan, C.S., Deng, C.C., Angelini, T.E., Sumerlin, B.S.: Photoreversible covalent hydrogels for soft-matter additive manufacturing. ACS Appl. Mater. Interfaces. **10**(19), 16793–16801 (2018)
19. El-Husseiny, H.M., Mady, E.A., Hamabe, L., Abugomaa, A., Shimada, K., Yoshida, T., Tanaka, T., Yokoi, A., Elbadawy, M., Tanaka, R.: Smart/stimuli-responsive hydrogels: cutting-edge platforms for tissue engineering and other biomedical applications. Mater. Today Bio. **13**, 100186 (2022)
20. Häring, M., Schiller, J., Mayr, J., Grijalvo, S., Eritja, R., Díaz Díaz, D.: Magnetic gel composites for hyperthermia cancer therapy. Gels. **1**(2), 135–161 (2015)
21. Hu, S.H., Liu, T.Y., Liu, D.M., Chen, S.Y.: Controlled pulsatile drug release from a ferrogel by a high-frequency magnetic field. Macromolecules. **40**(19), 6786–6788 (2007)
22. Lim, S.L., Ooi, C.W., Tan, W.S., Chan, E.S., Ho, K.L., Tey, B.T.: Biosensing of hepatitis B antigen with poly (acrylic acid) hydrogel immobilized with antigens and antibodies. Sensors Actuators B Chem. **252**, 409–417 (2017)
23. Dong, R., Pang, Y., Su, Y., Zhu, X.: Supramolecular hydrogels: synthesis, properties and their biomedical applications. Biomater. Sci. **3**(7), 937–954 (2015)
24. Gao, Y., Shi, J., Yuan, D., Xu, B.: Imaging enzyme-triggered self-assembly of small molecules inside live cells. Nat. Commun. **3**(1), 1033 (2012)
25. Giuri, D., Jurković, L., Fermani, S., Kralj, D., Falini, G., Tomasini, C.: Supramolecular hydrogels with properties tunable by calcium ions: a bio-inspired chemical system. ACS Appl. Biomater. **2**(12), 5819–5828 (2019)
26. Taylor, D.L., In Het Panhuis, M.: Self-healing hydrogels. Adv. Mater. **28**(41), 9060–9093 (2016)
27. Deng, L., Wang, B., Li, W., Han, Z., Chen, S., Wang, H.: Bacterial cellulose reinforced chitosan-based hydrogel with highly efficient self-healing and enhanced antibacterial activity for wound healing. Int. J. Biol. Macromol. **217**, 77–87 (2022)
28. Ahmed, A., Nath, J., Baruah, K., Rather, M.A., Mandal, M., Dolui, S.K.: Development of mussel mimetic gelatin based adhesive hydrogel for wet surfaces with self-healing and reversible properties. Int. J. Biol. Macromol. **228**, 68–77 (2023)
29. Ghobashy, M.M., El-Sattar, N.E.A.: Radiation synthesis of rapidly self-healing hydrogel derived from poly (acrylic acid) with good mechanical strength. Macromol. Chem. Phys. **221**(19), 2000218 (2020)

30. Zhao, J., Fang, C., Zhu, Y., He, G., Pan, F., Jiang, Z., Zhang, P., Cao, X., Wang, B.: Manipulating the interfacial interactions of composite membranes via a mussel-inspired approach for enhanced separation selectivity. J. Mater. Chem. A. **3**(39), 19980–19988 (2015)
31. Guo, W., Lu, C.H., Orbach, R., Wang, F., Qi, X.J., Cecconello, A., Seliktar, D., Willner, I.: pH-stimulated DNA hydrogels exhibiting shape-memory properties. Adv. Mater. **27**(1), 73–78 (2015)
32. Guvendiren, M., Lu, H.D., Burdick, J.A.: Shear-thinning hydrogels for biomedical applications. Soft Matter. **8**(2), 260–272 (2012)
33. Cyster, L.A., Grant, D.M., Howdle, S.M., Rose, F.R.A.J., Irvine, D.J., Freeman, D., Scotchford, C.A., Shakesheff, K.M.: The influence of dispersant concentration on the pore morphology of hydroxyapatite ceramics for bone tissue engineering. Biomaterials. **26**(7), 697–702 (2005)
34. Zhuang, J., Gordon, M.R., Ventura, J., Li, L., Thayumanavan, S.: Multi-stimuli responsive macromolecules and their assemblies. Chem. Soc. Rev. **42**(17), 7421–7435 (2013)
35. Chen, Y., Peng, C., Lu, Y., Liu, W., Xu, W.: Responsiveness and release characteristic of semi-IPN hydrogels consisting of nano-sized clay crosslinked poly (dimethylaminoethyl methacrylate) and linear carboxymethyl chitosan. J. Nanosci. Nanotechnol. **15**(1), 164–171 (2015)
36. Cheng, C., Xia, D., Zhang, X., Chen, L., Zhang, Q.: Biocompatible poly (N-isopropylacrylamide)-g-carboxymethyl chitosan hydrogels as carriers for sustained release of cisplatin. J. Mater. Sci. **50**, 4914–4925 (2015)
37. Meng, F.D., Ni, Y.X., Ji, S.F., Fu, X.H., Wei, Y.H., Sun, J., Li, Z.B.: Dual thermal-and pH-responsive polypeptide-based hydrogels. Chin. J. Polym. Sci. **35**(10), 1243–1252 (2017)
38. Chen, Y., Song, G., Yu, J., Wang, Y., Zhu, J., Hu, Z.: Mechanically strong dual responsive nanocomposite double network hydrogel for controlled drug release of asprin. J. Mech. Behav. Biomed. Mater. **82**, 61–69 (2018)
39. Mantha, S., Pillai, S., Khayambashi, P., Upadhyay, A., Zhang, Y., Tao, O., Pham, H.M., Tran, S.D.: Smart hydrogels in tissue engineering and regenerative medicine. Materials. **12**(20), 3323 (2019)
40. Hu, C., Liu, W., Long, L., Wang, Z., Zhang, W., He, S., Lu, L., Fan, H., Yang, L., Wang, Y.: Regeneration of infarcted hearts by myocardial infarction-responsive injectable hydrogels with combined anti-apoptosis, anti-inflammatory and pro-angiogenesis properties. Biomaterials. **290**, 121849 (2022)
41. Tohidi, H., Maleki-Jirsaraei, N., Simchi, A., Mohandes, F., Emami, Z., Fassina, L., Naro, F., Conti, B., Barbagallo, F.: An electroconductive, thermosensitive, and injectable chitosan/pluronic/gold-decorated cellulose nanofiber hydrogel as an efficient carrier for regeneration of cardiac tissue. Materials. **15**(15), 5122 (2022)
42. Díaz-Payno, P.J., Kalogeropoulou, M., Muntz, I., Kingma, E., Kops, N., D'Este, M., Koenderink, G.H., Fratila-Apachitei, L.E., van Osch, G.J.V.M., Zadpoor, A.A.: Swelling-dependent shape-based transformation of a human mesenchymal stromal cells-laden 4D bioprinted construct for cartilage tissue engineering. Adv. Healthc. Mater. **12**(2), 2201891 (2023)
43. Gilpin, A., Zeng, Y., Hoque, J., Ryu, J.H., Yang, Y., Zauscher, S., Eward, W., Varghese, S.: Self-healing of hyaluronic acid to improve in vivo retention and function. Adv. Healthc. Mater. **10**(23), 2100777 (2021)
44. Nafee, N., Zewail, M., Boraie, N.: Alendronate-loaded, biodegradable smart hydrogel: a promising injectable depot formulation for osteoporosis. J. Drug Target. **26**(7), 563–575 (2018)
45. Shen, X., Li, S., Zhao, X., Han, J., Chen, J., Rao, Z., Zhang, K., Quan, D., Yuan, J., Bai, Y.: Dual-crosslinked regenerative hydrogel for sutureless long-term repair of corneal defect. Bioact. Mater. **20**, 434–448 (2023)
46. Merceron, T.K., Burt, M., Seol, Y.J., Kang, H.W., Lee, S.J., Yoo, J.J., Atala, A.: A 3D bioprinted complex structure for engineering the muscle–tendon unit. Biofabrication. **7**(3), 035003 (2015)

Chapter 9
Injectable Hydrogels

Aysel Kızıltay

Abstract Due to the similarity of the physical properties of hydrogels to those of living tissues, various hydrogels have been studied over the past 20 years as implantable biomaterials that demonstrate good compatibility with soft tissues. Despite their unique properties, the applications of traditionally prepared and pre-shaped hydrogels are limited. Since they lack injectable fluidity, implanting them subcutaneously can cause significant discomfort for patients. To address this issue, injectable polymer materials have been developed that exist as a fluid solution at room temperature but transition from a sol to a gel structure after being injected into living tissues. These types of hydrogels are attractive systems for regenerative medicine applications due to their minimally invasive delivery via a syringe, ease of application, the ability to easily and homogeneously incorporate cells and bioactive molecules before gelation, minimal toxic effects, and enhanced patient comfort levels.

Keywords Injectable · Crosslinked · Hydrophobic interactions · Click chemistry · Schiff base reaction · Drug delivery · Wound healing

9.1 Preparation of Injectable Hydrogels

Injectable hydrogels can be prepared through chemical or physical crosslinking of polymers. While physically crosslinked hydrogels are considered safer due to the absence of chemical agents, chemically crosslinked hydrogels are expected to have stronger mechanical properties. Injectable hydrogels generally exhibit lower performance against deformation compared to in situ formed gels, as the injection process can affect the structure and properties of the resultant gel. Injectable hydrogels can

A. Kızıltay (✉)
Central Laboratory, Middle East Technical University, Ankara, Türkiye
e-mail: kiziltay@metu.edu.tr

N. Hasirci, V. Hasirci (eds.), *Hydrogels and Bioinks in Tissue Engineering*, Biological and Medical Physics, Biomedical Engineering,
https://doi.org/10.1007/978-3-031-90966-5_9

fill deep tissue wounds and lower the risk of infection while causing less pain and scarring than standard hydrogels. The mechanical characteristics of gel systems are directly impacted by the rate at which hydrogels gel and the transition to the gel phase can reduce the amount of soluble material that seeps into the surrounding tissues. Consequently, the chosen polymers, their molecular weight, concentration, and the kind of crosslinking all affect the crosslinking (gelation) rate.

Injectable hydrogels have been demonstrated to be promising materials for a range of biomedical applications, including controlled drug release, tissue engineering, and tissue regeneration, following loading with various active substances (drugs, proteins, genes, viruses, cells, etc.). These polymer solutions which have the ability to form gels in the body, can serve as a drug delivery system (DDS) that continuously releases drugs and/or bioactive agents, or function as scaffolds that mimic extracellular matrix (ECM) and tissue structures to facilitate the regeneration of damaged tissues. In general, the mechanical properties of hydrogel structures are much weaker than those of hard tissues in the body, which means that in most cases, hydrogels can be used not as load-bearing scaffolds, but rather for mimicking soft tissues or providing therapeutic drugs to tissues. Additionally, they can be applied in various biomedical fields such as drug delivery, anti-adhesive materials after surgical operations, biodegradable adhesives, and vascular embolization agents. For example, fibrin is a natural injectable hydrogel used for hemostatic purposes to stop bleeding [1]. Many physical and chemical crosslinking techniques, such as click chemistry, disulfide crosslinking, enzyme-mediated crosslinking, Michael-type addition reactions, Schiff base crosslinking, photo-crosslinking, ionic crosslinking, and reactions that are sensitive to heat or pH, can be used to create injectable hydrogels. It is possible to impart different functionalities to injectable hydrogels by modifying polymer design or employing new crosslinking methods.

9.2 Important Factors for Injectable Hydrogels

Injectable hydrogels are essentially a type of in situ forming gel. However, in addition to the ability to gel in situ, the term "injectable" implies that the hydrogel must be delivered to a target area for gelation via an injection device (Fig. 9.1). For clinical applications, an injectable hydrogel must be biocompatible and biodegradable, non-toxic, and possess the necessary mechanical, viscosity, and stability properties. It is important for the hydrogel to release drugs or support cell growth and differentiation without causing toxicity or immune reactions in the body. Additionally, the degradation products that result from the hydrogel's breakdown should not induce any adverse or chronic physiological or inflammatory responses.

Most natural polymers are more biocompatible than synthetic polymers because they structurally resemble the natural extracellular matrix (ECM) [2]. The viscosity of the polymer solution is a critical factor for ensuring easy injectability. The viscosity should be low enough to allow for the homogeneous distribution of drugs and/or cells within the matrix before gelation. Moreover, the gel should form very quickly

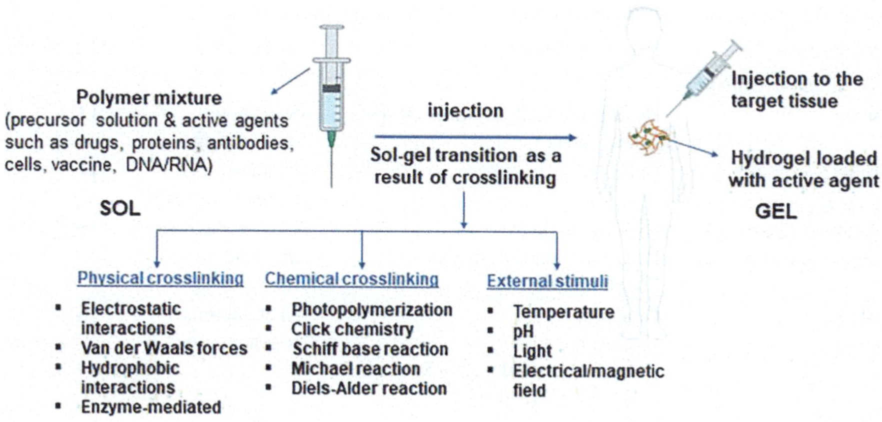

Fig. 9.1 Preparation techniques and applications of injectable hydrogels

after application to prevent unwanted leakage of the loaded agents or their movement to undesired locations.

After gelation, the hydrogel matrix must encapsulate the relevant drug molecules or cells and maintain a stable structure for cellular behaviors such as adhesion, migration, proliferation, and differentiation. Gelation kinetics is another important characteristic to consider in injectable hydrogels for tissue engineering and regenerative medicine applications. The kinetics of gelation can vary according to different gelation methods and should be sufficient for the targeted tissue. Other significant factors to consider include the solvents used to dissolve the polymer, the type and ratio of crosslinking agents, the time required for polymer crosslinking, the intensity of stimuli necessary for gelation, the concentration of initiators, and the functionality of the material.

9.3 Methods for Preparing Injectable Hydrogels

Hydrogels can be classified based on their responsiveness or the types of bonds they contain. Hydrogel formation can occur through physical or chemical crosslinking. Physical crosslinked hydrogels depend on fragile and reversible chemical bonds, which are influenced by the environmental conditions such as acid-base status, temperature, or the presence of ions. In contrast, chemically crosslinked hydrogels rely on covalent bonds formed using potentially toxic crosslinking agents or enzymes to create an irreversible polymer network.

Many physical and chemical crosslinking methods have been utilized to prepare injectable hydrogels (Table 9.1). The underlying crosslinking mechanism that leads to the solidification of a hydrogel determines its suitability as an injectable material. The physical or chemical crosslinking properties of precursor polymers in response

Table 9.1 Examples of methods of preparation of injectable hydrogels

Method	Polymeric material	Usage area	Reference
Physical crosslinking			
Ionic	Hyaluronic acid, gelatin, cellulose nanocrystals	Soft tissue engineering	[3]
Ionic	Chitosan	Immunization against viruses	[4]
Gelation due to temperature change	Hyaluronic acid Hekzametilendiizosiyanat Poloxamer 407	Diabetic wound dressing	[5]
Magnetic field	Gellan gum Hyaluronic acid Collagen coupled with iron oxide Magnetic nanoparticles	Skeletal muscle regeneration	[6]
Gelation due to temperature change	PNIPAAm Jelatin	Cardiac tissue engineering	[7]
Gelation due to temperature change	Akrileştirilmişhyaluronikasit NIPAAm	Ocular drug delivery	[8]
Gelation due to pH and temperature change	PNIAAm-co-IA Kitosan	Drug delivery for cancer treatment	[9]
Chemical crosslinking			
EDC/NHS	Gelatin 3-amino-4-methoxybenzoic acid,	Wound healing	[10]
Michael addition reaction	Collagen Hyaluronan	Tissue regeneration	[11]
Schiff base and PF127 crosslinking	Quaternized chitosan PEO99-b-PPO65-b-PEO99	Wound dressing	[12]
Enzymatic	Carboxymethyl cellulose (CMC), sulfated CMC, gelatin	Cartilage tissue regeneration	[13]

to various stimuli are fundamental characteristics governing the formation of an injectable hydrogel. Physical stimuli, such as temperature or ionic concentration, can induce crosslinking between polymers through non-covalent interactions like hydrogen bonds, hydrophobic interactions, and ionic interactions.

9.3.1 Physically Crosslinked Hydrogels

Physically crosslinked injectable hydrogels are three-dimensional networks in which polymer chains are held together by non-covalent interactions (such as environmental triggers like varying pH, temperature, ionic strength, as well as electrostatic, hydrogen bonding, and hydrophobic interactions). The network structures in physically crosslinked hydrogels are reversible. Due to their widespread applications in biomedical and pharmaceutical fields, these hydrogels have been extensively studied. The gels produced using this method offer significant advantages in terms of biocompatibility and toxicity, as no toxic chemicals are used for crosslinking. However, their mechanical strength is relatively weak. These fluid hydrogels, which exhibit decreased viscosity under shear stress, can be easily injected using a fine needle and can gel immediately after injection, effectively filling the targeted area [14].

Compared to chemical hydrogels, the preparation procedures for physically crosslinked hydrogels are simpler and do not require harsh production conditions, such as organic solvents or heat, making them safer materials for biomedical applications. Additionally, unlike the stable structures of chemical hydrogels, the reversible crosslinks allow these hydrogels to easily break down under physiological conditions, which can be advantageous for certain clinical applications. The high permeability of the hydrogel matrix permits the free diffusion of drug molecules, leading to rapid and immediate drug release. The diffusion rate and solubility of molecules inside the hydrogel matrix are directly impacted by the control of hydrogel properties such as pore size, hydrophilicity, or crosslinking density. Physical interactions such as hydrogen bonds, hydrophobic interactions, and/or supramolecular interactions can enhance drug release from polysaccharide-based hydrogels. To treat infections in avascular or necrotic tissues, for instance, Hoque et al. created a dual-purpose injectable hydrogel that contains *N*-(2-hydroxypropyl)-3-trimethyl ammonium chitosan and polydextran aldehyde loaded with vancomycin. Extended, pH-dependent drug release from the gel was made possible by the reversible imine linkages that were created between vancomycin and polydextran aldehyde [15].

9.3.1.1 Ionically Crosslinked Hydrogels

Ionic interactions allow hydrogels to be crosslinked under mild conditions, at room temperature and physiological pH. Ionizable polymers like alginate and pectin, or proteins like collagen that have been crosslinked with oppositely charged substances or metal ions like Fe^{3+}, Zn^{2+}, Ca^{2+}, and $AuCl_4$, are the main sources of ionic crosslinked hydrogels. Depending on the chosen ion, the hydrogel's permeability, degree of crosslinking, or drug release kinetics can be altered.

The main disadvantage of ionic crosslinked hydrogels is their short-term stability under physiological conditions. These hydrogels respond to changes in ionic strength in the body, causing the hydrogel structure to degrade due to the release of ions from the matrix. This property can be beneficial for controlling the release of chemotherapy drugs locally injected into tumor tissues. For instance, mucoadhesive and pH-sensitive hydrogels made by mixing chitosan-grafted-dihydrocaffeic acid with oxidized pullulan (OP) have been shown to continuously release the anticancer drug doxorubicin for the treatment of colon cancer, demonstrating lethal effects on colon tumor cells (HCT116). Additionally, these hydrogels, when applied subcutaneously in rats, exhibited good injectability, tissue adhesion, and rapid gelation. The grafting of dihydrocaffeic acid onto chitosan (CS-DA) enhances the poor solubility of chitosan. CS-DA undergoes crosslinking between polysaccharides through the conversion of catechol to o-quinone at physiological pH levels (7.4), subsequently reacting with surrounding molecules that contain catechols and amines. However, the oxidation of dihydrocaffeic acid occurs slowly under acidic conditions, reducing the density of crosslinking within the hydrogel structure and leading to increased swelling. Considering that tumor tissues exhibit a more acidic pH compared to normal tissues, the CS-DA injectable hydrogel system is effective as a low pH-sensitive drug release system. The mucoadhesive properties of CS-DA further enhance the hydrogel's adhesion to tissues, limiting drug diffusion to other tissues and resulting in fewer side effects [16].

9.3.1.2 Hydrophobic Interactions

Hydrophobic interactions are one of the most commonly used physical methods for injectable hydrogels, and these interactions are typically reversible. Hybrid hydrogels crosslinked through hydrophobic interactions are derived from amphiphilic polymers, which contain both hydrophilic and hydrophobic components. Changes in temperature or pH can alter the hydrophobic interactions between polymer chains, leading to hydrogel gelation. These interactions cause the polymer to swell and encapsulate water within the hydrogel structure. For instance, Wang et al. used hydrophobic interactions to create an injectable hydrogel based on chitosan that has the ability to crosslink quickly and function as a tissue adhesive. A mussel adhesion protein was included into the chitosan matrix to improve the chitosan's solubility and tissue adherence. Catechol-modified chitosan (Cat-CS), a thermoresponsive polymer (PNAM), and $NaHCO_3$ for dual ionic crosslinking were used to create an injectable bioadhesive chitosan-based hydrogel. Using sodium bicarbonate as a crosslinking agent and functional polymers including the thermoresponsive monomer *N*-isopropylacrylamide, the amino groups of chitosan molecular chains electrostatically interacted to form the injectable hydrogel. With a gelation time of 50 s at 37 °C, this sticky hydrogel showed exceptional adhesive qualities resembling the functional groups and electric charges of mussel proteins. Furthermore, it was reported that the hydrogel adhesive could be injected at room temperature and quickly transformed into a gel state at body temperature [17].

Poloxamer triblock copolymers are an example of a synthetic substance utilized in this application. Triblock copolymers called poloxymers are made up of two hydrophilic polyethylene oxide (PEO) blocks around a core hydrophobic polypropylene oxide (PPO) block. For instance, concentrated solutions of P407 (also known as Pluronic® F127) undergo gelation due to hydrophobic interactions between polymer chains resulting from temperature changes [18]. P407 has been utilized as an injectable hydrogel in some studies aimed at enhancing cell adhesion and chronic wound healing. Bercea et al. noted that Pullulan/P407 mixtures remain in a liquid state at low temperatures (where hydrophilic interactions dominate) and spontaneously transform into gel at physiological temperatures (where hydrophobic interactions are stronger). These thermoresponsive gels were prepared by adding poly(ethylene glycol) (PEG) and glycerin, enhancing biological properties with egg white to improve gel stability. P407 transitions from a micellar liquid structure to an injection-friendly viscous gel structure at concentrations above 15% due to physical interactions. As the temperature increases, intermolecular interactions between pullulan chains increase, resulting in gelation of the structure [19].

9.3.2 *Stimuli-Responsive Hydrogels*

In reaction to environmental stimuli like temperature, pH, light, ionic strength, and electric or magnetic fields, stimuli-responsive hydrogels show sudden changes in their swelling characteristics, network topologies, permeability, and mechanical strength. They are frequently called "responsive" or "smart" hydrogels. Notably, because of their customizable swelling qualities, injectable smart hydrogels that are sensitive to pH and temperature can regulate medication release in the tissues where they are injected.

9.3.2.1 Thermosensitive Crosslinked Hydrogels

Thermosensitive hydrogels possess the ability to transition from a free-flowing liquid at room temperature to a viscous gel at body temperature. These types of hydrogels offer advantages in terms of reduced cytotoxicity and shorter gelation times. Temperature is the easiest stimulus to manipulate in environmentally sensitive hydrogels, making thermosensitive systems the most studied group among injectable smart gels. The sol-gel transition of thermosensitive hydrogels is typically triggered by an increase or decrease in temperature. After injection, these materials form a gel in the applied area due to temperature-dependent phase transitions. The gel acts like a liquid when the outside temperature is higher or lower than body temperature, making it injectable outside the body. After injection, the system becomes a gel with mechanical characteristics akin to those of natural tissues due to the extensive covalent and non-covalent three-dimensional crosslinking that is induced by physiological temperatures. Because thermosensitive crosslinked hydrogels may be easily applied to temperature stimuli without the need for complicated equipment or external triggers, they have been the subject of much research. A $(mPEG\text{-}PCL\text{-}Se)_2$ copolymer solution, for example, has been shown to change from an injectable fluid state at ambient temperature to a gel at temperatures as high as 37 °C [20].

In a study, a thermosensitive hydroxybutyl chitosan (HBC) hydrogel was prepared with the aim of being used as an embolic material to occlude selected blood vessels for controlling bleeding. The hydroxybutyl groups added to the chitosan main chain provided the HBC hydrogel with a rapid sol-gel transition feature depending on the surrounding temperature changes. It was demonstrated that the HBC solution could be injected into the desired vascular region, and the in situ formed HBC hydrogel could effectively embolize blood vessels to control bleeding [21].

Most thermosensitive hydrogels designed for biomedical and tissue engineering applications are based on changes between body temperature and ambient temperature. These gels can be injected into tissue areas without blocking the syringe needle and gel rapidly at body temperature.

Natural, synthetic, or hybrid materials have been used to create a wide variety of thermosensitive hydrogel systems. Natural polymers that are frequently investigated include chitosan, dextran, cellulose derivatives, hyaluronic acid, heparin, and alginate. A cationic substance with nitrogen in its molecular structure, chitosan has the ability to generate in situ gels and polyelectrolyte complexes, among other useful characteristics. Chitosan's cationic composition facilitates the production of carboxylate salts, including citrate, acetate, and lactate, which dissolve the hydrogel structure in water. The main reason limiting the use of chitosan is its poor solubility in water. The solubility of chitosan can be enhanced through simple chemical modifications, such as the protonation of amine ($-NH_2$) groups in D-glucosamine units [22]. In particular, chitosan-based injectable gels containing various polyols like glycerol, ethylene glycol, and sorbitol are important for the release of specific drugs in targeted tissues [23]. These polyols help maintain the chitosan solution in a liquid state. They create a hydration shell around chitosan chains in acidic solutions, preserving chitosan's solubility at lower temperatures and higher pH values.

Glycerol phosphate (β-GP) is another polymer that contains chitosan and is utilized to produce the sol-gel transition at physiological pH and temperature. β-GP has attracted a lot of interest in the creation of gel delivery systems based on thermosensitive chitosan. For example, adding β-GP to a chitosan solution raises the pH, which decreases electrostatic repulsion and enhances hydrogen bonding between chitosan chains [24]. Furthermore, increasing the temperature strengthens the hydrophobic contacts between the chitosan chains by releasing hydrogen bonds between water molecules and the chains.

Methylcellulose is a modified cellulose that dissolves in water at low temperatures. The hydrophobic methyl groups in its structure confer thermosensitive crosslinking behavior at temperatures between 40 and 50 °C, depending on its molecular weight and degree of substitution. Increasing the degree of substitution and the concentration of methylcellulose, as well as adding salts like NaCl to the solution, lowers the gelation temperature [25]. The effectiveness of injectable hydrogels based on methylcellulose containing PEG and chitosan sulfate as barriers to reduce postoperative adhesions was evaluated using a rat cecal abrasion model. The results demonstrated that injectable methylcellulose hydrogels were effective in reducing adhesion formation and reopening adhesions [26].

Collagen is the source of the protein gelatin, which is another naturally occurring polymer with thermosensitive properties. In contrast to other thermosensitive polymers, gelatin liquefies at temperatures above 30 °C and gels at temperatures below 25 °C. Gelatin's special quality has led to its combination with synthetic polymers like PEG and NIPAAm, as well as other natural polymers like chitosan and alginate, to produce thermosensitive hydrogels for tissue engineering and medication delivery. For example, in a study, a PNIPAAm-g-PEG-based hydrogel injected into rats to repair spinal cord injuries resulted in improved motor functions and sensitivity, with nerve axon growth observed within the hydrogel [7].

The most researched thermosensitive synthetic polymer is poly(*N*-isopropylacrylamide) (PNIPAAm), which is favored for biomedical applications due to its sharp sol-gel transition at 32 °C, or near body temperature. It dissolves in aquatic settings below its lower critical solution temperature (LCST), but it forms a hydrogel beyond this temperature. Heat helps break the hydrogen bonds between NIPAAm molecules and water, increasing interactions between the hydrophobic portions of the polymer and causing the sol-gel transition. To produce hydrogels with variable gel temperatures, mechanical characteristics, and biodegradability for various purposes, PNIPAAm can be crosslinked with other natural polymers including chitosan, collagen, gelatin, and hyaluronic acid, or with synthetic polymers like polyethylene glycol. Navaei et al. developed a thermosensitive bi-hybrid injectable hydrogel based on poly(*N*-isopropylacrylamide) PNIPAAm-gelatin with suitable mechanical strength for damaged myocardium repair and regeneration. In this study, the researchers conducted extensive in vitro analyses on cardiac function before proceeding to in vivo experiments [27]. The PNIPAAm polymer can be modified by copolymerization with different monomers for various purposes, such as controlled degradation, increased biocompatibility, and rapid response characteristics.

Tan et al. developed a thermosensitive injectable hydrogel for soft tissue engineering by grafting PNIPAAm-COOH onto amine-modified hyaluronic acid chains. It was reported that as the degree of PNIPAAm grafting increased, both enzymatic degradation and swelling values decreased. Moreover, the encapsulated model cells, derived from adipose stem cells, maintained their viability and proliferated within the copolymer hydrogel, demonstrating biocompatibility when injected into mice [28].

Triblock copolymers including PEO-PPO-PEO (Pluronic®), PLGA-PEG-PLGA, PEG-PLLA-PEG, PCLA-PEG-PCLA, and PEG-PCL-PEG are an important class of synthetic thermosensitive polymers. In aqueous solutions, these amphiphilic copolymers combine to form gels at high temperatures and change into micellar structures at low temperatures. These thermosensitive polymers' gelation characteristics are influenced by their molecular weight, content, and solution concentration. Poloxamers are non-ionic triblock copolymers made up of two hydrophilic polyoxyethylene chains encircling a core hydrophobic polyoxypropylene chain [29]. They are also marketed under the Pluronics® brand. Because of hydrophobic interactions, the aqueous solutions of thermosensitive polymers, like Pluronics® and poloxamers, show sol-gel transitions at higher temperatures, which promote the self-assembly of polymer chains within a particular concentration range. For example, in one study, a thermosensitive Pluronic (F127)-hyaluronic acid-based hydrogel containing umbilical cord mesenchymal stem cells was injected into rat uterine wounds to treat cesarean section injuries. Examination of the uterine tissue on day 28 post-injection revealed good cell adhesion, proliferation, and vascularization.

Additionally, it was reported that a 28% degradation rate observed in the presence of lysozyme at the end of day 14 was sufficient for tissue repair [30].

9.3.2.2 pH Sensitive Crosslinked Hydrogels

pH is one of the commonly used physical stimuli in the preparation of injectable hydrogels. In reaction to changes in the pH of their surroundings, pH-sensitive polymers can either take or release protons thanks to their basic or acidic functional groups. Depending on the pH of the solution, these polymers—known as polyelectrolytes because of their ionizable groups—can create reversible hydrogels. Ionization and deionization are two ways that a polymer might go through a phase shift. It is possible to create ion-induced complex hydrogels by electrostatically interacting oppositely charged ions within the polymers. Due to problems with injection needle obstruction, absence of ionic interactions in protein/drug/gene delivery, polymer solubility, and material storage, pH-sensitive or pH/temperature-sensitive hydrogels have been developed for numerous thermosensitive hydrogel applications. Every human system has a distinct physiological pH environment, as is well established (e.g., stomach pH 1.35–3.5, liver pH 7.6–8.8, gut pH 6–7.5, tumor areas pH 6.4–7.0, body pH 7.4, and vaginal pH 3.8–4.5) [31]. This variation is crucial for ensuring that drugs and bioactive agents delivered via pH-sensitive systems are effective in the desired regions. For instance, pH-sensitive injectable hydrogels serve as smart carriers for anticancer drug delivery, releasing the drug effectively in the targeted acidic tumor environment while retaining it in normal physiological conditions.

pH-sensitive polymers are considered weak polyelectrolytes due to the presence of acidic groups, such as sulfonamides, carboxylic acids, or basic amine groups that ionize at low or high pH levels. Examples of pH-sensitive synthetic polymer-based injectable hydrogels used in medicine include PMA-PEG, NIPPAm-AAc, NIPPAm-DMAEMA, 2-(dimethylamino)ethyl methacrylate, and Pluronic® F127-DMAEMA. For example, one study developed biodegradable thermosensitive and pH-sensitive hydrogel systems based on polyacrylic acid (PAA) as carriers for stem cells aimed at cardiac therapy. The hydrogels were synthesized by copolymerizing NIPAAm, PAA, methacrylate poly(ethylene glycol) methyl ether (MA-PEG), and a biologically degradable macromer, 2-hydroxyethyl methacrylate-co-oligo(trimethylene carbonate) (HEMA-oTMC). These hydrogels were reported to be injectable through catheters at a solution pH of 8.0 and could form solid gels in an acidic environment (pH 6.5) similar to that of infarcted cardiac tissue. Additionally, it was reported that stem cells cultured in the hydrogel maintained their viability for up to 7 days. The findings imply that the created hydrogels might make appropriate injectable cell carriers for cardiac cell treatment [32].

Biopolymers like alginate, gelatin, dextran, chitosan, and albumin are examples of injectable natural polymers that react to changes in pH. Since cellulose's many hydroxyl groups are insensitive to changes in acidity, it cannot be employed directly to create pH-sensitive injectable hydrogels. However, cellulose can be made pH-sensitive by functionalizing its chains with amines, carboxylic acid groups, or other ionizable groups. Zhang et al. prepared a pH-sensitive hydrogel based on oxidized dextran (Dex-CHO) and achieved a dual-antibiotic hydrogel by grafting sulfanilamide (an antifungal) and tobramycin (an antibiotic) onto oxidized dextran. The pH difference between the infected area and normal tissue provides a specific target. The dual-antibiotic injectable hydrogel releases drugs in a pH-sensitive manner in mildly acidic environments. In order to generate smart hydrogels for the topical administration of antibiotics and the innovative treatment of localized bacterial infections, this work offers a straightforward and promising approach [33].

9.3.3 Chemically Crosslinked Hydrogels

Chemical crosslinked hydrogels are a group of hydrogels that transition from a liquid state to a gel state (sol-gel transition) due to specific chemical reactions forming new covalent bonds within the polymer network. Various methods have been developed to create chemical crosslinked hydrogels using natural or synthetic polymers, including click chemistry, light-sensitive polymerization, Schiff base formation, enzyme-catalyzed reactions, and Michael addition reactions. Generally, the synthesis of chemical crosslinked hydrogels exhibits slow gelation kinetics.

9.3.3.1 Photo-Crosslinked Hydrogels

Photo-crosslinked hydrogels are produced by incorporating light-sensitive molecules or polymerization catalysts into polymer solutions. Two main mechanisms can be identified based on the characteristics of the photosensitive materials: the first involves adding the material to a thermosensitive gel, where it transforms light energy into heat energy and raises the gel's temperature to its phase transition temperature; the second involves directly integrating the photosensitive materials into the gel structure, where exposure to UV/visible light causes polymerization and the production of free radicals. For example, methacrylate (acrylate groups) undergoes rapid polymerization when exposed to a light source. Advantages of photopolymerization include controlled gelation kinetics, the ability to design hydrogel structures for drug delivery, low energy requirements, the absence of overheating and regional toxicity, and rapid gelation at body temperature [34].

Typically, polymers such as chitosan and gelatin are modified to become photoactive. For instance, GelMA is a methacrylated derivative of gelatin and possesses photo-crosslinkable properties. Similarly, it is possible to synthesize methacrylated derivatives of hyaluronic acid and chitosan. Recently, a study developed a GelMA-HAMA/nHAP composite hydrogel drug delivery system aimed at bone regeneration using GelMA, hyaluronic acid methacrylate (HAMA), and nano-hydroxyapatite (nHAP). Exosomes derived from human urine stem cells (human USCEXOs) were loaded into the hydrogel solution to enhance bone regeneration. After injection into cranial bone defects created in rats, gelation was achieved through photo-crosslinking at 405 nm. Histological results showed improved osteogenesis and angiogenesis in bone tissue in the presence of USCEXOs [35].

Yu et al. designed a composite hydrogel comprising a methacrylated decellularized dermal extracellular matrix (ddECMMA), thiolated chitosan, polycaprolactone nanofibers, and poly(ethylene glycol) diacrylate for skin regeneration. The prepared hydrogel solution was injected into wound defects created on the backs of rats and subsequently exposed to UV light for gelation (100 mW/cm^2, 100 s). It was reported that the injected hydrogel accelerated skin tissue regeneration in the rat wound models [36].

9.3.3.2 Hydrogels Obtained via Click Chemistry Reactions

It is also possible to prepare injectable chemical crosslinked hydrogels without initiators. These hydrogels are created when complementary chemical groups react, as in Schiff base reactions and click chemistry. The following are examples of click chemical reactions: radical-mediated thiol-Michael reactions, Diels-Alder cycloaddition, non-catalyzed alkyne-azide coupling reactions, and Cu(I)-catalyzed alkyne-azide cycloaddition. One common illustration of click chemistry is the interaction between alkyne and azide groups. High efficiency and chemo-selectivity under physiological settings are two major benefits of click chemistry, a novel approach to creating injectable hydrogel scaffolds in tissue engineering. The Cu(I)-catalyzed azide-alkyne (CuAAC) click reaction has attracted growing attention in chemical synthesis, drug discovery, bioconjugation, and biochemistry since the groundbreaking research by Sharpless and Meldal in 2002 because of its biocompatibility, quick reaction rate, high efficiency, superior regional selectivity, and bioorthogonal qualities [37].

The click reaction has become a valuable bioorthogonal method for preparing chemically crosslinked biopolymer-based hydrogels over the past 15 years. Under physiological conditions, CuAAC's reactions are quick, usually allowing gelation to occur in a matter of seconds to minutes. The first click hydrogels were made with poly(vinyl alcohol) functionalized with azide and alkyne [38]. Recent advancements in copper-free click chemistry have allowed for hydrogel development without potential toxic catalysts or immunogenic enzymes. Examples of copper-free click reactions include azide-alkyne cycloaddition, radical-mediated thiol-ene chemistry, Diels-Alder reactions, tetrazole-alkene photo-click chemistry, and oxime reactions. Notably, click chemistry, unlike other chemical methods, allows for the encapsulation of live cells and bioactive molecules (such as proteins and drugs) into gels without damage, enabling the preparation of multifunctional hydrogels. Consequently, injectable hydrogels developed through click chemistry have applications in controlled drug delivery, tissue engineering, and regenerative medicine.

Diels-Alder (DA) click reactions are becoming more and more popular since they don't require copper, making them a novel approach to creating injectable hydrogel scaffolds for tissue engineering. Additionally, reactions between functional groups and proteins in organisms occur infrequently, making DA click chemistry applicable both in vitro and in vivo. However, when DA click chemistry is the sole crosslinking method used for preparing injectable hydrogels in vivo, the gelation time tends to be longer. The most effective solution to this issue is to implement the DA reaction in conjunction with another rapid gelation system. For instance, Bai et al. used the DA click reaction to create a dual crosslinked injectable hydrogel system with a low polymer content. Using the DA click reaction, F127@ChS (F127 crosslinked chondroitin sulfate) was produced. PEG-(maleimido terminated polyethylene glycol) and F127@ChS were then used to construct dual crosslinked hydrogels. While the in situ DA click reaction between F127@ChS and PEG produced covalent crosslinking for the hydrogel system, the physical crosslinking of F127@ChS allowed the hydrogel to form quickly. The dual crosslinked injectable hydrogels were loaded with BMP-4 and then used as scaffold constructs to heal cranial defects in rats. The development of new bone tissue in the defect locations was demonstrated by histological investigation and X-ray imaging 12 weeks later. According to the findings, dual crosslinked injectable hydrogel offers a promising alternative for cranial bone tissue engineering [39].

9.3.3.3 Hydrogels Obtained via Schiff Base Reactions

Schiff bases are crosslinked chemical compounds that are usually defined by a double bond between carbon and nitrogen. They are created when the electrophilic carbon atoms of aldehydes or ketones combine with nucleophilic amines or hydrazines. Under some biological circumstances, Schiff base reactions can take place without the aid of chemicals or catalysts. For example, an injectable gel can be created via a Schiff base reaction between the amino groups of glycol chitosan and oxidized sodium hyaluronate that has a high abundance of aldehyde groups. One study developed injectable oxidized hyaluronate/glycol chitosan gels for cartilage regeneration. Using Schiff base pathways, hyaluronate was combined with glycol chitosan to form a gel after being partially oxidized with sodium periodate to produce aldehyde groups throughout the polymer chains. In vitro studies were conducted to examine the impact of oxidation degree, polymer concentration, and polymer composition on the mechanical characteristics of oxidized hyaluronate/glycol chitosan hydrogels. It was observed that this hydrogel system could be used as an injectable solution for cartilage regeneration and other tissue engineering applications [40].

9.3.3.4 Thiol-Based Michael Reactions

The Michael reaction is a kind of click reaction when a nucleophile is added to an unsaturated carbonyl molecule that is either α or β. A thiol-based Michael reaction occurs when the nucleophilic components include compounds that contain thiols and amines, which normally react with unsaturated carbonyl components such as vinyl sulfone groups and acrylate/methacrylate. This reaction method's regulated reaction speeds, non-reactivity with biomolecules, and high chemical yields make it popular in biomedicine, optoelectronics, pharmaceuticals, composites, coatings, and adhesives.

Thiol-based Michael reactions are employed in biomedical applications for the release of biomolecules (e.g., proteins, anticancer agents) and the production of in situ crosslinked injectable hydrogels for tissue engineering. Vinyl sulfone-functionalized β-cyclodextrins can crosslink amino derivatives of hyaluronic acid via this method. The azo-type Michael reaction between amino groups and vinyl groups can trigger a rapid and spontaneous crosslinking process in an aqueous environment at 37 °C, forming a gel. Such hydrogels can be utilized for delivering anticancer agents to tumor sites. Fiorica et al. developed an in situ forming injectable hydrogel for solid tumor treatment, obtained through the azo-type Michael reaction between amino derivatives of hyaluronic acid and vinyl sulfone-functionalized β-cyclodextrins. Hydrogels loaded with the anticancer drug doxorubicin were injected near the tumor mass in model mice, demonstrating a reduction in tumor mass without toxic effects on cardiac tissue [41].

Overall, click reactions such as Michael reactions eliminate disadvantages that limit the clinical use of typical hydrogels, such as excessive swelling, complex and potentially toxic crosslinking processes, or the lack of slip thinning properties. To enable click reactions for crosslinking, natural polymers such as polysaccharides and proteins need to be modified. Thiolated collagen can be converted into an injectable hydrogel via crosslinking with multifunctional PEG-maleimide through a Michael addition reaction [42]. The gelation ability and crosslinking density increase with the concentration of functional groups.

9.3.3.5 Enzymatically Catalyzed Reactions to Produce Hydrogels

Enzymatically catalyzed crosslinking reactions occur in the presence of enzymes, where substrate specificity is crucial. Maintaining a neutral pH and appropriate temperature is essential to prevent damage to the enzyme, and the solvent used must be water. Enzymes such as transglutaminase, horseradish peroxidase, phosphopantetheinyl transferase, tyrosinase, and lysyl oxidase have been utilized to prepare hydrogel systems from natural polymers like gelatin, dextran, hyaluronic acid, and chitosan, particularly in tissue engineering.

For example, hyaluronic acid-tyramine (HA-TA) and chondroitin sulfate-tyramine (CS-TA) were combined to create an injectable enzyme-mediated cross-linkable hydrogel. Bone marrow-derived mesenchymal stem cells (BMSCs) and bone morphogenetic protein-2 (BMP2) were simultaneously encapsulated in this hydrogel, which was created in 15 s with hydrogen peroxide (H_2O_2) and catalyzed horseradish peroxidase (HRP). The gelation rate and mechanical strength could be easily adjusted by varying the concentrations of H_2O_2 and HRP. The BMSC and BMP2-loaded hydrogel system was injected into defects created in the femurs of rats. In vitro and in vivo studies showed that enzyme-mediated crosslinking resulted in hydrogel formation with pore sizes ranging from 100 to 300 μm within 15 s, indicating its potential to promote BMSC proliferation and osteogenic differentiation [43].

Injectable hydrogels with improved mechanical properties and controlled gelation and degradation rates can be produced by adding phenolic groups to thermosensitive carboxymethyl chitin chains [44]. Gelatin modified with tannic acid can be crosslinked using transglutaminase, which catalyzes the acyl-transfer reaction between the γ-carboxamide of glutamine and the ε-amino group of lysine, leading to covalent bonding with relevant amino acids in tissues. The hydrogels obtained in this manner can be utilized in applications such as hemostatic agents for tissue wounds or as bio-glues for tissue repair.

9.4 Biomedical Applications of Injectable Hydrogels

Injectable hydrogels possess soft tissue characteristics due to their high-water content, minimizing mechanical damage to surrounding tissues (e.g., through friction). They exhibit low adhesion to mucosal surfaces and tissues, which opens up many opportunities for applications in medicine due to their similarity to living tissues.

9.4.1 Drug/Biomolecule Delivery

Traditional drug delivery approaches, such as systemic and topical applications, often face challenges like undesired drug absorption, rapid metabolism, frequent dosing, systemic toxicity, and easy degradation under specific physiological conditions. Injectable hydrogels offer promising smart drug delivery systems capable of overcoming these drawbacks through extended and controlled drug release and adjustable physical properties. Their biocompatibility, minimally invasive characteristics, injectability, and responsiveness to environmental or enzymatic stimuli make these hydrogels attractive as smart therapeutic systems. Macromolecular compounds such as genes, peptides, proteins, and even live cells can be encapsulated in such hydrogels.

Stimulus-responsive polypeptides for drug/protein delivery, cancer treatment, and the creation of artificial protein-based systems have been the subject of recent research [45]. Different crosslinking techniques allow for customization of the molecular weights and chemical structures of polypeptide hydrogels. These changes (reversible, hydrophilic, hydrophobic, etc.) can be brought about by external stimuli like light, heat, and temperature, or by endogenous reactions like pH shifts, enzymes, and reductions. Several drug carriers, such as nanogels, hydrogels, and micelles for various delivery systems, are made using stimulus-responsive polypeptides [46].

9.4.2 Tissue Engineering

In soft tissue applications, injectable hydrogels are important, especially for cartilage regeneration and repair. Tissue engineering is said to be a quick method that combines bioactive compounds, cells, and scaffolds. The hydrogel structure may cover anatomical voids where cells can absorb, grow, and reorganize to generate completely functional tissues because of its predetermined, porous, and biocompatible 3D architecture. Commonly used polymeric materials for creating injectable hydrogel scaffolds include various proteoglycans, collagen, hyaluronic acid, alginate-based substrates, and chitosan [47, 48].

9.4.3 Wound Healing

The development of new wound dressings with antibacterial properties and accelerated wound healing has become a focus of current research. Wound infection is a primary cause of delayed healing, making the development of wound dressings with strong antibacterial properties essential. Hydrogels, with high water swelling capacity and oxygen permeability, provide a moist environment conducive to skin wound healing. Incorporating antibacterial polymers or loading antibacterial agents into the hydrogel structure can help combat bacteria-induced infections. Recent advancements have led to the development of self-healing hydrogels that can repair tissue damage by forming dynamic covalent and non-covalent bonds, thereby extending their stability [49].

9.5 Challenges in Injectable Hydrogel Formulations

Developing injectable hydrogel formulations involves significant challenges, particularly in maintaining low viscosity while ensuring adequate elasticity in situ. Since these hydrogels are administered via a syringe, concerns regarding in situ gelation and repeated dosing arise.

It is crucial that hydrogels protect biomolecules like DNA, peptides, proteins, and drugs from enzymatic degradation while allowing fragile biomolecules and cells to undergo a compatible crosslinking process, ensuring healthy cell growth within the hydrogel environment. Additionally, it is vital to prevent cytotoxicity and inflammatory responses when the injected hydrogel interacts with surrounding tissues. Immunological responses can adversely affect both the patient's health and the performance of the hydrogel, making the minimization of immune responses during in situ gel transitions a critical biological design parameter.

When developing hydrogel structures, factors such as the gelation mechanism, viscosity during injection, mechanical strength post-gelation, degradation time, and release profiles of bioactive agents should be carefully considered. Furthermore, injectable hydrogels should be specifically designed to suit the pathological and physical structure of the target tissue.

The effective release of bioactive molecules within tissues is determined by factors such as cargo-gel size, affinity, and interactions. Currently, injectable hydrogels used in clinical applications often include microparticle-based depot systems. Among the most commonly approved depot formulations are injectable hydrogels based on hyaluronic acid containing lidocaine (an anesthetic) for wrinkle filling. In applications such as wound dressings, the continuous and extended release of drug molecules from hydrogel formulations is critical. To prolong the release duration in such formulations, the hydrogel network's pore size must be reduced through physical or chemical crosslinking.

In tissue regeneration applications, it is essential for the hydrogel to be bioactive. Hydrogel bioactivity depends on cells or growth factors sticking to natural or artificial adhesives like gelatin, fibrin, or hyaluronic acid. Some gelatin-based hydrogel formulations have been developed for this purpose and are currently in clinical bioactivity study stages. Another challenge related to hydrogels is their rapid degradation times, highlighting the importance of developing polymer formulations that degrade gradually over weeks or months.

9.6 Conclusion

Injectable hydrogels have demonstrated significant potential for various biomedical applications; however, certain challenges continue to hinder their widespread use. The mechanical properties and gelation rates of these hydrogels can be significantly influenced by the complex internal environment of the body. Therefore, it is essential to recognize that injectable hydrogels may exhibit markedly different behaviors in in vivo and in vitro settings. The high solid content of physically crosslinked injectable hydrogels can increase production costs and potentially lead to discomfort in patients. Consequently, it is crucial to develop polymer solutions with low solid content, low viscosity prior to injection, and rapid gelation post-injection. Maintaining a very low viscosity can prevent needle clogging, while rapid gelation can minimize the risk of the polymer solution migrating to unintended sites. As the release of bioactive agents from these systems is largely dependent on the degradation rate of the hydrogels, efforts must be directed toward the modulation of these degradation rates. Additionally, there is a need for a variety of stimuli that can trigger rapid gelation of polymer solutions. Another significant consideration is the storage stability of polymer solutions, ensuring that the structural integrity is maintained during storage. Finally, the incorporation of additional functional groups into injectable hydrogels could expand their range of applications. Research in this direction is likely to yield innovative and effective injectable hydrogels that will hold considerable importance in health applications.

References

1. Shaygani, H., Mofrad, Y.M., Rezaei Demneh, S.M., Hafezi, S., Almasi-Jaf, A., Shamloo, A.: Cartilage and bone injectable hydrogels: a review of injectability methods and treatment strategies for repair in tissue engineering. Int. J. Biol. Macromol. **282**, 136689 (2024)
2. Tan, Z., Tian, L., Luo, Y., Ai, K., Zhang, X., Yuan, H., Zhou, J., Ye, G., Yang, S., Zhong, M., Li, G., Wang, Y.: Preventing postsurgical colorectal cancer relapse: a hemostatic hydrogel loaded with METTL3 inhibitor for CAR-NK cell therapy. Bioact. Mater. **44**, 236–255 (2025)
3. Warwar, C., Martin, P., Vasilyev, G., Vilensky, R., Fadul, R., Redenski, I., Samer, S., Zussman, E.: Injectable hydrogels based on inter-polyelectrolyte interactions between hyaluronic acid, gelatin, and cationic cellulose nanocrystals. Biomacromolecules. **23**(8), 3222–3234 (2022)
4. Hao, H., Wu, S., Lin, J., Zheng, Z., Zhou, Y., Zhang, Y., et al.: Immunization against Zika by entrapping live virus in a subcutaneous self-adjuvanting hydrogel. Nat. Biomed. Eng. **7**, 928–942 (2023)
5. Lin, Y.J., Chien, B.Y.C., Lee, Y.H.: Injectable and thermoresponsive hybrid hydrogel with antibacterial, anti-inflammatory, oxygen transport, and enhanced cell growth activities for improved diabetic wound healing. Eur. Polym. J. **175**, 111364 (2022)
6. Rossi, A., Bassi, G., Cunha, C., Baldisserri, C., Ravaglia, N., Gardini, D., et al.: Magnetically induced anisotropic structure in an injectable hydrogel for skeletal muscle regeneration. J. Colloid Interface Sci. **678**(Part C), 334–345 (2025)
7. Navaei, A., Truong, D., Heffernan, J., Cutts, J., Brafman, D., Sirianni, R.W., Vernon, B., Nikkhah, M.: PNIPAAm-based biohybrid injectable hydrogel for cardiac tissue engineering. Acta Biomater. **32**, 10–23 (2016)
8. Awwad, S., Abubakre, A., Angkawinitwong, U., Khaw, P.T., Brocchini, S.: In situ antibody-loaded hydrogel for intravitreal delivery. Eur. J. Pharm. Sci. **137**, 104993 (2019)
9. Fathi, M., Alami-Milani, M., Geranmayeh, M.H., Barar, J., Erfan-Niya, H., Omidi, Y.: Dual thermo-and pH-sensitive injectable hydrogels of chitosan/(poly (N-isopropylacrylamide-co-itaconic acid)) for doxorubicin delivery in breast cancer. Int. J. Biol. Macromol. **128**, 957–964 (2019)
10. Wu, Y., Wang, Y., Li, W., Li, D., Song, P., Kang, Y., et al.: Construction of piezoelectric, conductive and injectable hydrogels to promote wound healing through electrical stimulation. Acta Biomater. **191**, 205–215 (2025)
11. Li, R., Cai, Z., Li, Z., Zhang, Q., Zhang, S., Deng, L., Li, L., Zhou, C.: Synthesis of in-situ formable hydrogels with collagen and hyaluronan through facile Michael addition. Mater. Sci. Eng. C. **77**, 1035–1043 (2017)
12. Qu, J., Zhao, X., Liang, Y., Zhang, T., Ma, P.X., Guo, B.: Antibacterial adhesive injectable hydrogels with rapid self-healing, extensibility and compressibility as wound dressing for joints skin wound healing. Biomaterials. **183**, 185–199 (2018)
13. Arora, A., Mahajan, A., Katti, D.S.: TGF-β1 presenting enzymatically cross-linked injectable hydrogels for improved chondrogenesis. Colloids Surf. B: Biointerfaces. **159**, 838–848 (2017)
14. Nguyen, M.K., Lee, D.S.: Injectable biodegradable hydrogels. Macromol. Biosci. **10**(6), 563–579 (2010)
15. Hoque, J., Bhattacharjee, B., Prakash, R.G., Paramanandham, K., Haldar, J.: Dual function injectable hydrogel for controlled release of antibiotic and local antibacterial therapy. Biomacromolecules. **19**(2), 267–278 (2018)
16. Liang, Y., Zhao, X., Ma, P.X., Guo, B., Du, Y., Han, X.: pH-responsive injectable hydrogels with mucosal adhesiveness based on chitosan-grafted-dihydrocaffeic acid and oxidized pullulan for localized drug delivery. J. Colloid Interface Sci. **536**, 224–234 (2019)
17. Wang, Y., Zhao, J., Wang, X., Zhang, R., Liang, F.: Mussel-mimetic chitosan based injectable hydrogel with fast-crosslinking and water-resistance as tissue adhesive. Int. J. Adhes. **124**, 103382 (2023)

18. Bercea, M., Constantin, M., Plugariu, I.A., Daraba, M.O., Ichim, D.L.: Thermosensitive gels of pullulan and poloxamer 407 as potential injectable biomaterials. J. Mol. Liq. **362**, 119717 (2022)
19. Zhang, Y., Xu, Y., Wei, C., Zhang, Y., Yang, L., Song, Z., Lang, M.: Diselenide-containing poly (ε-caprolactone)-based thermo-responsive hydrogels with oxidation and reduction-triggered degradation. Mater. Today Chem. **4**, 172–179 (2017)
20. Sun, G., Feng, C., Jiang, C., Zhang, T., Bao, Z., Zuo, Y., Kong, M., Cheng, X., Liu, Y., Chen, X.: Thermo-responsive hydroxybutyl chitosan hydrogel as artery intervention embolic agent for hemorrhage control. Int. J. Biol. Macromol. **105**, 566–574 (2017)
21. Singha, I., Basu, A.: Chitosan based injectable hydrogels for smart drug delivery applications. Sens. Int. **3**, 100168 (2022)
22. Wu, J., Wei, W., Wang, L.Y., Su, Z.G., Ma, G.H.: A thermosensitive hydrogel based on quaternized chitosan and poly (ethylene glycol) for nasal drug delivery system. Biomaterials. **28**(13), 2220–2232 (2007)
23. Berger, J., Reist, M., Mayer, J.M., Felt, O., Peppas, N.A., Gurny, R.: Structure and interactions in covalently and ionically crosslinked chitosan hydrogels for biomedical applications. Eur. J. Pharm. Biopharm. **57**(1), 19–34 (2004)
24. Liu, Z., Yao, P.: Injectable thermo-responsive hydrogel composed of xanthan gum and methylcellulose double networks with shear-thinning property. Carbohydr. Polym. **132**, 490–498 (2015)
25. Zhang, Y., Gao, C., Li, X., Xu, C., Zhang, Y., Sun, Z., Gao, J.: Thermosensitive methyl cellulose-based injectable hydrogels for post-operation anti-adhesion. Carbohydr. Polym. **101**, 171–178 (2014)
26. Bonnet, M., Trimaille, T., Brezun, J.M., Feron, F., Gigmes, D., Marqueste, T., Decherchi, P.: Motor and sensitive recovery after injection of a physically cross-linked PNIPAAm-g-PEG hydrogel in rat hemisectioned spinal cord. Mater. Sci. Eng. C. **107**, 110354 (2020)
27. Tan, H., Ramirez, C.M., Miljkovic, N., Li, H., Rubin, J.P., Marra, K.G.: Thermosensitive injectable hyaluronic acid hydrogel for adipose tissue engineering. Biomaterials. **30**(36), 6844–6853 (2009)
28. Jiang, Z., You, Y., Deng, X., Hao, J.: Injectable hydrogels of poly (ε-caprolactone-co-glycolide)–poly (ethylene glycol)–poly (ε-caprolactone-co-glycolide) triblock copolymer aqueous solutions. Polymer. **48**(16), 4786–4792 (2007)
29. Hu, Q., Xie, N., Liao, K., Huang, J., Yang, Q., Zhou, Y., Deng, K.: An injectable thermosensitive Pluronic F127/hyaluronic acid hydrogel loaded with human umbilical cord mesenchymal stem cells and asiaticoside microspheres for uterine scar repair. Int. J. Biol. Macromol. **219**, 96–108 (2022)
30. Mathew, A.P., Uthaman, S., Cho, K.H., Cho, C.S., Park, I.K.: Injectable hydrogels for delivering biotherapeutic molecules. Int. J. Biol. Macromol. **110**, 17–29 (2018)
31. Chen, N., Wang, H., Ling, C., Vermerris, W., Wang, B., Tong, Z.: Cellulose-based injectable hydrogel composite for pH-responsive and controllable drug delivery. Carbohydr. Polym. **225**, 115207 (2019)
32. Li, Z., Fan, Z., Xu, Y., Lo, W., Wang, X., Niu, H., Khan, M., Guan, J.: pH-sensitive and thermosensitive hydrogels as stem-cell carriers for cardiac therapy. ACS Appl. Mater. Interfaces. **8**(17), 10752–10760 (2016)
33. Zhang, M., Chen, G., Lei, M., Lei, J., Li, D., Zheng, H.: A pH-sensitive oxidized-dextran based double drug-loaded hydrogel with high antibacterial properties. Int. J. Biol. Macromol. **182**, 385–393 (2021)
34. Tan, H., Marra, K.G.: Injectable, biodegradable hydrogels for tissue engineering applications. Materials. **3**(3), 1746–1767 (2010)
35. Lu, W., Zeng, M., Liu, W., Ma, T., Fan, X., Li, H., Hu, Y., Xie, J.: Human urine-derived stem cell exosomes delivered via injectable GelMA templated hydrogel accelerate bone regeneration. Mater. Today Bio. **19**, 100569 (2023)

36. Yu, F., Zheng, H., Li, X., Mohamed, E.N., Hany, E.H., Morsi, Y., Li, J., Wu, J., Mo, X.: A photocrosslinking antibacterial decellularized matrix hydrogel with nanofiber for cutaneous wound healing. Colloids Surf. B. Biointerfaces. **217**, 112691 (2022)
37. Jiang, Y., Chen, J., Deng, C., Suuronen, E.J., Zhong, Z.: Click hydrogels, microgels and nanogels: emerging platforms for drug delivery and tissue engineering. Biomaterials. **35**(18), 4969–4985 (2014)
38. Ossipov, D.A., Hilborn, J.: Poly (vinyl alcohol)-based hydrogels formed by "click chemistry". Macromolecules. **39**(5), 1709–1718 (2006)
39. Bai, X., Lü, S., Cao, Z., Ni, B., Wang, X., Ning, P., Ma, D., Wei, H., Liu, M.: Dual crosslinked chondroitin sulfate injectable hydrogel formed via continuous Diels-Alder (DA) click chemistry for bone repair. Carbohydr. Polym. **166**, 123–130 (2017)
40. Park, H., Kim, S.W., Lee, J.W., Lee, K.Y.: Injectable hydrogels prepared from partially oxidized hyaluronate and glycol chitosan for chondrocyte encapsulation. Carbohydr. Polym. **157**, 1281–1287 (2017)
41. Fiorica, C., Palumbo, F.S., Pitarresi, G., Puleio, R., Condorelli, L., Collura, G., Giammona, G.: A hyaluronic acid/cyclodextrin based injectable hydrogel for local doxorubicin delivery to solid tumors. Int. J. Pharm. **589**, 119879 (2020)
42. Pupkaite, J., Rosenquist, J., Hilborn, J., Samanta, A.: Injectable shape-holding collagen hydrogel for cell encapsulation and delivery cross-linked using thiol-michael addition click reaction. Biomacromolecules. **20**(9), 3475–3484 (2019)
43. Zhang, Y., Chen, H., Zhang, T., Zan, Y., Ni, T., Cao, Y., Wang, J., Liu, M., Pei, R.: Injectable hydrogels from enzyme-catalyzed crosslinking as BMSCs-laden scaffold for bone repair and regeneration. Mater. Sci. Eng. C. **96**, 841–849 (2019)
44. Bi, B., Liu, H., Kang, W., Zhuo, R., Jiang, X.: An injectable enzymatically crosslinked tyramine-modified carboxymethyl chitin hydrogel for biomedical applications. Colloids Surf. B. Biointerfaces. **175**, 614–624 (2019)
45. Asai, D., Xu, D., Liu, W., Quiroz, F.G., Callahan, D.J., Zalutsky, M.R., Craig, S.L., Chilkoti, A.: Protein polymer hydrogels by in situ, rapid and reversible self-gelation. Biomaterials. **33**(21), 5451–5458 (2012)
46. Jin, H., Zhao, G., Hu, J., Ren, Q., Yang, K., Wan, C., Chen, J., Zou, Z.: Melittin-containing hybrid peptide hydrogels for enhanced photothermal therapy of glioblastoma. ACS Appl. Mater. Interfaces. **9**(31), 25755–25766 (2017)
47. Liu, Y., Fan, D.: Novel hyaluronic acid-tyrosine/collagen-based injectable hydrogels as soft filler for tissue engineering. Int. J. Biol. Macromol. **141**, 700–712 (2019)
48. Gilarska, A., Lewandowska-Łańcucka, J., Guzdek-Zając, K., Karewicz, A., Horak, W., Lach, R., Wójcik, K., Nowakowska, M.: Bioactive yet antimicrobial structurally stable collagen/chitosan/lysine functionalized hyaluronic acid–based injectable hydrogels for potential bone tissue engineering applications. Int. J. Biol. Macromol. **155**, 938–950 (2020)
49. Zheng, B.D., Ye, J., Yang, Y.C., Huang, Y.Y., Xiao, M.T.: Self-healing polysaccharide-based injectable hydrogels with antibacterial activity for wound healing. Carbohydr. Polym. **275**, 118770 (2022)

Chapter 10
Conductive Hydrogels

Dilara Göksu Tamay

Abstract Biomaterials research has been progressing toward the development of "smart" products with advanced features to better meet the current needs of the health industry. Biocompatible and biodegradable conductive hydrogels with good electrical and optical properties, which can provide electrical, electromechanical, and electrochemical control to cells, tissues, and biomaterial function, have started to gain importance in recent years in tissue engineering, controlled drug delivery, and biosensor applications. Considering the electrical currents that provide communication between our cells, introducing conductivity to hydrogels that can mimic the mechanical properties and microstructure of the extracellular matrix is especially important for applications involving neural and cardiovascular tissue engineering, where conductivity directly affects the natural physiology and function of the tissues.

Keywords Poly(pyrrole) · Poly(3,4-ethylenedioxythiophene) · Poly(aniline) · Poly(phenylene vinylene) · Carbon nanotubes · Graphene · Fullerene · Gold nanoparticles · Silver nanoparticles · Platinum nanoparticles

D. G. Tamay (✉)
BIOMATEN Center of Excellence in Biomaterials and Tissue Engineering, Middle East Technical University, Ankara, Türkiye

Department of Biotechnology, Graduate School of Natural and Applied Sciences, Middle East Technical University, Ankara, Türkiye
e-mail: dtamay@metu.edu.tr

N. Hasirci, V. Hasirci (eds.), *Hydrogels and Bioinks in Tissue Engineering*, Biological and Medical Physics, Biomedical Engineering,
https://doi.org/10.1007/978-3-031-90966-5_10

10.1 Importance of Conductive Hydrogels

Conductive hydrogels (CHs) have great potential in the medical field as they amalgamate the advantageous properties of polymers (e.g., easy processability, biocompatibility, and biodegradability) with electrical conductivity. Communication between healthy and damaged cells and tissues can be provided using tissue scaffolds and implants made of such hydrogels, while creating an environment that supports electrophysiological functions such as heartbeat, transepithelial potential generation, neurotransmission, axon elongation, and contraction/relaxation in electroactive tissues like nerve, skin, and muscle. Conductive hydrogels can also help improve cellular behaviors such as adhesion, proliferation, and differentiation in stem cells of various origins (e.g., adipose-derived, neural, mesenchymal, etc.), and in neurons, cardiomyocytes, and myoblasts. Hydrogels developed for biomedical applications often utilize conductive polymers (e.g., poly(pyrrole), PPy; poly(3,4-ethylenedioxythiophene), PEDOT; poly(aniline), PANI) or blends of these polymers with proteins or polysaccharides (e.g., gelatin, gelatin methacrylate, collagen, chitosan, alginate, etc.). In hydrogels prepared with conductive polymers, electrical conductivity is based on charge transfer between the polymer chains forming the hydrogel network or charge transfer along a single polymer chain. In order for polymers to gain a high degree of electrical conductivity, "doping" is usually performed. Dopants added to the polymer chain structure can be ions (e.g., Cl−, SO_3^-, PO_4^-) or polymers that contribute to the oxidation reaction. In addition to conductive polymers, carbon-based nanomaterials and their derivatives (e.g., graphene; graphene oxide, GO; carbon nanotubes, CNTs) and metal nanoparticles (e.g., gold, silver, platinum or metal oxides), which are naturally conductive, can be entrapped in the hydrogel network to impart conductivity to the material. Since hydrogels exhibit swelling (water retention) behavior in ionic aqueous solutions, ions diffusing into the cross-linked hydrogel structure also contribute to the conductivity of hydrogels when soaked in such solutions. In addition, it is possible to form CHs based on ion mobility using polyelectrolytes of natural or synthetic origin that contain high amounts of ionic groups in their chain structures, or by using special salts called ionic liquids.

10.2 Imparting Conductivity to Hydrogels

Studies in the literature on conductive hydrogels suggest four different approaches to impart conductivity to the hydrogel structure: (1) in situ polymerization, (2) post-polymerization processing, (3) addition of conductive materials, and (4) ionic conductivity (Table 10.1).

Table 10.1 Comparison of methods used to impart conductivity to hydrogels

Method	Advantages	Disadvantages
In situ polymerization	– One-step production – Improved mechanical properties – Strong and stable electrical conductivity	– Use of monomers, oxidizers and solvents that may cause toxicity – Design and optimization of chemical synthesis – Low reproducibility
Post-polymerization processing	– Use of ready-made hydrogels – Material diversity – Ability to create conductive coating	– Use of monomers, oxidizers, and solvents that may cause toxicity – Non-homogeneous, low-stability electrical conductivity – Additional polymerization/processing step – Failure to maintain coating integrity in conductive coatings
Addition of conductive materials	– Easy, one-step production – Material diversity – Conductivity level controllable by changing the amount of additive – No use of toxic monomers and oxidizers – High reproducibility	– Inhomogeneous electrical conductivity – Decrease in gelation capacity – Use of micro-nano particles that may cause toxicity
Ionic conductivity	– Easy, one-step production – Use of ready-made hydrogels – No use of toxic monomers and oxidizers – Ability to use isotonic ionic solutions – Improved gelation, elasticity, and mechanical strength – Ability to provide stimuli-responsive "smart" system feature with ion addition	– Conductivity level and stability depending on diffusion limits – Low reproducibility

10.2.1 In Situ Polymerization

In *in situ* polymerization, a conductive polymer (CP) phase is constructed inside the hydrogel network of another polymer. To achieve this biphasic hydrogel, the monomers of the conductive polymer and an oxidizing molecule (e.g., sodium persulfate, SPS; ammonium persulfate, APS; iron (III) chloride, $FeCl_3$, etc.) are put into a one-step reaction within a solution of the nonconductive hydrogel polymer and its cross-linker (Fig. 10.1a). This way, an interpenetrating network (IPN) of conductive and nonconductive polymer chains is formed in tandem. Monomers of the CPs are usually grafted to polymers of natural origin (e.g., gelatin, chitosan, collagen, hyaluronic acid, etc.) via *in situ* polymerization and used in biomedical applications [1, 2].

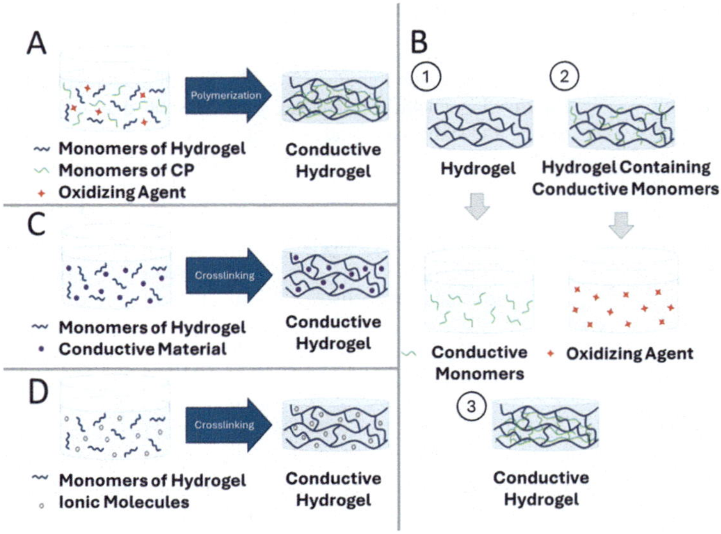

Fig. 10.1 Schematic representation of methods used to impart conductivity to hydrogels. (**a**) *In situ* polymerization. (**b**) Post-polymerization processing. (**c**) Adding conductive material. (**d**) Ionic conductivity

A disadvantage of this method is the aggregation of CPs in the hydrogel network—as most CPs are water insoluble—resulting in an inhomogeneous structure. To enhance the homogeneity and stability of these hydrogels, chemicals to form non-covalent weak bonds in the hydrogel network can be added to the reaction solution, or the CP can be chemically modified to increase its solubility in water [3]. Moreover, the nonconductive polymer chains forming the hydrogel network may adversely affect the inter-chain charge transfer within the CP phase. Therefore, hydrogels produced by *in situ* polymerization may have limited conductive capacity compared to hydrogels produced using only conductive polymers.

It is extremely important to remove unreacted monomers and oxidizing molecules from the hydrogel structures formed through *in situ* polymerization. Leaching of these molecules from the porous structures of hydrogels, which expand as they swell by retaining water, can decrease the overall conductivity or induce toxicity in the cells and tissues they come into contact with. Therefore, hydrogels formed via *in situ* polymerization are subjected to intensive washing steps post-production.

10.2.2 Post-Polymerization Processing

In the post-polymerization processing method, pre-cross-linked hydrogels are successively placed in the conductive monomer and oxidizer solutions, forming a CP phase within the hydrogel network structure (Fig. 10.1b) [4, 5]. This method is also used to create a conductive coating on the surfaces of hydrogels and other materials.

Post-polymerization processing is advantageous in that it allows the use of many different CPs and oxidizing molecules [6]. However, as it is a diffusion-based method, the diffusion of monomers or oxidizing molecules into the water-retaining hydrogel network may be restricted, and the formation of a homogeneous conductive phase may be negatively affected. Therefore, it is imperative to optimize the production parameters according to the diffusion limits [5]. As with *in situ* polymerization, CHs formed by this method should be subjected to extensive washing to remove unreacted toxic monomers and oxidizing molecules.

10.2.3 Addition of Conductive Materials

Hydrogels, consisting of polymer chains linked to each other by covalent bonds, allow small-sized molecules to be trapped in their structure [7]. Addition of micro/nano-sized conductive materials to a hydrogel solution will therefore result in their entrapment within the chain network after cross-linking, forming a kind of conductive particle network and rendering the whole structure conductive (Fig. 10.1c).

Nanoparticles with a high surface area-to-volume ratio (e.g., carbon nanotubes, CNTs) are often used to create an efficient conductive particle network. Metal nanoparticles, graphene, or short-chain conductive polymers are also conductive additives used in this method. The electrical conductivity of hydrogels formed with such naturally conductive materials is higher compared to hydrogels formed by other methods. Moreover, the extent of conductivity of the hydrogel can be tuned by varying the amount of conductive material added. However, achieving a stable electrical conductivity is dependent on the homogeneous distribution of the conductive material in the hydrogel network, and also on the biodegradability of the carrier hydrogel.

Since monomers or oxidizing solutions that can cause toxicity in cells and tissues are not used in this method, there is no need for extensive washing after production. The resulting hydrogel can be used directly in *in vitro* studies [8, 9]. However, before clinical applications, it must be demonstrated that micro/nano-sized conductive materials that may be released from the hydrogel—through diffusion or time-dependent degradation of the hydrogel—do not cause cellular or systemic toxicity [10].

10.2.4 Ionic Conductivity

Hydrogels consist of cross-linked polymer chains and can swell and retain 90% or more of their weight in water in aqueous solutions. Hydrogels are usually produced using ionic solutions such as phosphate-buffered saline (PBS) for biomedical applications. This way, the ions dispersed in the hydrogel network via diffusion during water retention give the hydrogels electrical conductivity (Fig. 10.1d). Considering *in vivo* conditions (pH 7.4), physiological fluids rich in ions (e.g., Ca^{2+}, Cl^-, Na^+, PO^{4-}, etc.) also increase the electrical conductivity of hydrogels that are in contact with them via diffusion [11]. In addition, CHs based on ionic strength and mobility can be formed by adding organic salt solutions with low melting temperatures known as "ionic liquids" or biocompatible polyelectrolyte materials to the hydrogel structure.

CHs based on ionic density do not require any additional conductive phase such as PPy, PANI, CNT, or metal nanoparticles. Therefore, non-cytotoxic, biocompatible, high optical transmittance hydrogels can be obtained by this method [12]. At the same time, ionic molecules in the structure may improve hydrogel properties such as gelation rate, elasticity, and mechanical strength or the physical or chemical structure of the hydrogel can become stimuli-responsive [13–15]. However, the maintenance of a stable electrical conductivity depends on the diffusion limits and the ion retention capacity of these hydrogels [16].

10.3 Materials Used to Impart Conductivity to Hydrogels

The materials used to impart electrical conductivity to conductive hydrogels can be listed in four different categories: (1) conductive polymers, (2) carbon-based nanomaterials, (3) metal nanoparticles, and (4) ionic materials.

10.3.1 Conductive Polymers

Conductive polymers (CPs) are materials that combine the useful properties of metals and polymers, such as electrical conductivity, easy synthesis and processing, and good optical properties [17]. CPs can be produced in many different forms (e.g., foam, electrospun mat, fiber, hydrogel, etc.), and their physical and chemical properties can be altered to suit a specific application. Pioneering studies on conductive polymers began when the electrical conductivity of poly(acetylene), a semiconductive polymer, increased ten-million-fold after being oxidized in iodine vapor [18]. This process, which is necessary for polymers to gain strong electrical conductivity, was named "doping." During doping, oxidizing agents called "dopants" remove or add electrons from/to the polymer chains.

Oxidation of poly(acetylene) in iodine vapor removes an electron from a carbon atom in the polymer, forming a radical cation (Fig. 10.2). This radical cation (i.e., polaron) can move and carry an electrical charge along the poly(acetylene) chain. A second polaron can be formed if another electron is removed from the already oxidized regions of the polymer, and a bipolaron can be formed if the unpaired electron is removed from the first polaron.

As poly(acetylene) is an unstable polymer in air and its synthesis is difficult, the search for alternative conductive polymers had begun, resulting in the emergence of polyheterocyclic polymers, a CP family with strong and stable electrical conductivity. Poly(pyrrole) (PPy), poly(aniline) (PANI), and poly(phenylenevinylene) (PPV) are well-known examples of the members of the polyheterocyclic polymer family (Fig. 10.3) [17]. Many different conductive polymer systems have been developed over the years after their discovery (Table 10.2).

CPs such as PPy, PEDOT, PANI, and PPV are conjugated polymers that have single and double bonds alternating in their chains [17]. Aromatic structures in their backbone render them capable of inter-chain π-electron transfer. Thus, they possess conductivity equal to that of semiconductors (10^{-3}–10^{5} S/cm). CHs made of conductive polymers are more elastic than composite CHs produced using carbon or metal nanoparticles [19]. Considering their advantageous electrical, chemical, and mechanical properties, CP hydrogels hold significant potential in the tissue engineering field (Table 10.3).

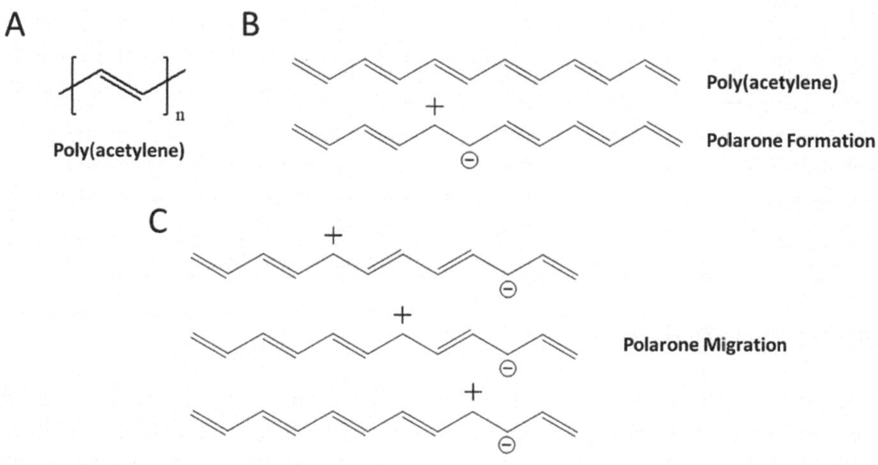

Fig. 10.2 Poly(acetylene) doping. (**a**) Chemical structure of poly(acetylene). (**b**) Polaron formation during the doping process. (**c**) Migration of polaron along the polymer chain

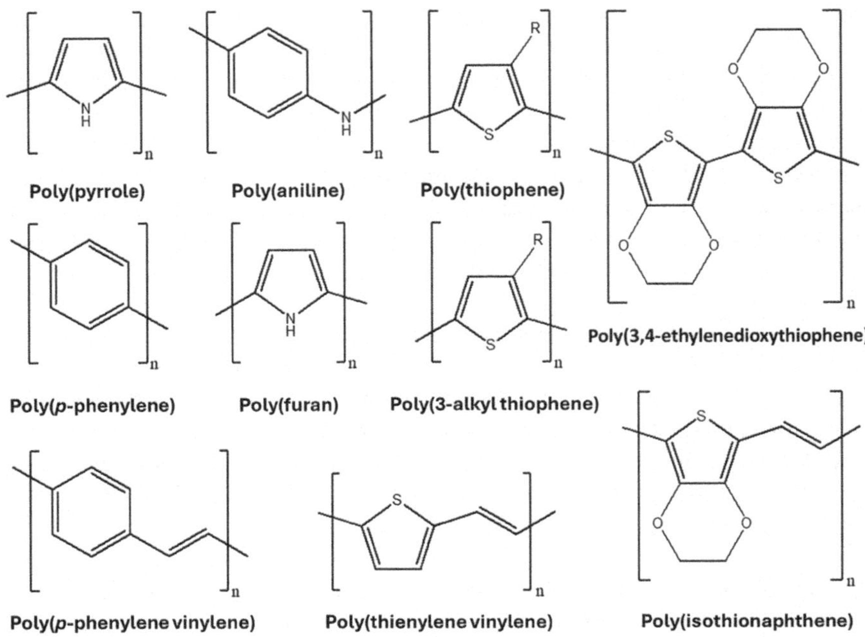

Fig. 10.3 Chemical structures of some conductive polymers of the polyheterocyclic polymer family

Table 10.2 Conductive polymers and their abbreviations

Poly(acetylene)	PAc	Poly(3,4-ethylenedioxythiophene)	PEDOT
Poly(pyrrole)	PPy	Poly(thiophene vinylene)	PTh-V
Poly(aniline)	PANI	Poly(2,5-thienylene vinylene)	PTV
Poly(thiophene)	PTh	Poly(2,5-furylene vinylene)	PFV
Poly(*p*-phenylene)	PPP	Poly(3-alkyl thiophene)	PAT
Poly(azulene)	PAz	Poly(*p*-phenylene sulfide)	PPS
Poly(furan)	PFu	Poly(*p*-phenylene vinylene)	PPV
Poly(isoprene)	PIP	Poly(*p*-phenylene terephthalamide)	PPTA
Poly(butadiene)	PBD	Poly(α-naphthylamine)	PNA
Poly(isothianaphthene)	PITN	Poly(3-octylthiophene-3-methylthiophene)	POTMT

10.3.1.1 Poly(Pyrrole)

Poly(pyrrole) (PPy), the most extensively studied member of the polyheterocyclic polymer family (Fig. 10.3), is an important "smart" biomaterial due to its cytocompatibility, biocompatibility, electrical conductivity, and stability under *in vivo* conditions, as well as its stimuli-responsive properties [20–22]. PPy can be synthesized easily under normal conditions in high quantities and with the desired properties, using many different solvents including water [23]. It can be produced in various

Table 10.3 Properties of various conductive polymers

Conductive polymer	Conductivity(mS/cm)	Advantages	Disadvantages	Application areas
Poly(pyrrole) (PPy)	10^3–5×10^4	– Strong conductivity – High stability – Biocompatibility – High mechanical strength	– Fragile – Prone to irreversible oxidation – Insoluble in water	– Biosensors – Antioxidants – Drug delivery – Beural prostheses – Tissue engineering
Poly(aniline)(PANI)	10^2–10^8	– Strong conductivity – High stability – Solubility in water	– Lack of plasticity – Poor cell adhesion and growth – Low solubility in water	– Biosensors – Antioxidants – Drug release – Bioactuators – Food industry – Tissue engineering
Poly(thiophene)(PTh)	10^1–10^4	– Good optical properties – Biocompatibility – Acquires various functions according to reactions	– Poor conductivity – Low stability – Low solubility in water	– Biosensors – Food industry – Tissue engineering
Poly(3,4-ethylenedioxythiophene) (PEDOT)	3×10^5–5×10^5	– Strong conductivity – High stability – Biocompatibility – High mechanical strength – Water solubility (when doped with PSS)	– Low mechanical strength	– Antioxidants – Drug delivery – Neural prostheses
Poly(*p*-phenylene vinylene)(PPV)	1–1×10^5	– High stability – Good optical properties – Precursor molecules can be manipulated in aqueous solutions	– Insoluble in water – Doping is necessary to increase conductivity	– Biosensors – Light-emitting diodes (LEDs) – Photovoltaic devices

degrees of porosity with a large surface area and can be grafted with different bioactive molecules for use in biomedical applications [24, 25]. Due to its stimuli-responsiveness, dynamic control of its properties is possible when an electric potential is applied to the polymer [26]. In various *in vitro* studies with endothelial cells, neurons, dorsal root ganglion cells, and rat pheochromocytoma cell line PC12, PPy was shown to induce focal adhesion and proliferation. However, PPy is difficult to process post-synthesis due to its molecular structure, which renders it crystalline, rigid, brittle, non-degradable, non-thermoplastic, and insoluble in water [22, 27–29]. To overcome these disadvantages, research on additives to ameliorate the structure of PPy and its analogs is ongoing [30]. PPy is utilized in fuel cells, anti-wear systems, computer monitors, microsurgical instruments, biosensors, drug delivery systems, and tissue engineering products [26, 31–35].

10.3.1.2 Poly(3,4-Ethylenedioxythiophene)

Poly(3,4-ethylenedioxythiophene) (PEDOT) (Fig. 10.3), a poly(thiophene) (PTh) derivative, is formed by polymerization of the bicyclic monomer 3,4-ethylenedioxythiophene [33, 36]. PEDOT contains a dioxyalkene bridging group at the third and fourth positions of its heterocyclic ring, which improves the properties of the polymer by reducing its bandwidth and oxidation/reduction potential. This chemical structure also gives the polymer better electrical, chemical, thermal, and environmental stability, and stronger electrical conductivity compared to PPy [37]. PEDOT's biocompatibility is similar to the natural biological molecules melanin and poly(thiophene) derivatives, and its hydrophilic nature allows it to be mixed with different materials to produce synthetic aqueous polymer systems with desired properties [38–41]. PEDOT is currently used in biosensor and bioengineering applications such as neural electrodes, nerve grafts, and muscle patches [36, 42].

10.3.1.3 Poly(Aniline)

Poly(aniline) (PANI) (Fig. 10.3), or aniline black, is the most extensively studied conductive polymer after PPy [33, 43]. PANI exists in various forms according to its degree of oxidation: fully oxidized *pernigraniline*, semi-oxidized *emeraldine*, and fully reduced *leucoemeraldine*. Among these forms, emeraldine is the most stable and conductive. PANI is easy to synthesize at a low cost, has good environmental stability, and has the ability to switch between electrically conductive and resistive states [44–46]. However, the polymer's lack of flexibility and biodegradability, difficulties in its processing, and its potential to induce chronic inflammation when implanted limit its biological applications [47]. Despite these drawbacks, there are ongoing studies investigating the utilization of PANI in biosensors, controlled drug-delivery systems, and tissue engineering applications [44, 48].

10.3.1.4 Poly(*p*-Phenylene Vinylene)

Poly(phenylene vinylene) (PPV) (Fig. 10.3) is a conductive material that can be processed into a highly ordered crystalline film. Well-doped PPV can be synthesized with good electrical conductivity as well as nonlinear optical properties and electroluminescence. PPV is insoluble in water as a polymer, but its precursor molecules can be dissolved in water to react in an aqueous solution. Since the density of PPV-based hydrogels can be controlled by varying the strength of the electric field applied to them, they are often used in drug targeting and controlled drug-delivery systems [49].

10.3.2 Carbon-Based Nanomaterials

Carbon-based nanomaterials such as carbon nanotubes (CNTs), graphene, and fullerene have long been utilized in industrial and medical fields owing to their advantageous chemical, electrical, thermal, and mechanical properties [50]. These materials are considered specific conjugated condensed aromatic polymers with large, delocalized electron planes (Fig. 10.4). They exhibit strong electrical conductivity as a result of π-orbit clustering, providing easy π-electron transfer. However, due to their low water solubility, it is very difficult to ensure their homogeneous distribution in hydrophilic hydrogels [51]. Although chemical modifications can be performed to graphene and carbon nanotubes to increase their water solubility, disruption of the conjugated structure can cause a decrease in electrical conductivity. Another disadvantage of carbon-based conductive nanomaterials is that they do not have good biodegradability *in vivo*, which may cause systemic toxicity under physiological conditions if they leach from the hydrogel structure they are incorporated into. Despite this, carbon-based nanomaterials are frequently used to construct CHs for tissue engineering applications due to their strong electrical conductivity and stable structure [52] (Table 10.4).

10.3.2.1 Graphene

Graphene has a two-dimensional, planar structure of hexagonally arranged sp^2-hybridized carbon atoms (Fig. 10.4) and is produced by various methods such as chemical vapor deposition (CVD), liquid-phase exfoliation (LPE), and mechanical exfoliation. Graphene and its derivatives (e.g., graphene oxide, GO; reduced graphene oxide, rGO; graphene nanomats) are often involved in research and application areas such as biological fuel cells, electrochemical biosensors, and tissue engineering because of their strong thermal and electrical conductivity and superior mechanical and optical properties [53]. They have a larger surface area compared to carbon nanotubes and can be dispersed more homogeneously in solvents [54]. When used as additives in the production of conductive hydrogels, graphene

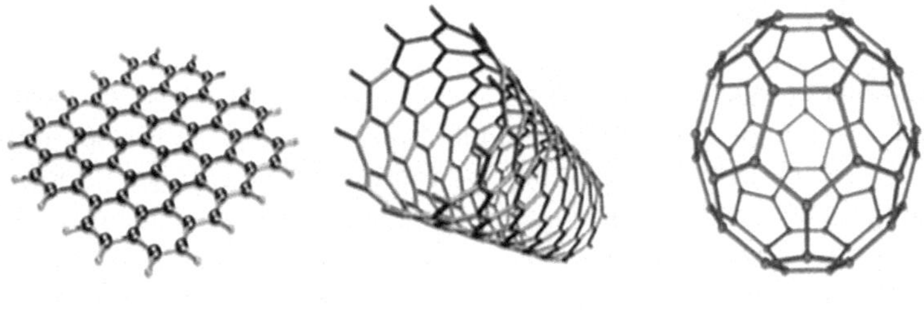

Fig. 10.4 Structures of carbon-based nanomaterials

Table 10.4 Properties of carbon-based nanomaterials

Carbon-based nanomaterial	Conductivity (mS/cm)	Advantages	Disadvantages
Graphene	10^8–10^9	– High mechanical strength – Strong electrical conductivity – Easy synthesis	– Oxidative stress effect – Toxic effect – High aggregation – Hydrophobicity
Carbon nanotube (CNT)	10^{-1}–10^{-5}	– Hydrophilicity – Controllable electrical/optical characteristics – Interactivity with various inorganic and organic materials	– Poor electrical conductivity – Sensitivity to moisture and oxidation – Low mechanical strength
Fullerene	10^7–10^8	– High mechanical strength – Strong electrical conductivity – Magnetic property	– Oxidative stress effect – Toxic effect – Hydrophobicity – Additional synthesis steps

molecules impart conductivity as well as increase elasticity and mechanical strength [55]. Although graphene and its derivatives provide various improvements in many biomedical systems, they should be subjected to extensive toxicological testing prior to *in vivo* application [56, 57].

10.3.2.2 Carbon Nanotubes

Carbon nanotubes (CNTs) can be considered as the cylindrical form of graphene sheets (Fig. 10.4). CNTs have unique electronic, magnetic, and chemical properties as a result of their strong interatomic bonding and high aspect ratio [58, 59]. According to their wall structure, CNTs are divided into two categories: single-walled carbon nanotubes (SWCNTs) and multi-walled carbon nanotubes (MWCNTs) [60]. Single-walled CNTs consist of a single graphene layer with a diameter of 1–2 nm, whereas multi-walled CNTs consist of multiple nesting graphene layers with diameters increasing from 3 nm to 30 nm [61]. Carbon nanotubes can be produced by chemical vapor deposition (CVD), arc discharge, and laser vaporization methods [62, 63]. They have high tensile strength, are ultra-light weight, and exhibit strong chemical and thermal stability, antibacterial and antifungal activity, and electrical conductivity. These superior chemical and physical properties of CNTs are exploited in the development of drug delivery systems, analytical and diagnostic biosensors, and tissue engineering products [64]. Their size and geometry, as well as their potential to induce oxidative stress, may cause toxicity in cells and tissues, therefore, CNTs must undergo extensive toxicological testing prior to *in vivo* applications [65–67].

10.3.2.3 Fullerene

Fullerenes are highly symmetrical spherical lattice nanomaterials consisting only of carbon atoms (Fig. 10.4). Since the first synthesis of the 60-carbon fullerene C_{60}, many different types of fullerenes have been obtained by various chemical modifications, such as C_{70} and water-soluble fullerenols. These materials can be produced via pressure, charge transfer, electron-beam irradiation, plasma polymerization, and photopolymerization using metal catalysts [68–70]. Fullerene molecules have superior structural, magnetic, superconductive, electrochemical, and photophysical properties, and are used in many fields of materials science and biotechnology [70, 71].

10.3.3 Metal Nanoparticles

Metal nanoparticles are nanomaterials whose optical properties and electrical conductivity depend on their type, geometry, and size [6]. Metal nanoparticles (e.g., Gold, Au; Silver, Ag; Platinum, Pt) and metal oxide nanoparticles (e.g., Iron Oxide, Fe_2O_3; Zinc Oxide, ZnO; Zirconium, ZrO_2) can impart strong electrical conductivity to hydrogels through thermodynamic diffusion and random collisions between nanoparticles, when entrapped in the chain network [72–75]. They can be produced in various other geometries such as nanorods, nanowires, and nanorosettes. These materials have a low percolation threshold as a result of their high aspect ratio; thus, they can provide high levels of electrical conductivity, even at low concentrations. Metal nanoparticles are advantageous materials for the production of conductive hydrogel-based tissue engineering implants due to their strong electrical conductivity, good biocompatibility, antibacterial activity, and chemical stability (Table 10.5).

Table 10.5 Properties of frequently studied metal nanoparticles

Metal nanoparticle	Size (nm)	Geometry	Advantages	Disadvantages	Application areas
Gold nanoparticles (AuNP)	1–60	Sphere Rod Polygonal Flower-like	– High stability – Low cytotoxic effect in the short term – Can be manufactured in high quantities	– Weak optical signal – Cytotoxic in the long term – High cost	– Immunomarking and biological imaging – Diagnostic and treatment kits – Catalytic agent – Cell therapy for cancer
Silver nanoparticles (AgNP)	4–120	Sphere Rod Wire Oval Polygonal	– Antimicrobial activity – Strong optical signal	– Cytotoxic effect – Low stability without surface modification – High cost	– Antimicrobial agent – Gas/vapor sensors – Water treatment – Cell therapy for cancer
Platinum nanoparticles (PtNP)	10–100	Sphere Cubic Flower-like	– High stability – Strong optical signal – Catalytic activity	– Cytotoxic effect – High cost	– Biosensor – Bone tissue strengthening – Cancerous cell screening
Iron oxide nanoparticles	4–45	Sphere Tube Cluster	– Low cytotoxicity – Super-paramagnetic property – Low cost	– Low mechanical strength – Low stability – Toxic solvent requirement	– Gas sensor – Magnetic resonance imaging (MRI)
Zinc oxide nanoparticles	20–600	Rod Wire Flower-like Mat	– Antimicrobial activity – Strong optical signal – Piezoelectric and pyroelectric properties – Broadband UV absorption – Low cost	– Cytotoxic effect – Low stability – Toxic solvent requirement	– Biosensors – Gas sensors – Photocatalytic agent – UV radiation absorber

10.3.3.1 Gold Nanoparticles

Gold nanoparticles (AuNPs) are nanomaterials that are highly stable under physiological conditions and can be easily modified with various organic and biological molecules [76]. Gold nanoparticles of various sizes (1–60 nm) and geometries can be produced by nanoscale lithography and chemical, electrochemical, photochemical, or thermal reduction methods. These nanomaterials have a biocompatible profile in short-term applications but may induce toxicity in cells and tissues in long-term applications [77]. AuNPs are currently used in bioimaging, photothermal therapy, antigen screening, radioactive immunolabeling, and various tissue engineering applications [72, 78–81].

10.3.3.2 Silver Nanoparticles

Silver nanoparticles (AgNPs) are frequently used in biomedical applications due to their unique optical, electronic, and antimicrobial properties. AgNPs can be produced by laser cutting, gamma irradiation, electron-beam irradiation, chemical reduction, photochemical methods, microwave processing, and biological synthesis [82]. The antimicrobial activity and cytotoxic effects of AgNPs can be controlled by changing their size and extent of aggregation during production [83–85]. Due to their good optical and photothermal properties, AgNPs are used in bioimaging and diagnostic applications. When AgNPs are used in CH production, they can transfer their advantageous antimicrobial, optical, and thermal properties to the hydrogel, in addition to their electrical conductivity. However, excessive antimicrobial activity or aggregation of AgNPs may cause adverse effects and toxicity *in vitro* and *in vivo*.

10.3.3.3 Platinum Nanoparticles

Platinum nanoparticles (PtNPs) are versatile metal nanomaterials that are commonly used as catalysts in industrial applications [86–88]. PtNPs ranging in size from 10 to 100 nm can be produced by chemical and physical synthesis methods such as electron beam evaporation, chemical vapor deposition, ion implantation, sol–gel process, and chemical reduction [86, 89]. Due to their catalytic activity, their ability to be produced in various geometries (e.g., nanospheres, nanotubes, nanofibers), and their good optical, thermal, electrical, and antibacterial properties, PtNPs are used as antibacterial coatings for surgical instruments, antioxidant agents in cosmetics, and contrast agents in medical imaging, as well as in electrochemical biosensor production, photothermal therapy, chemotherapy against cancer, and various other industrial and medical applications [86–92]. Despite their success in these fields, the application and research of PtNPs in the biomedical field are limited due to their cytotoxic effect and high production costs.

10.3.3.4 Metal Oxide Nanoparticles

Metal oxide nanoparticles are nanomaterials that can exist in different geometries and exhibit various levels of functional biocompatibility and catalytic activity [93]. Due to their high electron transfer kinetics and strong adsorption ability, these materials allow the immobilization of various biomolecules on their surfaces, improving their electron transfer and biosensing capacity. Iron oxide nanoparticles are used in magnetic resonance imaging (MRI) (as contrast enhancers), tissue regeneration, immunoassays, liquid purification systems, and cell sorting applications [94, 95]. Zinc oxide nanoparticles are used in biosensors, gas sensors, and as catalytic or UV absorbing agents due to their good optical, conductive, and piezoelectric properties, chemical sensing capabilities, and antimicrobial activity [96–98]. Titanium dioxide nanoparticles are commonly used in cosmetic products, paint, and food additives and have catalytic and antimicrobial properties [99, 100].

10.3.4 Ionic Materials

Apart from electron transfer within the hydrogel network structure, another factor that determines electrical conductivity is ionic strength and ion mobility. Ionic CHs are preferred in applications requiring high optical transmittance or flexibility due to their transparency and high elasticity. In addition, they can be used as electrolytes in batteries and accumulators (as bilayer structures with different levels of electron density) or exhibit high water retention or anti-freeze properties depending on their ionic composition [101]. Ionic CHs have better biocompatibility compared to other types of CHs since their production does not involve toxic organic solvents or a secondary conductive phase that can leach out of the hydrogel structure. Ionic conductive hydrogels are used in various applications such as biosensors, tissue engineering products, purification systems, surface coatings, wearable technologies, and energy storage systems. Biocompatible polyelectrolytes such as poly(acrylic acid) (PAA) and organic salts with melting temperatures below 100 °C called "ionic liquids" (IL) (e.g., 1-ethyl-3-methylimidazolium chloride) can be used to form ionic hydrogels [102–104] (Table 10.6).

Table 10.6 Properties of the materials used to impart ionic electrical conductivity to hydrogels

Ionic material	Advantages	Application areas
Polyelectrolytes	– Biocompatibility – Easy synthesis – Tunable properties – Stimuli-responsiveness – Solubility in water – Optical transmittance	– Controlled drug-delivery systems – Antithrombogenic agent – Energy storage – Tissue engineering products
Ionic liquids	– Biocompatibility – Strong and stable electrical conductivity – Reversibility – Stimuli-responsiveness – Good mechanical and thermal properties – Optical transmittance	– Catalytic agent – Energy storage – CO_2 capturing – Tissue engineering products

10.3.4.1 Polyelectrolytes

Polyelectrolytes are polymers that contain high concentrations of ionic groups in their backbone and thus show sensitivity to pH and electrical stimuli. They are classified into three groups: *polyanion*, *polycation*, and *amphophilic polyelectrolytes* [105, 106]. Polyelectrolytes used in CH production can be of natural (e.g., hyaluronic acid, chondroitin sulfate, calcium alginate) or synthetic (acrylate and methacrylate derivatives) origin [107]. Synthetic polyelectrolytes can be formed by introducing ionic groups into polymer chains by copolymerization or chemical modification [106]. The copolymerization method is preferred due to the diversity and tunability of the desired properties.

Anionic monomers (e.g., acrylic acid, methacrylic acid), cationic monomers (e.g., diallyl quaternary ammonium salts), and amphophilic monomers (e.g., vinyl-containing aminimides) are used in the synthesis of polyelectrolytes. The production of polyelectrolytes by chemical modification is realized through sulfonation, addition, substitution, hydrolysis, or oxidation reactions. Depending on their ionic strength and interactions between ions, polyelectrolytes can affect the viscosity and light scattering properties of the system they are in, and these effects can be controlled by factors such as polyelectrolyte concentration and pH. Since polyelectrolytes have better solubility in water than non-ionic polymers, they can be more easily incorporated into hydrogel systems. The use of polyelectrolytes in the biomedical field is usually based on the interaction of these molecules with surfaces or molecules that carry opposite charges to them. For example, cationic polyelectrolytes are used to prevent hemolysis of red blood cells by silica, which has a negative surface charge. Polyelectrolytes are also used in surface coatings, controlled drug-delivery systems, and as antithrombogenic agents. Due to their biocompatibility and ionic strength, polyelectrolytes have significant potential in the development of tissue-engineered CHs.

10.3.4.2 Ionic Liquids

Ionic liquids (ILs) are salt solutions with strong ionic conductivity and stability and melting temperatures below 100 °C [108, 109]. ILs are classified as sustainable (green) solutions because they are non-flammable, non-volatile, and recyclable [103, 110]. Since CHs produced with ILs exhibit good mechanical and thermal properties and optical transmittance in addition to their ionic strength, they are used in electrolyte-coated transistors, gas separator membranes, and energy storage [111, 112]. In addition, "smart" ionic hydrogels with stimuli-responsive properties are utilized in the development of chemical and biological sensors. ILs are divided into four groups according to their cationic components: alkylammonium-, dialkylimidazolium-, phosphonium-, and N-alkylpyridinium-based ILs [113, 114]. Ammonium-based ILs are used as electrolyte solutions in high-energy electrochemical devices due to their good cathodic protection properties, low melting temperatures, and high fluidity. Imidazolium-based ILs, the most studied and easily synthesized of all IL types, have high chemical stability and fluidity and are used as catalytic agents in organic reactions due to their chemical selectivity. Studies on the stability, reactivity, and ionic and catalytic activities of pyridinium-based ILs, a more innovative type compared to imidazolium-based ILs, are ongoing. Due to their high thermal stability, phosphonium-based ILs are used as catalytic agents in reactions at high temperatures and in carbon dioxide capturing.

10.4 Applications of Conductive Hydrogels in Tissue Engineering and Biomedical Fields

In addition to being physically, chemically, and mechanically similar to the natural tissue, conductive hydrogels may require additional properties specific to the tissue they will replace or work with such as antimicrobial activity or electrical conductivity. In this context, conductive hydrogels find use in neural, cardiovascular, skin, bone, and cartilage tissue engineering, and in the production of biosensors and wearable technologies (Table 10.7).

10.4.1 Nerve Tissue Engineering

The nervous system, which consists of two components, the central and peripheral nervous system, is responsible for carrying out complex vital functions such as signal transmission to organs and limbs to perform physiological functions and control of senses and motor skills. Injuries to the nervous system cause dysfunctions and permanent disabilities in organs and limbs, reducing the quality of life and creating a great economic and social burden on individuals. The extent of functional tissue regeneration and recovery is rather low in traditional treatment methods of nerve injuries; therefore, innovative tissue engineering products are being developed to address this problem. Conductive hydrogels have great potential in the regeneration of nerve tissue where electrical conductivity is of central importance (Table 10.7).

Guarino et al. combined the conductive polymer PANI with poly(ethylene glycol) diacrylate (PEGDA) hydrogel network structure through *in situ* polymerization method, and after sodium chloride was added to the structure and UV cross-linking was performed, a CH with interconnected pores ranging in size (136 to 158 μm) was obtained [115]. *In vitro* studies of the resultant CH with rat pheochromocytoma cell line PC12 and human mesenchymal stem cells (hMSCs) showed improvement in the biological functions of these cells. In another study conducted by Heo et al. using photopolymerization-based 3D printing techniques, poly(ethylene glycol)

Table 10.7 Electrically conductive hydrogel studies in various tissue engineering fields

Application area	Method of imparting conductivity	Hydrogel matrix	Conductive phase	Reference
Nerve tissue engineering	In situ polymerization	Poly(Ethylene Glycol) Diacrylate	Poly(Aniline) (PANI)	[115]
		Poly(Ethylene Glycol) Diacrylate	Poly(3,4-Dioxyethylenethiophene)/Poly(styrene sulfonate) (PEDOT/PSS)	[116]
	Post-polymerization processing	Alginate	Poly(Pyrrole) (PPy)	[5]
		Carboxymethyl chitosan	Poly(Pyrrole) (PPy)	[117]
	Addition of conductive materials	Type I collagen-based Matrigel™	Single-walled carbon nanotube (SWCNT)	[119]
		Poly(acrylamide)/Poly(ethylene glycol) (PA/PEG)	Multi-walled carbon nanotube (MWCNT)	[120]
		Chitosan	Graphene oxide	[170]
		Poly(acrylamide)/gelatin/sodium alginate(PA/gel/SA)	Graphene oxide	[122]
	Ionic conductivity	Gelatin methacrylate (GelMA)	Choline acrylate	[123]
	Hybrid approaches	Catechol-modified hyaluronic acid (HA-CA)	Poly(pyrrole) (PPy) and/or single-walled carbon nanotube (SWCNT) (in situ polymerization and addition of conductive material)	[118]
		Oligo poly(ethylene glycol) fumarate	Carbon nanotube and Graphen oxide (in situ polymerization and addition of conductive material)	[121]
Bone tissue engineering	In situ polymerization	Oxidized polysaccharides and gelatin	Poly(ε-Caprolactone)/poly(aniline) (PCL/PANI)	[149]
	Addition of conductive materials	Cellulose Esther	Graphene	[150]
		Gelatin methacrylate (GelMA)	Gold nanoparticles	[151]
		Silk fibroin/Nanohydroxyapatite (SF/nHAp)	Gold and silver nanoparticles (AuNP and AgNP)	[152]

(continued)

Table 10.7 (continued)

Application area	Method of imparting conductivity	Hydrogel matrix	Conductive phase	Reference
Cardiovascular tissue engineering	In situ polymerization	Chitosan (CS)	Poly(pyrrole) (PPy)	[135]
		Poly(lactic acid)/poly(ethylene)/poly(lactic acid) (PLA/PEG/PLA)	Carboxyl-functionalized Tetraaniline (CTA)	[136]
		Gelatin methacrylate/poly(ethylene glycol) Diacrylate (GelMA/PEGDA)	Poly(pyrrole) nanoparticles	[137]
		Collagen/silk fibroin/hyaluronic acid/poly(ethylene glycol)	Poly(aniline)/poly(Vinylpyrrolidone) (PANI/PVP)	[138]
		Dibenzaldehyde-functionalized poly(ethylene glycol) (PEG-DA) and chitosan (CS)	Aniline tetramer (AT)	[139]
	Addition of conductive materials	Gelatin methacrylate (GelMA)	Carbon nanotube (CNT)	[124]
		Gelatin methacrylate (GelMA)	Graphene oxide (GO)	[125]
		Gelatin methacrylate (GelMA)	Reduced graphene oxide (rGO)	[126]
		Chitosan/gelatin	Single-walled carbon nanotube (SWCNT)	[127]
		Type I collagen	Carbon nanotube (CNT)	[128]
		Collagen methacrylate (ColMA) and alginate	Carboxyl-modified carbon nanotube (CNT)	[131]
		Decellularized pericardial tissue	Carbodihydrazide-modified carbon nanotube (CDH-CNT)	[129]
		Decellularized pericardial tissue	Hydrazide-modified carbon nanotube (CNT)	[130]
		Poly(2-hydroxyethyl methacrylate) (pHEMA)	Rosette carbon nanotubes (RNTs) and carbon nanofibers (CNF)	[132]
		Thiol-2-hydroxyethyl methacrylate/2-hydroxyethyl methacrylate (thiol-HEMA/HEMA)	Silver ions	[133]
		Gelatin methacrylate (GelMA)	Gold Nanorods	[134]
		Chitosan (CS)	Gold nanoparticles (gold NPs)	[72]

Skin tissue engineering	In situ polymerization	Quaternarized chitosan (QCS) And benzaldehyde-modified poly(ethylene glycol)/poly(glycerol Sebacate) (PEGS)	Poly(aniline) (PANI)	[140]
		N-Carboxyethyl chitosan (CEC) and hyaluronic acid (HA)	Aniline tetramer (TA)	[1]
		Guar gum	Poly(3,4-Dioxyethylenethiophene)/poly(styrene sulfonate) (PEDOT/PSS)	[141]
	Post-polymerization processing	Poly(acrylamide)/chitosan (PA/CS)	Poly(pyrrole) (PPy)	[142]
		Poly(2-hydroxyethyl methacrylate) (pHEMA)	Poly(pyrrole) (PPy)	[143]
	Addition of conductive materials	Gelatin or trypsin containing poly(acrylamide)/Poli(ethylene glycol) (PA/PEG)	Graphene oxide (GO)	[144]
		Poly(vinyl alcohol)/sodium alginate (PVA/SA)	Reduced graphene oxide (rGO) and terbium ions (Tb^{3+})	[145]
		Dopamine grafted gelatin and chitosan (GT-DA/CS)	Poly(dopamine) coated carbon nanotube (CNT-PD)	[146]
		Poly(vinyl alcohol)/N-Carboxyethyl chitosan/agarose (PVA/CEC/AGA)	Silver nanowires (AgNT)	[147]
	Ionic conductivity	Poly(acrylamide) (PA)	Lithium chloride (LiCl)	[148]
Cartilage tissue engineering	In situ polymerization	Oxidized alginate-gelatin (ADA-GEL)	Poly(pyrrole)/poly(styrene sulfonate) (PPy/PSS)	[16]
	Addition of conductive materials	Poly(D,L-lactic acid)/poly(ethylene glycol) (PDLLA/PEG)	Graphene oxide (GO)	[161]
		Poly(vinyl alcohol) (PVA)	Sodium Phytate (PANa)	[162]
		Chitosan/ß-Glycerophosphate (CS/BGP)	Pyrrole oligomers (OPy)	[163]

(continued)

Table 10.7 (continued)

Application area	Method of imparting conductivity	Hydrogel matrix	Conductive phase	Reference
Biosensors and wearable technologies	In situ polymerization	Poly(vinyl alcohol) (PVA)	Poly(acrylamide) (PA)	[165]
	Post-polymerization processing	Poly(acrylamide)/(chitosan) (PA/CS)	Poly(pyrrole) (PPy)	[142]
	Addition of conductive materials	Cellulose	Poly(dopamine) reduced graphene oxide (PGO)	[166]
	Hybrid approaches	Poly(vinyl alcohol) (PVA)	Poly(aniline) (PANI) and poly(dopamine)-modified silver nanoparticles (PDA-AgNP) (in situ polymerization and addition of conductive material)	[167]
		Poly(vinyl alcohol) (PVA) and chitosan (CS)	Poly(pyrrole) (PPy) and zinc (Zn) (in situ polymerization and addition of conductive material)	[168]
		Chitosan (CS)	Silicotungstic acid (SiW) and poly(acrylamide) (PA) (ionic conductivity and in situ polymerization)	[169]

diacrylate (PEGDA) hydrogels containing poly(3,4-ethylenedioxythiophene)/poly(styrene sulfonate) (PEDOT/PSS) were produced and the electrical conductivity of the hydrogel increased with PEDOT/PSS addition [116]. These hydrogels were interacted with dorsal root ganglion cells encapsulated in GelMA hydrogels and a significant increase in the expression of neural differentiation markers BDNF, NT-3, and erbB2 was observed when the cells were exposed to electrical stimuli.

Using a different approach to CH production, Bu et al. designed a peripheral nerve guide by coating carboxymethyl chitosan hydrogels with conductive poly(pyrrole) particles [117]. During *in vitro* tests, PC12 cell line and bone marrow-derived mesenchymal stem cells showed a greater tendency to adhere and proliferate on PPy-containing hydrogels compared to those without PPy, while *in vivo* tests with adult Sprague Dawley (SD) rats demonstrated the biocompatibility of the material and its positive effect on nerve regeneration. In another study using post-polymerization processing method, Yang et al. immersed ionically cross-linked alginate hydrogels in a pyrrole monomer solution and chemically cross-linked the pyrrole monomers to form a conductive PPy phase in the hydrogel [5]. It was observed that expressions of neural cell markers such as Tuj1 and MAP2 increased in human mesenchymal stem cells cultured on these hydrogels.

In a study where conductive polymers and carbon-based conductive nanomaterials were combined, Shin et al. doped catechol-modified hyaluronic acid hydrogels with PPy and single-walled carbon nanotubes (CNTs) individually and in combination and observed positive effects of these CHs on the differentiation of human neural precursor cells and neural stem cells *in vitro* [118]. Combining single-walled CNT with type I collagen-based hydrogels, Koppes et al. obtained CHs where nanomaterial addition did not affect hydrogel microstructure and elasticity negatively [119]. In a similar study, Imaninezhad et al. added multi-walled CNTs to poly(acrylamide)/poly(ethylene glycol) (PA/PEG) hydrogels to produce CHs [120]. Using two different carbon-based materials, Liu et al. synthesized CHs with acrylate-functionalized graphene oxide and carbon nanotube-functionalized poly(ethylene glycol) [121].

In another study, Zhao et al. obtained composite conductive hydrogels via *in situ* free radical polymerization method by adding different ratios of gelatin (Gel) and sodium alginate (SA) to graphene oxide (GO) doped poly(acrylamide) (PA) hydrogels [122]. In these studies, an increase in orientation, axon extension, number of neural extensions, and neuron-related gene expression were observed for cells cultured on the CHs.

Neuman et al. achieved ionic conductivity by modifying gelatin methacrylate (GelMA) chains with choline acrylate (ChoA) and developed a photopolymerizable conductive hydrogel with good optical transmittance [123]. Dorsal root ganglion cells and Schwann cells encapsulated in the hydrogels were cultured for 5 and 7 days, respectively, and were able to maintain their viability. However, no positive effect on axon extension for dorsal root ganglions was observed.

10.4.2 Cardiovascular Tissue Engineering

Many cardiovascular diseases, such as myocardial infarction (i.e., heart attack), are caused by abnormal electrical activity in the cardiac (heart) muscle (myocardium) due to the necrosis of cardiac muscle cells. The regenerative capacity of myocardium is very limited compared to tissues such as skin and bone, and after injury, a fibrotic scar tissue forms in the area leading to permanent tissue loss. The development of cardiac muscle patches and tissue mimics to replace the lost tissue is, therefore, an important area of research. In the literature, there are studies where the growth and migration behavior of cardiomyocytes and related precursor cells are enhanced by electrophysiological stimulation. As such, investigating the use of conductive hydrogels in cardiovascular tissue applications to mimic the electroactive nature of the cardiovascular cell environment has become a popular research area (Table 10.7).

In their study to develop a cardiac patch, Shin et al. doped gelatin methacrylate (GelMA) hydrogels with CNTs and developed a hybrid scaffold that combines the high porosity, biocompatibility, and biodegradability of hydrogels with the mechanical strength and electrical conductivity of CNTs [124]. With the adhesion and differentiation of myocardial cells cultured on the hydrogel, the frequency of intercellular electrical communication and spontaneous synchronized contraction increased threefold, and it was determined that the excitation threshold of cells decreased by 85%. The hydrogels were also found to protect cells from the effects of cardiotoxic (doxorubicin) and cardioinhibitory (heptanol) molecules. In another study, the group investigated the mechanical properties and cell–material interaction of graphene oxide (GO)-doped GelMA hydrogels [125]. GelMA/GO hydrogel supported the distribution and orientation of fibroblasts entrapped within the three-dimensional microenvironment and enhanced cell viability and proliferative capacity. In another study where electrical conductivity was obtained by mixing reduced graphene oxide (rGO) nanoparticles, another form of graphene, with GelMA, a positive effect on cardiomyocyte cell viability, proliferation, and differentiation was reported, and it was shown that primary cardiomyocytes performed strong and spontaneous synchronized contraction when cultured on these gels [126]. Using carbon-based nanomaterials, Pok et al. added single-walled CNTs to chitosan/GelMA hydrogels [127]. The composite hydrogel bridged signal transduction between cardiomyocytes and improved synchronized contraction behavior. Yu et al. combined CNT with collagen and investigated the mechanical properties, electrical conductivity, and cardiomyocyte behavior on the hydrogel [128]. Culturing primary neonatal rat cardiomyocytes on hydrogels, the group reported an increase in cell population and rhythmic contraction, and the surface area where these contractions were seen was 3 times larger in composite hydrogels compared to unmodified hydrogels.

In a study using decellularized sheep pericardial extracellular matrix as an injectable hydrogel base material, Roshanbinfar et al. used carbodihydrazide (CDH)-modified multi-walled CNT as a conductive additive [129]. An increase in cell viability, Cx34-like cardiac marker gene expression, and synchronized rhythmic contraction was observed in HL-1 cardiomyocytes cultured on the material. In another study, the group examined the effect of this material on the differentiation of human induced pluripotent stem cells (hiPSC) and observed that differentiated cardiomyocytes were oriented, and their contractility was increased in the absence of arrhythmia [130].

Utilizing three-dimensional bioprinting methods, Izadifar et al. produced a tissue scaffold for a myocardial patch [131]. The electrical conductivity, mechanical strength, elasticity, and biological activity of collagen methacrylate (ColMA)/carboxyl modified carbon nanotube (CNT) and alginate/CNT hybrid hydrogel scaffolds were investigated, as well as their potential to support neovascularization. The proliferation, migration, and differentiation behavior of endothelial cells were greatly enhanced on CNT-doped hydrogels, indicating that these materials can be used in the production of pre-vascularized, electrically conductive, and high mechanical strength cardiac muscle patches that can be printed in complex three-dimensional structures. In a study using rosette nanotubes (RNT) and carbon nanofibers (CNF) in combination with poly(2-Hydroxyethyl methacrylate) (pHEMA) hydrogels, Meng et al. produced injectable biomimetic conductive hydrogels [132]. *In vitro* cell–material interaction tests using cardiomyocytes showed that the samples with the highest nanomaterial concentration were also the ones exhibiting the highest density of cultured cells.

In order to form IHs with silver nanoparticles (AgNP) and pHEMA, You et al. utilized thiol functionalization and obtained a biocompatible and conductive hydrogel [133]. The elastic modulus of the composite hydrogel under tensile load was found to be dependent on the thiol-HEMA concentration, and the resultant hydrogels were found to be 1 to 4 times softer compared to those obtained from PPy or PANI. In addition, the electrical conductivity of the AgNP hydrogels increased up to 30% compared to non-composite hydrogels and exceeded the conductivity of intra- and extracellular tissues under physiological conditions. In a study where injectable and thermoresponsive IHs were obtained using gold nanoparticles (AuNPs), Baei et al. preferred chitosan to form the main hydrogel matrix [72]. The group obtained conductive hydrogels with gelation rates and conductivities tunable through AuNP concentration, which supported the viability, metabolism, migration, and proliferation of mesenchymal stem cells *in vitro*. AuNP addition enhanced cardiomyogenic differentiation marker expression in these cells, as well as improving mechanical properties of the hydrogel and the resistance to degradation *in vitro*. In another study using gold particles in nanorod form, Navaei et al. used UV-crosslinkable GelMA hydrogels [134]. This composite hydrogel structure, which affects the F-actin fiber organization of cardiomyocytes, provided more cell adhesion to the material surface, promoted synchronized rhythmic contractions at the tissue-level, and also increased the expression of different cardiac markers by the cells depending on the concentration of gold nanoparticles in the hydrogel.

Using conductive polymers, Mihic et al. obtained semiconductive poly(pyrrole)/chitosan (PPy/CS) hydrogels via *in situ* polymerization method [135]. PPy/CS hydrogels were able to provide electrical conduction between skeletal muscle tissues isolated from adult rats. The *in vivo* performance was evaluated in a rat heart attack model, and subjects injected with the hydrogel showed electrocardiogram (ECG) wave patterns similar to healthy control subjects injected with PBS solution. In another study, Cui et al. grafted carboxyl-functionalized tetraaniline (CTA) conductive polymer functional groups to poly(lactic acid)/poly(ethylene glycol)/poly(lactic acid) (PLA/PEG/PLA) copolymers through *in situ* polymerization and produced injectable electroactive hydrogels by cross-linking these copolymers [136]. *In vitro* cytocompatibility tests with fibroblasts, cardiomyocytes, and pre-osteoblasts showed that these hydrogels increased cell proliferation under electrical stimulation. Subcutaneous biocompatibility tests with Sprague-Dawley (SD) rats showed a transition from acute inflammation to mild chronic inflammation at the implant site within 2 weeks and a return to normal state with complete degradation of the hydrogel at the end of the fourth week.

In a study inspired by the strong adhesion of mussel byssus to many different surfaces and materials, Wang et al. added poly(pyrrole) (PPy) nanoparticles to gelatin methacrylate/poly(ethylene glycol) diacrylate (GelMA/PEGDA) hydrogels by cryogelation method, using dopamine as a cross-linker [137]. Strong adhesion and improvement in cardiac functions were observed in cardiomyocytes cultured on the hydrogels *in vitro*. When tested *in vivo* with a rat heart attack model, the hydrogel led to shrinkage of scar tissue and improved cardiac function. In order to mimic the complex bioactive extracellular matrix structure of cardiac tissue physically and chemically, Alves et al. added water-soluble poly(aniline)/poly(vinylpyrrolidone) (PANI/PVP) conductive copolymers to a hydrogel composed of varying concentrations of collagen, silk fibroin, hyaluronic acid, and poly(ethylene glycol) [138]. The composite hydrogel possessed semiconductor-level electrical conductivity, and an interconnected porous structure allowing oxygen and nutrient exchange. The physiomechanical and physiochemical properties of the hydrogels supported the viability and proliferation of cardiomyocytes *in vitro*, and it was suggested that the antioxidant activity of the hydrogels could protect the regenerating heart tissue from oxidative stress.

Hydrogels that are implanted to or work with the myocardium are exposed to a cyclic load as the myocardium is in a continuous contraction/relaxation cycle. This cyclic load may affect the integrity of hydrogels in time, by causing mechanical damage. Dong et al., focusing on this problem, produced self-healing electroactive hydrogels with the potential to release drugs into cardiac tissue by mixing aniline tetramer grafted chitosan copolymers with dibenzaldehyde-functionalized poly(ethylene glycol) [139]. These hydrogels were reported to have antibacterial activity as well as electrical conductivity similar to the natural myocardium tissue. In addition, the mechanical properties, gelation rate, and swelling capacity of the hydrogels were shown to be adjustable by changing the aniline tetramer concentration.

10.4.3 Skin Tissue Engineering

Skin is the largest organ in the body with functions such as protecting the human body against environmental factors, sensing the environment, and controlling the body temperature. The fact that it forms the outermost layer of the human body and has a soft and elastic structure makes the skin tissue vulnerable to damage. In addition to properties such as elasticity, mechanical strength, and high regenerative capacity, electrical conductivity also plays an important role in the function of skin tissue. When skin integrity is disrupted, the healing process is initiated by triggering the formation of electric fields in the range of 40–200 mV/mm at the damage site. Electrical impulses accelerate the healing process by reducing edema formation in the damaged area, increasing keratinocyte migration to the area, triggering dermal angiogenesis with epithelial tissue regeneration, and providing an antibacterial effect. Therefore, the production of conductive hydrogels that can transmit electrical impulses is frequently studied for the development of elastic and bioactive artificial skin and wound dressings (Table 10.7).

In order to develop antioxidant, antibacterial, self-renewable, and injectable conductive wound dressings, Zhao et al. grafted poly(aniline) (PANI) to quaternarized chitosan (QCS) polymers, then blended this material with poly(ethylene glycol)/(poly(glycerol sebacate) (PEGS) copolymer modified with benzaldehyde groups to form composite hydrogels with dynamic cross-linking [140]. These hydrogels supported cell viability and showed superior antibacterial effects by eliminating *E. coli* (Gram-negative) and *S. aureus* (Gram-positive) colonies with which they made contact. When the *in vivo* performance of the hydrogels as wound dressings was investigated, an effective reduction in wound surface area, a decrease in the inflammatory cell population migrating to the wound site, and an increase in the expression of growth factors and collagen production associated with wound healing and angiogenesis were observed. In another study, the group used N-carboxyethyl chitosan (CEC) and aniline tetramer grafted oxidized hyaluronic acid (OHA-AT) to produce multifunctional hydrogels, then added amoxicillin to the hydrogel structure and observed a positive effect of these hydrogels on *in vivo* wound healing [1].

Li et al., addressing areas where skin elasticity is important (i.e., elbows, wrists, knees, neck, and nape), produced elastic hydrogels with a conductivity similar to the human dermis layer by gelling guar gum and poly(3,4-ethylenedioxythiophene)/poly(styrene sulfonate) (PEDOT/PSS) conductive polymer system through acid–base reactions [141]. Gan et al. prepared electrically conductive hydrogels that can be used in wound dressing or biosensor applications by imparting conductivity to UV photopolymerizable poly(acrylamide)/chitosan (PA/CS) hydrogel structure using poly(pyrrole) (PPy) nanorods [142]. The hydrogel was capable of controlled drug release in response to electrical stimuli and provided rapid wound closure under *in vivo* conditions, and the thickness of epithelial tissue whose regeneration was triggered by these hydrogels in vivo was similar to the thickness of healthy epithelial tissue. Lu et al. developed pHEMA/PPy conductive hydrogels to replace the metal electrodes traditionally used in the treatment of chronic skin wounds with electrical stimulation [143]. Hydrogels exhibited cytocompatibility under *in vitro* conditions and effectively transmitted the electrical stimuli applied to them to cultured fibroblasts in scratch tests, triggering cell migration and accelerating the closure of the scratched area.

To develop electroactive smart wound dressings capable of drug release, di-Luca et al. designed graphene oxide (GO) doped poly(acrylamide)/poly(ethylene glycol dimethacrylate) (PA/PEG) hydrogels containing gelatin or trypsin [144]. The hydrogel was able to release antibacterial drugs in a controlled manner *in vitro*, in response to electrical stimulation. *In vitro* studies with human fetal fibroblasts showed a cell viability of 94% and higher, indicating that the hydrogels did not have a cytotoxic effect. Wang et al. aimed to produce antibacterial conductive hydrogels for the treatment of chronic skin injuries using the rare earth element terbium (Tb3+) and reduced graphene oxide (rGO). The group added these nanomaterials into poly(vinyl alcohol)/sodium alginate (PVA/SA) hydrogel structure [145]. PVA/SA/rGO/Tb hydrogels, which were found to exhibit antibacterial effects against Gram-negative and Gram-positive bacteria through their surface properties that inhibit bacterial adhesion, provided rapid healing by effectively suppressing inflammation in infected diabetic wound models. Liang et al. designed conductive composite hydrogels with dopamine-grafted gelatin (GT-DA), chitosan (CS), and poly(dopamine)-coated carbon nanotubes (CNT-PDA), and produced hydrogels with controllable swelling, drug release, and degradation rates [146]. The hydrogels, which were reported to have good cell compatibility and hemostatic capacity, showed photothermal antibacterial activity due to the CNT addition. Composite, antibiotic (doxycycline)-loaded hydrogels implanted in an infected wound tissue model contributed to denser collagen deposition and epithelial tissue formation in the wound area compared to commercially available wound dressings and accelerated wound healing and closure.

Zhao et al. combined silver nanowires (AgNWs) and poly(vinyl alcohol)/N--carboxyethyl chitosan/agarose (PVA/CEC/AGA) hydrogel systems to produce electrically conductive, antibacterial, and self-renewable hydrogels whose sol–gel transformation could be triggered by physical (thermal or ultrasonic) or chemical (pH or diol-containing molecules) stimuli, thus enabling complete dissolution of the hydrogel structure on demand [147]. During *in vivo* application in diabetic ulcer models, these hydrogels triggered angiogenesis, increased collagen deposition, and prevented bacterial infection, thus accelerating wound healing.

Developing ionic triboelectric nanogenerator (iTENG) wound dressings, Jeong et al. aimed to accelerate the healing of chronic wounds such as diabetes ulcers by converting the biophysical energy generated by body movements into an electrical potential that can be transferred to the wound area [148]. The group added lithium chloride (LiCl) to a poly(acrylamide) hydrogel solution and formed an elastic wound dressing by turning the hydrogel into a wound-fabric-like structure. The electric field generated by friction in the wound dressing in contact with the skin tissue caused an electric potential difference between the wound area and the wound dressing, providing electrical stimulation to the area. The electrical stimuli provided by the wound dressing increased migration, proliferation, and growth factor expression of fibroblasts *in vitro* and provided faster wound healing compared to the control groups in vivo.

10.4.4 Bone Tissue Engineering

Bone, a hard tissue, has a complex structure that requires a combination of many different physical and mechanical properties. Collagen and mineralized extracellular matrix give bone tissue high toughness, hardness, and elasticity. In addition to the sensitivity to mechanical stimuli for healing, piezoelectric and vaporization potential mechanisms make bone tissue responsive to electrical stimuli. Adhesion, extracellular matrix production, proliferation, mineralization, and intercellular communication via growth factors and transmembrane proteins can be triggered through electrical stimuli in bone cells, accelerating the healing processes in bone tissue. Therefore, conductive hydrogels with high mechanical and environmental resistance that can easily transfer electrical impulses have great potential in the development of "smart" bone tissue engineering products (Table 10.7).

Khorshidi et al. produced conductive hydrogels dynamically cross-linked by Schiff base reactions, using oxidized polysaccharides and gelatin in the main hydrogel network and poly(ε-caprolactone)/poly(aniline) (PCL/PANI) polymer blend in the conductive phase [149]. *In vitro* studies showed that the material improved the adhesion, proliferation, and morphology of osteoblast-like cells. Lu et al. produced multilayer graphene-doped cellulose ester membrane hydrogels [150], aiming to achieve directional regeneration of bone tissue while separating it from the rapidly produced connective tissue after injury. This membrane with high mechanical strength, elasticity, and selective permeability successfully isolated the damaged site and accelerated bone tissue regeneration by triggering early osteogenesis and increasing mineralization.

Based on the literature where gold nanoparticles trigger differentiation toward bone lineages in mesenchymal stem cells, Heo et al. combined gold nanoparticles with biodegradable UV photopolymerizable GelMA hydrogel for bone tissue engineering [151]. The conductive hydrogel increased the proliferation of adipose-derived stem cells and their differentiation into osteoblasts, enhanced alkaline phosphatase activity, and triggered a high degree of bone regeneration *in vivo*. To produce bone tissue scaffolds with antibacterial properties against the formation of antibiotic-resistant bacterial infection in the wound area, a common problem encountered in the treatment of bone injuries, Ribeiro et al. added gold and silver nanoparticles to osteoconductive silk fibroin/nanohydroxyapatite (SF/nHAp) hydrogels [152]. Nanoparticle concentrations that can exert an antibacterial effect without adversely affecting the growth and proliferation of osteoblast-like MG63 cell lines were determined, and it was concluded that these multi-effective hydrogels can be used in the treatment of bone tissue injuries.

10.4.5 Cartilage Tissue Engineering

Cartilage tissue is a complex, specialized connective tissue of great importance for joint mobility and health. Due to its dense extracellular matrix structure consisting of type II collagen fibrils and chondroitin sulfate proteoglycans intertwined, cartilage tissue is capable of water retention up to 80% of its weight [153]. This aqueous structure of the extracellular matrix not only provides lubrication to the joints but also delivers nutrients—via diffusion—to the chondrocytes (cartilage cells), as cartilage tissue lacks vasculature. Chondrocytes trapped in the extracellular matrix make up 2% of the tissue, and they are sensitive to mechanical, magnetic, and electrical stimuli [154–156]. Due to the low regenerative capacity of cartilage tissue, injury due to trauma or degenerative diseases leads to progressive degradation of the cartilage tissue and loss of joint function [157]. With cartilage tissue engineering, scaffolds with good viscoelasticity and porosity, appropriate mechanical strength, low friction, abrasion resistance, biocompatibility, biodegradability, and long-term durability can be produced for the treatment of these injuries [158]. Hydrogels have significant potential in cartilage tissue engineering due to their similar water retention capacity, physical and mechanical properties, and biodegradability.

Electrical stimulation is known to accelerate the healing of cartilage tissue damage in vivo by increasing the proliferation of chondrocytes and their production of extracellular matrix components [159]. In addition, mesenchymal stem cells exposed to electrical stimulation have been observed to produce more chondrogenic marker molecules such as type II collagen, aggrecan, and glycosaminoglycans [160]. Therefore, research on conductive hydrogels that can mimic the natural properties of cartilage tissue and have a positive effect on tissue healing with electrical stimuli has gained importance in the field of cartilage tissue engineering. Despite the scarcity of studies on electrically conductive hydrogels in the literature for cartilage tissue engineering, the results obtained indicate that these materials are an innovative and advanced alternative to traditional materials and approaches (Table 10.7).

Shen et al. investigated the chondrogenic potential of photopolymerizable poly(D,L-lactic acid)/poly(ethylene glycol) (PDLLA/PEG) nanocomposite hydrogels containing graphene oxide (GO) [161]. When bone marrow mesenchymal stem cells were cultured on these hydrogels, they produced chondrogenic marker molecules and the expression of transcriptional factors increased significantly. The hydrogels, which were also injectable, were reported to be biodegradable. In another study, Zhang et al. produced a hydrogel exhibiting electrical conductivity and good mechanical properties by adding sodium phytate (PANa) to the poly(vinyl alcohol) (PVA) hydrogel structure [162]. The hydrogel was found to have high elastic strength and an electrical conductivity similar to that of bovine cartilage tissue. In addition, the long-term stability of the hydrogel in aqueous media indicated that implant integrity could be maintained *in vivo*.

Utilizing additive manufacturing methods, Distler et al. produced 3D printable oxidized alginate-gelatin (ADA-GEL)/poly(pyrrole):poly(styrene sulfonate) (PPy:PSS) hydrogels [16]. The mechanical strength and electrical conductivity of the hydrogels were found to be similar to natural cartilage tissue. PPy modification caused a minimal decrease in the adhesion and proliferation of mouse teratocarcinoma precursor chondrogenic cell line ATDC5, and this effect was associated with an increase in hydrogel stiffness due to the modification. In order to improve the biodegradability of natural conductive polymers used in the production of conductive hydrogels, Kashi et al. synthesized pyrrole oligomers and added this conductive material to the chitosan/ß-glycerophosphate hydrogel network structure [163]. Synthesizing short pyrrole chains facilitated hydrogel production by increasing the water solubility of the polymer. It was determined that the hydrogels performed optimally in terms of biodegradability, biocompatibility, swelling, and electrical activity.

10.4.6 Biosensors and Wearable Technologies

The mechanical properties of hydrogel-based stable conductive materials, which are similar to soft tissues in the human body; and their electrical conductivity, which they can maintain while in contact with body fluids, form the basis for the design of bioelectronic devices such as biosensors and wearable technological devices used for signal monitoring, disease diagnosis, and treatment [164]. The development of hydrogels with suitable physical and chemical properties, biocompatibility, and stability for these applications is an important research area (Table 10.7).

Li et al. aimed to develop electrically conductive hydrogels that can be used as body sensors by improving the poor mechanical properties of poly(vinyl alcohol) (PVA) polymer, which is frequently used in wound dressings, and formed a poly(acrylamide) (PA) phase in PVA via *in situ* polymerization method [165]. The interpenetrating network (IPN) hydrogel was able to sense body movements, such as stretching and bending, and their directions and angles, repetitively through changes in electrical signal transmission. Yan et al. produced conductive hydrogels using the physical self-assembly behavior of poly(dopamine)-reduced graphene oxide (PGO) hybridized cellulose (PGC) bio-nanostructures to develop an "e-skin" that can be implanted subcutaneously or to other desired parts of the body [166]. When used as a diabetic wound dressing, the hydrogel was able to effectively transfer electrical impulses to the tissue, accelerating the wound healing. In addition, hydrogel was able to stably detect physiological signals when in contact with biological fluids.

Gan et al. formed a poly(pyrrole) (PPy) conductive phase in the poly(acrylamide)/chitosan (PA/CS) hydrogel structure by post-polymerization processing method and obtained a hydrogel with high mechanical strength, capable of releasing drugs in response to electrical stimuli and monitoring the movement of human limbs thanks to its conductive structure [142]. These hydrogels with good biocompatibility also accelerated the diabetic wound healing process under electrical stimulation. Using hybrid approaches to impart conductivity to hydrogels, Zhao et al. developed composite hydrogels containing poly(dopamine) modified silver nanoparticles (PDA-AgNP), poly(aniline) (PANI), and poly(vinyl alcohol) (PVA) [167]. The hydrogels had good processability, electrochemical properties, self-renewal function, and antibacterial effect. Furthermore, with good adhesion to tissues, the hydrogels were able to track large-scale body movements in real time, indicating that they can be used in the development of epidermal sensors for the monitoring of reflex movements.

Aiming to achieve flexibility, conductivity, self-renewal, and biocompatibility with cost-effective chemicals and approaches, Zhang et al. obtained conductive hydrogels by cross-linking poly(pyrrole) (PPy) and zinc (Zn) functionalized chitosan molecules with poly(vinyl alcohol) (PVA) [168]. The dynamic bond structure of the hydrogels provided the ability to rapidly regenerate itself in the case of physical damage. In addition, the hydrogel could be effectively used as a sensor that can detect temperature changes and body movements. To design self-renewable, antibacterial conductive hydrogels with strong adhesion that can be used as biosensors, Wei et al. utilized silicotungstic acid (SiW), chitosan (CS), and poly(acrylamide) (PA) [169]. The hydrogel, which exhibited good adhesion to various surfaces such as Teflon, ceramic, stainless steel, wood, and biological tissues in a repetitive manner, was also able to regenerate itself with the same efficiency against repeated physical deformations. It was emphasized that the hydrogel, which also shows antibacterial properties thanks to its negatively charged chitosan chains and silicotungstic acid addition, has potential in the development of sensors that can perform real-time tracking of body movements, due to its conductivity that changes with the mechanical forces applied to it.

10.5 Conclusion

Conductive hydrogels can be prepared using many different materials and production methods and utilized in various diagnostic and therapeutic applications in the biomedical field. Conductive materials, which generally have low mechanical strength, processing difficulties, and poor biocompatibility and biodegradability, and hydrogels, which are frequently used in tissue engineering applications and have high biocompatibility, are combined to produce innovative and superior conductive hydrogels that bring together the advantages of both materials. For biomedical applications, uniform distribution of the conductive phase in the hydrogel network and stability of the material under physiological conditions are of utmost importance. This can only be achieved by carefully selecting the type and concentration of materials used, as well as the production methods. The presence of many naturally electroactive tissues such as nerves and heart muscles in the human body indicates that conductive hydrogels have a significant potential in the field of tissue engineering. In today's world where technological developments are rapidly increasing, wearable technologies and the production of advanced biosensors are also research areas where the advantageous properties of conductive hydrogels are being investigated. Despite many technical challenges such as maintaining good mechanical strength, stable electrical conductivity, and biocompatibility under physiological conditions, conductive hydrogels suitable for biomedical use remain an important area of research and development.

References

1. Qu, J., Zhao, X., Liang, Y., Xu, Y., Ma, P.X., Guo, B.: Degradable conductive injectable hydrogels as novel antibacterial, anti-oxidant wound dressings for wound healing. Chem. Eng. J. **362**, 548–560 (2019)
2. Zhao, X., Li, P., Guo, B., Ma, P.X.: Antibacterial and conductive injectable hydrogels based on quaternized chitosan-graft-polyaniline/oxidized dextran for tissue engineering. Acta Biomater. **26**, 236–248 (2015)
3. Pan, L., Yu, G., Zhai, D., Lee, H.R., Zhao, W., Liu, N., et al.: Hierarchical nanostructured conducting polymer hydrogel with high electrochemical activity. Proc. Natl. Acad. Sci. U. S. A. **109**(24), 9287–9292 (2012). https://doi.org/10.1073/pnas.1202636109
4. Wu, Y., Chen, Y.X., Yan, J., Yang, S., Dong, P., Soman, P.: Fabrication of conductive polyaniline hydrogel using porogen leaching and projection microstereolithography. J. Mater. Chem. B. **3**(26), 5352–5360 (2015)
5. Yang, S., Jang, L.K., Kim, S., Yang, J., Yang, K., Cho, S.W., et al.: Polypyrrole/alginate hybrid hydrogels: electrically conductive and soft biomaterials for human mesenchymal stem cell culture and potential neural tissue engineering applications. Macromol. Biosci. **16**(11), 1653–1661 (2016). https://doi.org/10.1002/mabi.201600148
6. Min, J.H., Patel, M., Koh, W.G.: Incorporation of conductive materials into hydrogels for tissue engineering applications. Polymers. **10**, 1078 (2018)
7. Hasirci, V., Hasirci, N.: Fundamentals of Biomaterials [Internet], 2nd edn. Springer Nature, Cham (2024). https://doi.org/10.1007/978-3-031-54046-2
8. Sayyar, S., Gambhir, S., Chung, J., Officer, D.L., Wallace, G.G.: 3D printable conducting hydrogels containing chemically converted graphene. Nanoscale. **9**(5), 2038–2050 (2017)
9. Lee, S.J., Zhu, W., Nowicki, M., Lee, G., Heo, D.N., Kim, J., et al.: 3D printing nano conductive multi-walled carbon nanotube scaffolds for nerve regeneration. J. Neural Eng. **15**(1), 016018 (2018)
10. Ray, P.C., Yu, H., Fu, P.P.: Toxicity and environmental risks of nanomaterials: challenges and future needs. J. Environ. Sci. Health C Environ. Carcinog. Ecotoxicol. Rev. **27**(1), 1–35 (2009). https://doi.org/10.1080/10590500802708267
11. Mawad, D., Lauto, A., Wallace, G.G., Mawad, D., Lauto, A., Wallace, G.G.: Conductive polymer hydrogels. In: Polymeric Hydrogels as Smart Biomaterials, pp. 19–44 (2016). https://doi.org/10.1007/978-3-319-25322-0_2
12. Tian, K., Bae, J., Bakarich, S.E., Yang, C., Gately, R.D., Spinks, G.M., et al.: 3D printing of transparent and conductive heterogeneous hydrogel–elastomer systems. Adv. Mater. **29**(10), 1604827 (2017). https://doi.org/10.1002/adma.201604827
13. Jia, Z., Zeng, Y., Tang, P., Gan, D., Xing, W., Hou, Y., et al.: Conductive, tough, transparent, and self-healing hydrogels based on catechol-metal ion dual self-catalysis. Chem. Mater. **31**(15), 5625–5632 (2019). https://doi.org/10.1021/acs.chemmater.9b01498
14. Odent, J., Wallin, T.J., Pan, W., Kruemplestaedter, K., Shepherd, R.F., Giannelis, E.P., et al.: Highly elastic, transparent, and conductive 3D-printed ionic composite hydrogels. Adv. Funct. Mater. **27**(33), 1701807 (2017). https://doi.org/10.1002/adfm.201701807
15. Zhou, Y., Wan, C., Yang, Y., Yang, H., Wang, S., Dai, Z., et al.: Highly stretchable, elastic, and ionic conductive hydrogel for artificial soft electronics. Adv. Funct. Mater. **29**(1), 1806220 (2019). https://doi.org/10.1002/adfm.201806220
16. Distler, T., Polley, C., Shi, F., Schneidereit, D., Ashton, M.D., Friedrich, O., et al.: Electrically conductive and 3D-printable oxidized alginate-gelatin Polypyrrole:PSS hydrogels for tissue engineering. Adv. Healthc. Mater. **10**(9), 2001876 (2021). https://doi.org/10.1002/adhm.202001876
17. Plieth, W.: Intrinsically conducting polymers. Electrochem. Mater. Sci. **1**, 323–363 (2008)
18. Shirakawa, H.: Nobel Lecture: The discovery of polyacetylene film—the dawning of an era of conducting polymers. Rev. Mod. Phys. **73**(3), 713 (2001). https://doi.org/10.1103/RevModPhys.73.713

19. Jiang, L., Wang, Y., Liu, Z., Ma, C., Yan, H., Xu, N., et al.: Three-dimensional printing and injectable conductive hydrogels for tissue engineering application. Tissue Eng. Part B Rev. **25**(5), 398–411 (2019). https://doi.org/10.1089/ten.teb.2019.0100
20. George, P.M., Lavan, D.A., Burdick, J.A., Chen, C.Y., Liang, E., Langer, R.: Electrically controlled drug delivery from biotin-doped conductive polypyrrole. Adv. Mater. **18**(5), 577–581 (2006). https://doi.org/10.1002/adma.200501242
21. Li, Y., Neoh, K.G., Kang, E.T.: Plasma protein adsorption and thrombus formation on surface functionalized polypyrrole with and without electrical stimulation. J. Colloid Interface Sci. **275**(2), 488–495 (2004)
22. Meng, S., Rouabhia, M., Shi, G., Zhang, Z.: Heparin dopant increases the electrical stability, cell adhesion, and growth of conducting polypyrrole/poly(L,L-lactide) composites. J. Biomed. Mater. Res. A. **87A**(2), 332–344 (2008). https://doi.org/10.1002/jbm.a.31735
23. Chronakis, I.S., Grapenson, S., Jakob, A.: Conductive polypyrrole nanofibers via electrospinning: electrical and morphological properties. Polymer. **47**(5), 1597–1603 (2006)
24. Gomez, N., Schmidt, C.E.: Nerve growth factor-immobilized polypyrrole: bioactive electrically conducting polymer for enhanced neurite extension. J. Biomed. Mater. Res. A. **81A**(1), 135–149 (2007). https://doi.org/10.1002/jbm.a.31047
25. Song, H.K., Toste, B., Ahmann, K., Hoffman-Kim, D., Palmore, G.T.R.: Micropatterns of positive guidance cues anchored to polypyrrole doped with polyglutamic acid: a new platform for characterizing neurite extension in complex environments. Biomaterials. **27**(3), 473–484 (2006)
26. Ferraz, N., Straømme, M., Fellström, B., Pradhan, S., Nyholm, L., Mihranyan, A.: In vitro and in vivo toxicity of rinsed and aged nanocellulose–polypyrrole composites. J. Biomed. Mater. Res. A. **100A**(8), 2128–2138 (2012). https://doi.org/10.1002/jbm.a.34070
27. Brahim, S., Guiseppi-Elie, A.: Electroconductive hydrogels: electrical and electrochemical properties of polypyrrole-poly(HEMA) composites. Electroanalysis. **17**(7), 556–570 (2005). https://doi.org/10.1002/elan.200403109
28. Walczak, R.M., Reynolds, J.R.: Poly(3,4-alkylenedioxypyrroles): the PXDOPs as versatile yet underutilized electroactive and conducting polymers. Adv. Mater. **18**(9), 1121–1131 (2006)
29. Zhang, Z., Rouabhia, M., Wang, Z., Roberge, C., Shi, G., Roche, P., et al.: Electrically conductive biodegradable polymer composite for nerve regeneration: electricity-stimulated neurite outgrowth and axon regeneration. Artif. Organs. **31**(1), 13–22 (2007). https://doi.org/10.1111/j.1525-1594.2007.00335.x
30. Thompson, B.C., Moulton, S.E., Richardson, R.T., Wallace, G.G.: Effect of the dopant anion in polypyrrole on nerve growth and release of a neurotrophic protein. Biomaterials. **32**(15), 3822–3831 (2011)
31. Bettinger, C.J., Bruggeman, J.P., Misra, A., Borenstein, J.T., Langer, R.: Biocompatibility of biodegradable semiconducting melanin films for nerve tissue engineering. Biomaterials. **30**(17), 3050–3057 (2009)
32. Castano, H., O'Rear, E.A., McFetridge, P.S., Sikavitsas, V.I.: Polypyrrole thin films formed by Admicellar polymerization support the osteogenic differentiation of mesenchymal stem cells. Macromol. Biosci. **4**(8), 785–794 (2004). https://doi.org/10.1002/mabi.200300123
33. Ghasemi-Mobarakeh, L., Prabhakaran, M.P., Morshed, M., Nasr-Esfahani, M.H., Baharvand, H., Kiani, S., et al.: Application of conductive polymers, scaffolds and electrical stimulation for nerve tissue engineering. J. Tissue Eng. Regen. Med. **5**(4), e17–e35 (2011). https://doi.org/10.1002/term.383
34. Gomez, N., Lee, J.Y., Nickels, J.D., Schmidt, C.E.: Micropatterned polypyrrole: a combination of electrical and topographical characteristics for the stimulation of cells. Adv. Funct. Mater. **17**(10), 1645–1653 (2007). https://doi.org/10.1002/adfm.200600669
35. Lee, J.W., Serna, F., Nickels, J., Schmidt, C.E.: Carboxylic acid-functionalized conductive polypyrrole as a bioactive platform for cell adhesion. Biomacromolecules. **7**(6), 1692–1695 (2006). https://doi.org/10.1021/bm060220q

36. Zhou, D.D., Cui, X.T., Hines, A., Greenberg, R.J.: Conducting Polymers in Neural Stimulation Applications, pp. 217–252 (2009). https://doi.org/10.1007/978-0-387-98120-8_8
37. Peramo, A., Urbanchek, M.G., Spanninga, S.A., Povlich, L.K., Cederna, P., Martin, D.C.: In situ polymerization of a conductive polymer in acellular muscle tissue constructs. Tissue Eng. Part A. **14**(3), 423–432 (2008). https://doi.org/10.1089/tea.2007.0123
38. Karagkiozaki, V., Karagiannidis, P.G., Gioti, M., Kavatzikidou, P., Georgiou, D., Georgaraki, E., et al.: Bioelectronics meets nanomedicine for cardiovascular implants: PEDOT-based nanocoatings for tissue regeneration. Biochim. Biophys. Acta Gen. Subj. **1830**(9), 4294–4304 (2013)
39. Luo, S.C., Ali, E.M., Tansil, N.C., Yu, H.H., Gao, S., Kantchev, E.A.B., et al.: Poly(3,4-ethylenedioxythiophene) (PEDOT) nanobiointerfaces: thin, ultrasmooth, and functionalized PEDOT films with in vitro and in vivo biocompatibility. Langmuir. **24**(15), 8071–8077 (2008). https://doi.org/10.1021/la800333g
40. Rad, A.T., Ali, N., Kotturi, H.S.R., Yazdimamaghani, M., Smay, J., Vashaee, D., et al.: Conducting scaffolds for liver tissue engineering. J. Biomed. Mater. Res. A. **102**(11), 4169–4181 (2014). https://doi.org/10.1002/jbm.a.35080
41. Strakosas, X., Wei, B., Martin, D.C., Owens, R.M.: Biofunctionalization of polydioxythiophene derivatives for biomedical applications. J. Mater. Chem. B. **4**(29), 4952–4968 (2016)
42. Kim, B.G., Lim, J.H., Kim, J.Y., Frances Glover, C., Watson, T., et al.: Toxicity evaluation of PEDOT/biomolecular composites intended for neural communication electrodes. Biomed. Mater. **4**(4), 045009 (2009). https://doi.org/10.1088/1748-6041/4/4/045009
43. Blinova, N.V., Stejskal, J., Trchová, M., Prokeš, J.: Control of polyaniline conductivity and contact angles by partial protonation. Polym. Int. **57**(1), 66–69 (2008). https://doi.org/10.1002/pi.2312
44. Borriello, A., Guarino, V., Schiavo, L., Alvarez-Perez, M.A., Ambrosio, L.: Optimizing PANi doped electroactive substrates as patches for the regeneration of cardiac muscle. J. Mater. Sci. Mater. Med. **22**(4), 1053–1062 (2011). https://doi.org/10.1007/s10856-011-4259-x
45. Cullen, D.K., Patel, A.R., Doorish, J.F., Smith, D.H., Pfister, B.J.: Developing a tissue-engineered neural-electrical relay using encapsulated neuronal constructs on conducting polymer fibers. J. Neural Eng. **5**(4), 374 (2008). https://doi.org/10.1088/1741-2560/5/4/002
46. Prabhakaran, M.P., Ghasemi-Mobarakeh, L., Jin, G., Ramakrishna, S.: Electrospun conducting polymer nanofibers and electrical stimulation of nerve stem cells. J. Biosci. Bioeng. **112**(5), 501–507 (2011)
47. Huang, L., Zhuang, X., Hu, J., Lang, L., Zhang, P., Wang, Y., et al.: Synthesis of biodegradable and electroactive multiblock polylactide and aniline pentamer copolymer for tissue engineering applications. Biomacromolecules. **9**(3), 850–858 (2008). https://doi.org/10.1021/bm7011828
48. Yu, Q.Z., Shi, M.M., Deng, M., Wang, M., Chen, H.Z.: Morphology and conductivity of polyaniline sub-micron fibers prepared by electrospinning. Mater. Sci. Eng. B. **150**(1), 70–76 (2008)
49. Niamlang, S., Buranut, T., Niansiri, A., Sirivat, A.: Controlled aloin release from crosslinked polyacrylamide hydrogels: effects of mesh size, electric field strength and a conductive polymer. Materials. **6**, 4787–4800 (2013)
50. Eivazzadeh-Keihan, R., Bahojb Noruzi, E., Chidar, E., Jafari, M., Davoodi, F., Kashtiaray, A., et al.: Applications of carbon-based conductive nanomaterials in biosensors. Chem. Eng. J. **442**, 136183 (2022)
51. Gao, C., Song, S., Lv, Y., Huang, J., Zhang, Z.: Recent development of conductive hydrogels for tissue engineering: review and perspective. Macromol. Biosci. **22**(8), 2200051 (2022). https://doi.org/10.1002/mabi.202200051
52. Singh, E., Nalwa, H.S.: In: Singh, E., Nalwa, H.S. (eds.) Nanomaterial-Based Flexible and Multifunctional Sensors, pp. 1–651 (2018)

53. Sanchez, V.C., Jachak, A., Hurt, R.H., Kane, A.B.: Biological interactions of graphene-family nanomaterials: an interdisciplinary review. Chem. Res. Toxicol. **25**(1), 15–34 (2012). https://doi.org/10.1021/tx200339h
54. Bussy, C., Ali-Boucetta, H., Kostarelos, K.: Safety considerations for graphene: lessons learnt from carbon nanotubes. Acc. Chem. Res. **46**(3), 692–701 (2013). https://doi.org/10.1021/ar300199e
55. Thompson, B.C., Murray, E., Wallace, G.G.: Graphite oxide to graphene. Biomaterials to bionics. Adv. Mater. **27**(46), 7563–7582 (2015). https://doi.org/10.1002/adma.201500411
56. Liu, S., Zeng, T.H., Hofmann, M., Burcombe, E., Wei, J., Jiang, R., et al.: Antibacterial activity of graphite, graphite oxide, graphene oxide, and reduced graphene oxide: membrane and oxidative stress. ACS Nano. **5**(9), 6971–6980 (2011). https://doi.org/10.1021/nn202451x
57. Seabra, A.B., Paula, A.J., De Lima, R., Alves, O.L., Durán, N.: Nanotoxicity of graphene and graphene oxide. Chem. Res. Toxicol. **27**(2), 159–168 (2014). https://doi.org/10.1021/tx400385x
58. Mohanta, D., Patnaik, S., Sood, S., Das, N.: Carbon nanotubes: evaluation of toxicity at biointerfaces. J. Pharm. Anal. **9**(5), 293–300 (2019)
59. Terrones, H., Terrones, M., Hernández, E., Grobert, N., Charlier, J.C., Ajayan, P.M.: New metallic allotropes of planar and tubular carbon. Phys. Rev. Lett. **84**(8), 1716 (2000). https://doi.org/10.1103/PhysRevLett.84.1716
60. He, H., Pham-Huy, L.A., Dramou, P., Xiao, D., Zuo, P., Pham-Huy, C.: Carbon nanotubes: Applications in pharmacy and medicine. Biomed. Res. Int. **2013**, 578290 (2013)
61. Hirlekar, R., Garse, H., Vij, M., Yamagar, M., Vilasrao, K.: Carbon nanotubes and its applications: a review. Carbon nanotubes and its applications: a review article. Asian J. Pharm. Clin. Res. **2**(4), 17–27 (2009)
62. Hou, P.X., Bai, S., Yang, Q.H., Liu, C., Cheng, H.M.: Multi-step purification of carbon nanotubes. Carbon. **40**(1), 81–85 (2002)
63. Eatemadi, A., Daraee, H., Karimkhanloo, H., Kouhi, M., Zarghami, N., Akbarzadeh, A., et al.: Carbon nanotubes: properties, synthesis, purification, and medical applications. Nanoscale Res. Lett. **9**(1), 1–13 (2014). https://doi.org/10.1186/1556-276x-9-393
64. Che, J., Çağin, T., Goddard, W.A.: Thermal conductivity of carbon nanotubes. Nanotechnology. **11**(2), 65 (2000). https://doi.org/10.1088/0957-4484/11/2/305
65. Kostarelos, K., Lacerda, L., Partidos, C.D., Prato, M., Bianco, A.: Carbon nanotube-mediated delivery of peptides and genes to cells: translating nanobiotechnology to therapeutics. J. Drug. Deliv. Sci. Technol. **15**(1), 41–47 (2005)
66. Malik, M.A., Wani, M.Y., Hashim, M.A., Nabi, F.: Nanotoxicity: dimensional and morphological concerns. Adv. Phys. Chem. **2011**, 450912 (2011)
67. Yan, L., Zhao, F., Li, S., Hu, Z., Zhao, Y.: Low-toxic and safe nanomaterials by surface-chemical design, carbon nanotubes, fullerenes, metallofullerenes, and graphenes. Nanoscale. **3**(2), 362–382 (2011)
68. Bogdanov, A.A., Deininger, D., Dyuzhev, G.A.: Development prospects of the commercial production of fullerenes. Tech. Phys. **45**(5), 521–527 (2000). https://doi.org/10.1134/1.1259670
69. Coro, J., Suárez, M., Silva, L.S.R., Eguiluz, K.I.B., Salazar-Banda, G.R.: Fullerene applications in fuel cells: a review. Int. J. Hydrogen Energy. **41**(40), 17944–17959 (2016)
70. Giacalone, F., Martín, N.: Fullerene polymers: synthesis and properties. Chem. Rev. **106**(12), 5136–5190 (2006). https://doi.org/10.1021/cr068389h
71. Bhakta, P., Barthunia, B.: Fullerene and its applications: a review. J. Indian Acad. Oral Med. Radiol. **32**(2), 159 (2020)
72. Baei, P., Jalili-Firoozinezhad, S., Rajabi-Zeleti, S., Tafazzoli-Shadpour, M., Baharvand, H., Aghdami, N.: Electrically conductive gold nanoparticle-chitosan thermosensitive hydrogels for cardiac tissue engineering. Mater. Sci. Eng. C. **63**, 131–141 (2016)
73. Paquet, C., De Haan, H.W., Leek, D.M., Lin, H.Y., Xiang, B., Tian, G., et al.: Clusters of superparamagnetic iron oxide nanoparticles encapsulated in a hydrogel: a particle architec-

ture generating a synergistic enhancement of the T2 relaxation. ACS Nano. **5**(4), 3104–3112 (2011). https://doi.org/10.1021/nn2002272
74. Peña, B., Maldonado, M., Bonham, A.J., Aguado, B.A., Dominguez-Alfaro, A., Laughter, M., et al.: Gold nanoparticle-functionalized reverse thermal gel for tissue engineering applications. ACS Appl. Mater. Interfaces. **11**(20), 18671–18680 (2019). https://doi.org/10.1021/acsami.9b00666
75. Zare, M., Ramezani, Z., Rahbar, N.: Development of zirconia nanoparticles-decorated calcium alginate hydrogel fibers for extraction of organophosphorous pesticides from water and juice samples: facile synthesis and application with elimination of matrix effects. J. Chromatogr. A. **1473**, 28–37 (2016)
76. Conde, J., Dias, J.T., Grazú, V., Moros, M., Baptista, P.V., de la Fuente, J.M.: Revisiting 30 years of biofunctionalization and surface chemistry of inorganic nanoparticles for nanomedicine. Front. Chem. **2**(JUL), 48 (2014)
77. Sani, A., Cao, C., Cui, D.: Toxicity of gold nanoparticles (AuNPs): a review. Biochem. Biophys. Rep. **26**, 100991 (2021)
78. Chung, U.S., Kim, J.H., Kim, B., Kim, E., Jang, W.D., Koh, W.G.: Dendrimer porphyrin-coated gold nanoshells for the synergistic combination of photodynamic and photothermal therapy. Chem. Commun. **52**(6), 1258–1261 (2016)
79. Jaiswal, L., Shankar, S., Rhim, J.W.: Applications of nanotechnology in food microbiology. Methods Microbiol. **46**, 43–60 (2019)
80. Nossier, A.I., Eissa, S., Ismail, M.F., Hamdy, M.A., Azzazy, H.M.E.S.: Direct detection of hyaluronidase in urine using cationic gold nanoparticles: a potential diagnostic test for bladder cancer. Biosens. Bioelectron. **54**, 7–14 (2014)
81. Rengan, A.K., Bukhari, A.B., Pradhan, A., Malhotra, R., Banerjee, R., Srivastava, R., et al.: In vivo analysis of biodegradable liposome gold nanoparticles as efficient agents for photothermal therapy of cancer. Nano Lett. **15**(2), 842–848 (2015). https://doi.org/10.1021/nl5045378
82. Iravani, S., Korbekandi, H., Mirmohammadi, S.V., Zolfaghari, B.: Synthesis of silver nanoparticles: chemical, physical and biological methods. Res Pharm Sci. **9**(6), 385 (2014)
83. Durán, N., Nakazato, G., Seabra, A.B.: Antimicrobial activity of biogenic silver nanoparticles, and silver chloride nanoparticles: an overview and comments. Appl. Microbiol. Biotechnol. **100**(15), 6555–6570 (2016). https://doi.org/10.1007/s00253-016-7657-7
84. Johnston, H.J., Hutchison, G., Christensen, F.M., Peters, S., Hankin, S., Stone, V.: A review of the in vivo and in vitro toxicity of silver and gold particulates: particle attributes and biological mechanisms responsible for the observed toxicity. Crit. Rev. Toxicol. **40**(4), 328–346 (2010). https://doi.org/10.3109/10408440903453074
85. Rai, M.K., Deshmukh, S.D., Ingle, A.P., Gade, A.K.: Silver nanoparticles: the powerful nanoweapon against multidrug-resistant bacteria. J. Appl. Microbiol. **112**(5), 841–852 (2012). https://doi.org/10.1111/j.1365-2672.2012.05253.x
86. Khan, M.A.R., Mamun, M.S., Ara, M.H.: Review on platinum nanoparticles: synthesis, characterization, and applications. Microchem. J. **171**, 106840 (2021)
87. Nash, K.M., Ahmed, S.: Nanomedicine in the ROS-mediated pathophysiology: applications and clinical advances. Nanomedicine. **11**(8), 2033–2040 (2015)
88. Li, S., Liu, Y., Wu, Y., Ren, L., Lu, Y., Yamaguchi, S., et al.: An outlook on platinum-based active ingredients for dermatologic and skincare applications. Nanomaterials. **14**(15), 1303 (2024)
89. Stepanov, A.L., Golubev, A.N., Nikitin, S.I., Osin Rev, Y.N.: A review on the fabrication and properties of platinum nanoparticles. Adv. Mater. Sci. **38**, 160–175 (2014)
90. Lee, S., Kwon, D., Yim, C., Jeon, S.: Facile detection of troponin i using dendritic platinum nanoparticles and capillary tube indicators. Anal. Chem. **87**(9), 5004–5008 (2015). https://doi.org/10.1021/acs.analchem.5b00921
91. Li, M., Bo, X., Mu, Z., Zhang, Y., Guo, L.: Electrodeposition of nickel oxide and platinum nanoparticles on electrochemically reduced graphene oxide film as a nonenzymatic glucose sensor. Sens. Actuators B Chem. **192**, 261–268 (2014)

92. Venu, R., Ramulu, T.S., Anandakumar, S., Rani, V.S., Kim, C.G.: Bio-directed synthesis of platinum nanoparticles using aqueous honey solutions and their catalytic applications. Colloids Surf. A Physicochem. Eng. Asp. **384**(1–3), 733–738 (2011)
93. Sengul, A.B., Asmatulu, E.: Toxicity of metal and metal oxide nanoparticles: a review. Environ. Chem. Lett. **18**(5), 1659–1683 (2020). https://doi.org/10.1007/s10311-020-01033-6
94. Xie, J., Liu, G., Eden, H.S., Ai, H., Chen, X.: Surface-engineered magnetic nanoparticle platforms for cancer imaging and therapy. Acc. Chem. Res. **44**(10), 883–892 (2011). https://doi.org/10.1021/ar200044b
95. Xu, Q., Li, J., Li, S., Pan, H.: A highly sensitive electrochemiluminescence immunosensor based on magnetic nanoparticles and its application in CA125 determination. J. Solid State Electrochem. **16**(9), 2891–2898 (2012). https://doi.org/10.1007/s10008-012-1719-2
96. Fan, Z., Lu, J.G.: Zinc oxide nanostructures: synthesis and properties. J. Nanosci. Nanotechnol. **5**(10), 1561–1573 (2005)
97. Li, M., Zhu, L., Lin, D.: Toxicity of ZnO nanoparticles to escherichia coli: mechanism and the influence of medium components. Environ. Sci. Technol. **45**(5), 1977–1983 (2011). https://doi.org/10.1021/es102624t
98. Zhang, H., Chen, B., Jiang, H., Wang, C., Wang, H., Wang, X.: A strategy for ZnO nanorod mediated multi-mode cancer treatment. Biomaterials. **32**(7), 1906–1914 (2011)
99. Shi, H., Magaye, R., Castranova, V., Zhao, J.: Titanium dioxide nanoparticles: a review of current toxicological data. Part. Fibre Toxicol. **10**(1), 1–33 (2013). https://doi.org/10.1186/1743-8977-10-15
100. Weir, A., Westerhoff, P., Fabricius, L., Hristovski, K., Von Goetz, N.: Titanium dioxide nanoparticles in food and personal care products. Environ. Sci. Technol. **46**(4), 2242–2250 (2012). https://doi.org/10.1021/es204168d
101. Peng, Q., Chen, J., Wang, T., Peng, X., Liu, J., Wang, X., et al.: Recent advances in designing conductive hydrogels for flexible electronics. InfoMat. **2**(5), 843–865 (2020). https://doi.org/10.1002/inf2.12113
102. Di, X., Ma, Q., Xu, Y., Yang, M., Wu, G., Sun, P.: High-performance ionic conductive poly(vinyl alcohol) hydrogels for flexible strain sensors based on a universal soaking strategy. Mater. Chem. Front. **5**(1), 315–323 (2021)
103. Gomes, J.M., Silva, S.S., Reis, R.L.: Biocompatible ionic liquids: fundamental behaviours and applications. Chem. Soc. Rev. **48**(15), 4317–4335 (2019)
104. Sui, X., Guo, H., Cai, C., Li, Q., Wen, C., Zhang, X., et al.: Ionic conductive hydrogels with long-lasting antifreezing, water retention and self-regeneration abilities. Chem. Eng. J. **419**, 129478 (2021)
105. Kudaibergenov, S.E., Sigitov, V.B.: Swelling, shrinking, deformation, and oscillation of polyampholyte gels based on vinyl 2-aminoethyl ether and sodium acrylate. Langmuir. **15**(12), 4230–4235 (1999). https://doi.org/10.1021/la981070a
106. Scranton, A.B., Rangarajan, B., Klier, J.: Biomedical applications of polyelectrolytes. Adv. Polym. Sci. **122**, 1–54 (1995). https://doi.org/10.1007/3540587888_13
107. Murdan, S.: Electro-responsive drug delivery from hydrogels. J. Control. Release. **92**(1–2), 1–17 (2003)
108. Ghorbanizamani, F., Moulahoum, H., Guler Celik, E., Timur, S.: Ionic liquids enhancement of hydrogels and impact on biosensing applications. J. Mol. Liq. **357**, 119075 (2022)
109. Lei, Z., Chen, B., Koo, Y.M., Macfarlane, D.R.: Introduction: ionic liquids. Chem. Rev. **117**(10), 6633–6635 (2017). https://doi.org/10.1021/acs.chemrev.7b00246
110. Hulsbosch, J., De Vos, D.E., Binnemans, K., Ameloot, R.: Biobased ionic liquids: solvents for a green processing industry? ACS Sustain. Chem. Eng. **4**(6), 2917–2931 (2016). https://doi.org/10.1021/acssuschemeng.6b00553
111. Bara, J.E., Lessmann, S., Gabriel, C.J., Hatakeyama, E.S., Noble, R.D., Gin, D.L.: Synthesis and performance of polymerizable room-temperature ionic liquids as gas separation membranes. Ind. Eng. Chem. Res. **46**(16), 5397–5404 (2007). https://doi.org/10.1021/ie0704492

112. Wishart, J.F., Castner, E.W.: The physical chemistry of ionic liquids. J. Phys. Chem. B. **111**(18), 4639–4640 (2007). https://doi.org/10.1021/jp072262u
113. Ghandi, K.: A review of ionic liquids, their limits and applications. Green Sustain. Chem. **2014**(1), 44–53 (2014)
114. Marsh, K.N., Boxall, J.A., Lichtenthaler, R.: Room temperature ionic liquids and their mixtures—a review. Fluid Phase Equilib. **219**(1), 93–98 (2004)
115. Guarino, V., Antonio Alvarez-Perez, M., Borriello, A., Napolitano, T., Ambrosio, L., Guarino, V., et al.: Conductive PANi/PEGDA macroporous hydrogels for nerve regeneration. Adv. Healthc. Mater. **2**(1), 218–227 (2013). https://doi.org/10.1002/adhm.201200152
116. Heo, D.N., Lee, S.J., Timsina, R., Qiu, X., Castro, N.J., Zhang, L.G.: Development of 3D printable conductive hydrogel with crystallized PEDOT:PSS for neural tissue engineering. Mater. Sci. Eng. C. **99**, 582–590 (2019)
117. Bu, Y., Xu, H.X., Li, X., Xu, W.J., Yin, Y.X., Dai, H.L., et al.: A conductive sodium alginate and carboxymethyl chitosan hydrogel doped with polypyrrole for peripheral nerve regeneration. RSC Adv. **8**(20), 10806–10817 (2018)
118. Shin, J., Choi, E.J., Cho, J.H., Cho, A.N., Jin, Y., Yang, K., et al.: Three-dimensional Electroconductive hyaluronic acid hydrogels incorporated with carbon nanotubes and Polypyrrole by catechol-mediated dispersion enhance neurogenesis of human neural stem cells. Biomacromolecules. **18**(10), 3060–3072 (2017). https://doi.org/10.1021/acs.biomac.7b00568
119. Koppes, A.N., Keating, K.W., McGregor, A.L., Koppes, R.A., Kearns, K.R., Ziemba, A.M., et al.: Robust neurite extension following exogenous electrical stimulation within single walled carbon nanotube-composite hydrogels. Acta Biomater. **39**, 34–43 (2016)
120. Imaninezhad, M., Pemberton, K., Xu, F., Kalinowski, K., Bera, R., Zustiak, S.P.: Directed and enhanced neurite outgrowth following exogenous electrical stimulation on carbon nanotube-hydrogel composites. J. Neural Eng. **15**(5), 056034 (2018). https://doi.org/10.1088/1741-2552/aad65b
121. Liu, X., Miller, A.L., Park, S., Waletzki, B.E., Zhou, Z., Terzic, A., et al.: Functionalized carbon nanotube and graphene oxide embedded electrically conductive hydrogel synergistically stimulates nerve cell differentiation. ACS Appl. Mater. Interfaces. **9**(17), 14677–14690 (2017). https://doi.org/10.1021/acsami.7b02072
122. Zhao, Y., Wang, Y., Niu, C., Zhang, L., Li, G., Yang, Y.: Construction of polyacrylamide/graphene oxide/gelatin/sodium alginate composite hydrogel with bioactivity for promoting Schwann cells growth. J. Biomed. Mater. Res. A. **106**(7), 1951–1964 (2018). https://doi.org/10.1002/jbm.a.36393
123. Neuman, K.E., Kenny, A., Shi, L., Koppes, A.N., Koppes, R.A.: Complex material properties of gel-Amin: a transparent and ionically conductive hydrogel for neural tissue engineering. Cells Tissues Organs. **212**, 45–63 (2022)
124. Shin, S.R., Jung, S.M., Zalabany, M., Kim, K., Zorlutuna, P., Kim, S.B., et al.: Carbon-nanotube-embedded hydrogel sheets for engineering cardiac constructs and bioactuators. ACS Nano. **7**(3), 2369–2380 (2013). https://doi.org/10.1021/nn305559j
125. Shin, S.R., Aghaei-Ghareh-Bolagh, B., Dang, T.T., Topkaya, S.N., Gao, X., Yang, S.Y., et al.: Cell-laden microengineered and mechanically tunable hybrid hydrogels of gelatin and graphene oxide. Adv. Mater. **25**(44), 6385–6391 (2013). https://doi.org/10.1002/adma.201301082
126. Shin, S.R., Zihlmann, C., Akbari, M., Assawes, P., Cheung, L., Zhang, K., et al.: Reduced graphene oxide-GelMA hybrid hydrogels as scaffolds for cardiac tissue engineering. Small. **12**(27), 3677–3689 (2016). https://doi.org/10.1002/smll.201600178
127. Pok, S., Vitale, F., Eichmann, S.L., Benavides, O.M., Pasquali, M., Jacot, J.G.: Biocompatible carbon nanotube-chitosan scaffold matching the electrical conductivity of the heart. ACS Nano. **8**(10), 9822–9832 (2014). https://doi.org/10.1021/nn503693h

128. Yu, H., Zhao, H., Huang, C., Du, Y.: Mechanically and electrically enhanced CNT-collagen hydrogels as potential scaffolds for engineered cardiac constructs. ACS Biomater Sci. Eng. **3**(11), 3017–3021 (2017). https://doi.org/10.1021/acsbiomaterials.6b00620
129. Roshanbinfar, K., Hilborn, J., Varghese, O.P., Oommen, O.P.: Injectable and thermoresponsive pericardial matrix derived conductive scaffold for cardiac tissue engineering. RSC Adv. **7**(51), 31980–31988 (2017)
130. Roshanbinfar, K., Mohammadi, Z., Sheikh-Mahdi Mesgar, A., Dehghan, M.M., Oommen, O.P., Hilborn, J., et al.: Carbon nanotube doped pericardial matrix derived electroconductive biohybrid hydrogel for cardiac tissue engineering. Biomater. Sci. **7**(9), 3906–3917 (2019)
131. Izadifar, M., Chapman, D., Babyn, P., Chen, X., Kelly, M.E.: UV-assisted 3D bioprinting of Nanoreinforced hybrid cardiac patch for myocardial tissue engineering. Tissue Eng. Part C Methods. **24**(2), 74–88 (2018). https://doi.org/10.1089/ten.tec.2017.0346
132. Meng, X., Stout, D.A., Sun, L., Beingessner, R.L., Fenniri, H., Webster, T.J.: Novel injectable biomimetic hydrogels with carbon nanofibers and self assembled rosette nanotubes for myocardial applications. J. Biomed. Mater. Res. A. **101A**(4), 1095–1102 (2013). https://doi.org/10.1002/jbm.a.34400
133. You, J.O., Rafat, M., Ye, G.J.C., Auguste, D.T.: Nanoengineering the heart: conductive scaffolds enhance connexin 43 expression. Nano Lett. **11**(9), 3643–3648 (2011). https://doi.org/10.1021/nl201514a
134. Navaei, A., Saini, H., Christenson, W., Sullivan, R.T., Ros, R., Nikkhah, M.: Gold nanorod-incorporated gelatin-based conductive hydrogels for engineering cardiac tissue constructs. Acta Biomater. **41**, 133–146 (2016)
135. Mihic, A., Cui, Z., Wu, J., Vlacic, G., Miyagi, Y., Li, S.H., et al.: A conductive polymer hydrogel supports cell electrical signaling and improves cardiac function after implantation into myocardial infarct. Circulation. **132**(8), 772–784 (2015). https://doi.org/10.1161/CIRCULATIONAHA.114.014937
136. Cui, H., Liu, Y., Cheng, Y., Zhang, Z., Zhang, P., Chen, X., et al.: In vitro study of electroactive tetraaniline-containing thermosensitive hydrogels for cardiac tissue engineering. Biomacromolecules. **15**(4), 1115–1123 (2014). https://doi.org/10.1021/bm4018963
137. Wang, L., Jiang, J., Hua, W., Darabi, A., Song, X., Song, C., et al.: Mussel-inspired conductive Cryogel as cardiac tissue patch to repair myocardial infarction by migration of conductive nanoparticles. Adv. Funct. Mater. **26**(24), 4293–4305 (2016). https://doi.org/10.1002/adfm.201505372
138. Alves, T., Souza, J., Amaral, V., Almeida, D., Grotto, D., Lima, R., et al.: Biomimetic dense lamellar scaffold based on a colloidal complex of the polyaniline (PANi) and biopolymers for electroactive and physiomechanical stimulation of the myocardial. Colloids Surf. A Physicochem. Eng. Asp. **579**, 123650 (2019)
139. Dong, R., Zhao, X., Guo, B., Ma, P.X.: Self-healing conductive injectable hydrogels with antibacterial activity as cell delivery carrier for cardiac cell therapy. ACS Appl. Mater. Interfaces. **8**(27), 17138–17150 (2016). https://doi.org/10.1021/acsami.6b04911
140. Zhao, X., Wu, H., Guo, B., Dong, R., Qiu, Y., Ma, P.X.: Antibacterial anti-oxidant electroactive injectable hydrogel as self-healing wound dressing with hemostasis and adhesiveness for cutaneous wound healing. Biomaterials. **122**, 34–47 (2017)
141. Li, S., Wang, L., Zheng, W., Yang, G., Jiang, X., Li, S., et al.: Rapid fabrication of self-healing, conductive, and injectable gel as dressings for healing wounds in stretchable parts of the body. Adv. Funct. Mater. **30**(31), 2002370 (2020). https://doi.org/10.1002/adfm.202002370
142. Gan, D., Han, L., Wang, M., Xing, W., Xu, T., Zhang, H., et al.: Conductive and tough hydrogels based on biopolymer molecular templates for controlling in situ formation of polypyrrole nanorods. ACS Appl. Mater. Interfaces. **10**(42), 36218–36228 (2018). https://doi.org/10.1021/acsami.8b10280
143. Lu, Y., Wang, Y., Zhang, J., Hu, X., Yang, Z., Guo, Y., et al.: In-situ doping of a conductive hydrogel with low protein absorption and bacterial adhesion for electrical stimulation of chronic wounds. Acta Biomater. **89**, 217–226 (2019)

144. di Luca, M., Vittorio, O., Cirillo, G., Curcio, M., Czuban, M., Voli, F., et al.: Electroresponsive graphene oxide hydrogels for skin bandages: the outcome of gelatin and trypsin immobilization. Int. J. Pharm. **546**(1–2), 50–60 (2018)
145. Wang, Y., Lu, Y., Zhang, J., Hu, X., Yang, Z., Guo, Y., et al.: A synergistic antibacterial effect between terbium ions and reduced graphene oxide in a poly(vinyl alcohol)–alginate hydrogel for treating infected chronic wounds. J. Mater. Chem. B. **7**(4), 538–547 (2019)
146. Liang, Y., Zhao, X., Hu, T., Han, Y., Guo, B.: Mussel-inspired, antibacterial, conductive, antioxidant, injectable composite hydrogel wound dressing to promote the regeneration of infected skin. J. Colloid Interface Sci. **556**, 514–528 (2019)
147. Zhao, Y., Li, Z., Li, Q., Yang, L., Liu, H., Yan, R., et al.: Transparent conductive supramolecular hydrogels with stimuli-responsive properties for on-demand dissolvable diabetic foot wound dressings. Macromol. Rapid Commun. **41**(24), 2000441 (2020). https://doi.org/10.1002/marc.202000441
148. Jeong, S.H., Lee, Y., Lee, M.G., Song, W.J., Park, J.U., Sun, J.Y.: Accelerated wound healing with an ionic patch assisted by a triboelectric nanogenerator. Nano Energy. **79**, 105463 (2021)
149. Khorshidi, S., Karkhaneh, A.: Hydrogel/fiber conductive scaffold for bone tissue engineering. J. Biomed. Mater. Res. A. **106**(3), 718–724 (2018). https://doi.org/10.1002/jbm.a.36282
150. Lu, J., Cheng, C., He, Y.S., Lyu, C., Wang, Y., Yu, J., et al.: Multilayered graphene hydrogel membranes for guided bone regeneration. Adv. Mater. **28**(21), 4025–4031 (2016). https://doi.org/10.1002/adma.201505375
151. Heo, D.N., Ko, W.K., Bae, M.S., Lee, J.B., Lee, D.W., Byun, W., et al.: Enhanced bone regeneration with a gold nanoparticle–hydrogel complex. J. Mater. Chem. B. **2**(11), 1584–1593 (2014)
152. Ribeiro, M., Ferraz, M.P., Monteiro, F.J., Fernandes, M.H., Beppu, M.M., Mantione, D., et al.: Antibacterial silk fibroin/nanohydroxyapatite hydrogels with silver and gold nanoparticles for bone regeneration. Nanomedicine. **13**(1), 231–239 (2017)
153. Sophia Fox, A.J., Bedi, A., Rodeo, S.A.: The basic science of articular cartilage: structure, composition, and function. Sports Health. **1**(6), 461–468 (2009). https://doi.org/10.1177/1941738109350438
154. Bhosale, A.M., Richardson, J.B.: Articular cartilage: structure, injuries and review of management. Br. Med. Bull. **87**(1), 77–95 (2008)
155. Farooqi, A.R., Bader, R., Van Rienen, U.: Numerical study on Electromechanics in cartilage tissue with respect to its electrical properties. Tissue Eng. Pt. B Rev. **25**(2), 152–166 (2019). https://doi.org/10.1089/ten.teb.2018.0214
156. Brady, M.A., Waldman, S.D., Ethier, C.R.: The application of multiple biophysical cues to engineer functional neocartilage for treatment of osteoarthritis. Part I: cellular response. Tissue Eng. Part B Rev. **21**(1), 1–19 (2015). https://doi.org/10.1089/ten.teb.2013.0757
157. Moreira-Teixeira, L.S., Georgi, N., Leijten, J., Wu, L., Karperien, M.: Cartilage tissue engineering. Endocr. Dev. **21**, 102–115 (2011)
158. Kwon, H.J., Yasuda, K., Gong, J.P., Ohmiya, Y.: Polyelectrolyte hydrogels for replacement and regeneration of biological tissues. Macromol. Res. **22**(3), 227–235 (2014). https://doi.org/10.1007/s13233-014-2045-6
159. Zuzzi, D.C., de Ciccone, C., Neves, L.M.G., Mendonça, J.S., Joazeiro, P.P., Esquisatto, M.A.M.: Evaluation of the effects of electrical stimulation on cartilage repair in adult male rats. Tissue Cell. **45**(4), 275–281 (2013)
160. Prasopthum, A., Deng, Z., Khan, I.M., Yin, Z., Guo, B., Yang, J.: Three dimensional printed degradable and conductive polymer scaffolds promote chondrogenic differentiation of chondroprogenitor cells. Biomater. Sci. **8**(15), 4287–4298 (2020)
161. Shen, H., Lin, H., Sun, A.X., Song, S., Zhang, Z., Dai, J., et al.: Chondroinductive factor-free chondrogenic differentiation of human mesenchymal stem cells in graphene oxide-incorporated hydrogels. J. Mater. Chem. B. **6**(6), 908–917 (2018)
162. Zhang, S., Li, Y., Zhang, H., Wang, G., Wei, H., Zhang, X., et al.: Bioinspired conductive hydrogel with ultrahigh toughness and stable antiswelling properties for articular cartilage replacement. ACS Mater. Lett. **3**, 807–814 (2021). https://doi.org/10.1021/acsmaterialslett.1c00203

163. Kashi, M., Baghbani, F., Moztarzadeh, F., Mobasheri, H., Kowsari, E.: Green synthesis of degradable conductive thermosensitive oligopyrrole/chitosan hydrogel intended for cartilage tissue engineering. Int. J. Biol. Macromol. **107**, 1567–1575 (2018)
164. Li, C.: Towards conductive hydrogels in e-skins: a review on rational design and recent developments. RSC Adv. **11**(54), 33835–33848 (2021)
165. Li, Z., Xu, W., Wang, X., Jiang, W., Ma, X., Wang, F., et al.: Fabrication of PVA/PAAm IPN hydrogel with high adhesion and enhanced mechanical properties for body sensors and antibacterial activity. Eur. Polym. J. **146**, 110253 (2021)
166. Yan, L., Zhou, T., Han, L., Zhu, M., Cheng, Z., Li, D., et al.: Conductive cellulose bionanosheets assembled biostable hydrogel for reliable bioelectronics. Adv. Funct. Mater. **31**(17), 2010465 (2021). https://doi.org/10.1002/adfm.202010465
167. Zhao, Y., Li, Z., Song, S., Yang, K., Liu, H., Yang, Z., et al.: Skin-inspired antibacterial conductive hydrogels for epidermal sensors and diabetic foot wound dressings. Adv. Funct. Mater. **29**(31), 1901474 (2019). https://doi.org/10.1002/adfm.201901474
168. Zhang, J., Wu, C., Xu, Y., Chen, J., Ning, N., Yang, Z., et al.: Highly stretchable and conductive self-healing hydrogels for temperature and strain sensing and chronic wound treatment. ACS Appl. Mater. Interfaces. **12**(37), 40990–40999 (2020). https://doi.org/10.1021/acsami.0c08291
169. Wei, X., Ma, K., Cheng, Y., Sun, L., Chen, D., Zhao, X., et al.: Adhesive, conductive, self-healing, and antibacterial hydrogel based on chitosan-polyoxometalate complexes for wearable strain sensor. ACS Appl. Polym. Mater. **2**(7), 2541–2549 (2020). https://doi.org/10.1021/acsapm.0c00150
170. Jafarkhani, M., Salehi, Z., Nematian, T.: Preparation and characterization of chitosan/graphene oxide composite hydrogels for nerve tissue engineering. Mater. Today Proc. **5**(7), 15620–15628 (2018)

Chapter 11
Hydrogels as Delivery Systems

Senem Büyüksungur

Abstract The therapeutic action of medications occurs when they bind to specific receptors in the tissues that they are intended to treat. The treatment's efficacy, however, declines if the medication does not reach to the intended location at the desired rate and required concentration. Bioavailability, the rate and extent to which the active substance reaches its site of action in the body, is essential for the success of the treatment. To achieve therapeutic efficacy, the concentration of the drug in the plasma must remain within the therapeutic dose range, i.e., between the "minimum effective concentration" and the "toxic level." In traditional drug applications, high doses or repeated doses are often administered at certain intervals to achieve the desired therapeutic effect. However, these methods, whether through injection or oral administration, can lead to patient discomfort. Furthermore, high doses can result in side effects or toxicity. The short half-life of certain bioactive substances, like peptides and proteins, in the bloodstream, along with their degradation before reaching the target tissue, further reduces therapeutic effectiveness. To address these drawbacks, controlled drug release systems are developed using membranes, nanoparticles, liposomes, and hydrogels.

Keywords Drug delivery · In situ gelling · Shear-thinning · Degradation · Release kinetics · Bioadhesion · Shape-memory hydrogels

11.1 Hydrogels in Drug Delivery Systems

Hydrogels are cross-linked, three-dimensional network structures. Hydrogels, with their network structure and high water retention capacity (70–99%), closely resemble biological tissues and are biocompatible. Hydrogels were first described as drug

S. Büyüksungur (✉)
BIOMATEN Center of Excellence in Biomaterials and Tissue Engineering, Middle East Technical University, Ankara, Türkiye

N. Hasirci, V. Hasirci (eds.), *Hydrogels and Bioinks in Tissue Engineering*, Biological and Medical Physics, Biomedical Engineering,
https://doi.org/10.1007/978-3-031-90966-5_11

delivery systems in 1960 [1]. There are four important factors that make hydrogels effective as drug carriers:

1. Hydrophilic drugs can be easily encapsulated within hydrogels.
2. The risk of drug denaturation and aggregation is minimal because hydrogels are formed in an aqueous solution and are not exposed to organic solvents.
3. The mechanical strength of hydrogels can be adjusted between 0.5 kPa and 5 MPa, which allows them to physically resemble soft tissues in the human body.
4. The cross-linked network structure prevents proteins/enzymes in the body from diffusing into the hydrogel, thus preserving the bioactive agent (active substance or drug) and preventing early degradation by enzymes.

When developing hydrogels as controlled drug delivery systems, it is crucial to ensure that the bioactivity of the drug is preserved during production, packaging, and transportation and that both the drug and the hydrogel remain physically and chemically stable. In addition, application-specific requirements vary depending on whether the application is short-term or long-term, the drug release profile (continuous or as-needed), and the physical and mechanical properties of the targeted tissue. Two important factors that determine the route and form of application in controlled drug delivery systems are the size of the hydrogel and bioadhesion, which refers to the ability of the hydrogel to adhere to biological tissues.

11.1.1 Importance of Hydrogel Size in Drug Delivery Systems

In hydrogels, significant interactions occur across multiple length scales, ranging from centimeters to nanometers (Fig. 11.1). These interactions affect the method of application and drug release. At the macro level, the design of the hydrogel determines how it will be applied to the body, while at the micron level, the presence of pores plays a crucial role in allowing water molecules to enter into the hydrogel and in facilitating drug transport (convective transport). As the scale shifts from microns to nanometers, the pore size in the network structure becomes more critical for drug diffusion. Another interaction at the molecular and atomic scale (nano and sub-nano levels) is the various chemical interactions between polymer chains and drug molecules. These chemical interactions can influence and control the drug release rate. The ability to modify the properties of the hydrogel at different scales, starting from the micron level, provides the opportunity to design modular systems tailored to the specific needs of controlled drug delivery.

Macro, micro, and nanogels are the three primary size-based categories into which hydrogel delivery systems belong. At the macro level, hydrogels can either have porous or nonporous structures. As the scale decreases to the micron level, the space between the polymer chains is referred to as the mesh structure, and mesh size can range from 5 to 100 nm. The nanoscale refers to molecular or atomic sizes, representing the physical or chemical interaction between drug molecules and polymer chains. Macroscopic hydrogels can be applied transepithelial or as patches

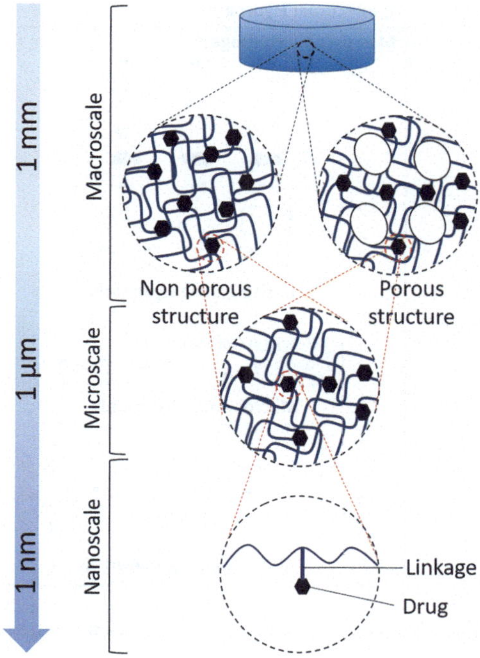

Fig. 11.1 Hydrogel interactions that change depending on their sizes

placed inside the body. On the other hand, microgels are used in oral and pulmonary drug delivery, while nanogels are employed for the systemic administration of drugs.

11.1.1.1 Macroscopic Hydrogels

Macroscopic hydrogels range in size from millimeters to centimeters. This category typically includes hydrogels that are surgically implanted into the body via surgery or used in direct contact with the body for transepithelial drug delivery (passing through or across epithelial cells). There are several successful clinical applications of implanted hydrogels used for drug delivery. For instance, the type 1 collagen gel in the INFUSE® bone graft, which is used to treat spinal disorders and long bone fractures, releases recombinant human bone morphogenetic protein-2 (BMP2) [2]. While epithelial barriers in the skin, small intestine, and mucosa allow drug permeability, they are impermeable to hydrogels, which makes them suitable for transdermal applications. Transdermal hydrogels are commonly used as wound dressings (to prevent wound infection) or for the delivery of proteins like insulin (to regulate blood sugar levels) or calcitonin (to regulate blood calcium levels) [3]. Examples of commercial antimicrobial wound dressings used in clinical practice include ALGICELL Ag, Prontosan, and Calamine-zinc gelatin, which are based on alginate, hydroxymethyl cellulose, and gelatin, respectively [4]. On the other hand, if the target tissue for drug delivery is located in deeper regions of the body or if there

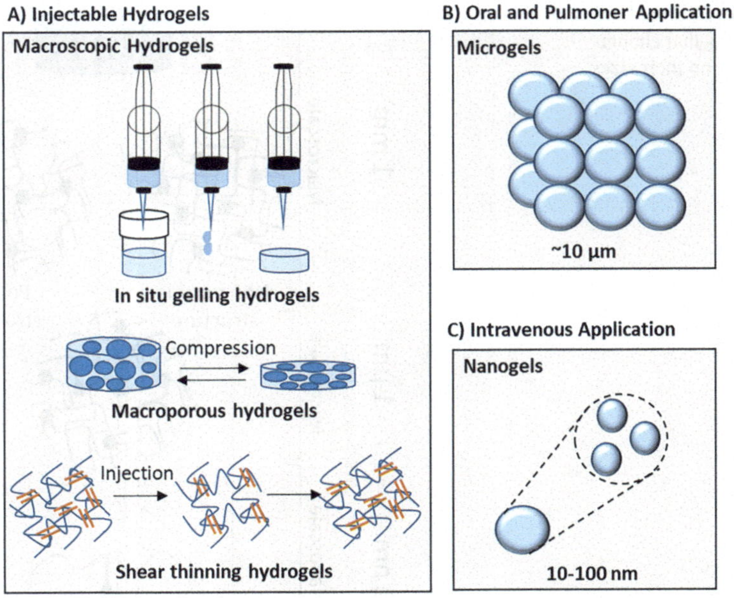

Fig. 11.2 Different applications of drug-carrying hydrogels

is a biological barrier with low permeability to the drug, the hydrogels must be implanted into the body. This allows the epithelial barriers to be bypassed, enabling drug release directly at the target site. However, implantation carries risks such as being an invasive procedure, reduced patient comfort, and the necessity of surgery. To overcome these disadvantages, injectable macroscopic hydrogels present a secondary and effective solution.

Injectable hydrogels are classified into three groups based on their gelation principles: "in situ-gelling hydrogels," "shear-thinning hydrogels," and "macroporous shape-memory hydrogels" (Fig. 11.2).

11.1.1.1.1 In Situ Gelling Hydrogels

In situ gelling hydrogels are injected into the body in liquid form and undergo a sol–gel transition to form a gel inside the body. The transition of these hydrogels from liquid to gel can be achieved using different strategies. One of these strategies involves the use of "slow-gelling systems." In these systems, gelation begins before injection, but the hardening process progresses slowly, which is enough to allow the injection to be completed before gelation is finalized. The kinetics of gelation in such systems must be slow enough to prevent needle blockage, but fast enough to prevent the pre-gel solution from being diluted by body fluids before gelation is complete. In situ hydrogels that gel slowly can undergo gelation through different mechanisms such as charge interaction, complexation, or Michael addition.

Alginate-based injectable hydrogels developed by Kalaf and colleagues for intervertebral disc tissue engineering can be given as an example of slow-gelling systems [5]. In this system, calcium carbonate is used as a calcium source to cross-link the alginate molecules, while glucono-δ-lactone (GDL) is used to slow down the gelation process.

Apart from slow-gelling systems, another gelation mechanism is temperature-induced gelation at body temperature. Temperature-sensitive hydrogels are produced using synthetic polymers. The most commonly used temperature-sensitive polymers are poly(N-isopropylacrylamide) (PNIPAm) and poly(ethylene oxide)-poly(propylene oxide)-poly(ethylene oxide) (PEO–PPO–PEO) [6, 7]. These types of hydrogels are typically classified as smart hydrogels.

11.1.1.1.2 Shear-Thinning Hydrogels

Shear-thinning hydrogels behave like a low-viscosity liquid during injection due to the applied shear force but become a gel once the shear force is removed inside the body. The shear-thinning property of these hydrogels is achieved through reversible physical cross-linking. These types of hydrogels gain their reversible nature through competing forces such as physical cross-linking, hydrophobic interactions, electrostatic interactions, or hydrogen bonding, which hold the gel together, and dissolving or electrostatic repulsion forces that pull it apart. In the fabrication of shear-thinning hydrogels, self-assembling peptides are commonly used because of the variety of amino acids and the simplicity of modifying peptides according to specific sequences [8]. Studies have also shown that alginate-based hydrogels exhibit shear-thinning properties. These alginate hydrogels can form a gel inside a syringe when combined with multivalent ions like zinc (Zn^{2+}) or calcium (Ca^{2+}) and can then be injected into the target area [9]. Another approach to the production of shear-thinning hydrogels involves the use of dynamic covalent bonds to obtain the gel. Common bonds used for this purpose include imine, hydrazone, and oxime bonds [10, 11].

11.1.1.1.3 Macroporous Shape-Memory Hydrogels

In the production of injectable hydrogels, another approach is the development of large hydrogels with interconnected porous structures that can shrink under mechanical stress and return to their original shape when the stress is removed. These hydrogels can be produced using methods such as porogen leaching, lyophilization, gas foaming, cryogelation, and microemulsion formation. In such systems, during injection, the liquid is forced out through the pores of the hydrogel, allowing the hydrogel to compress enough to pass through the needle. When the hydrogel gets out from the needle and the mechanical pressure from the needle walls is removed, the hydrogel expands back to its original shape. These types of hydrogels can undergo up to 90% volumetric changes without damaging their three-dimensional network structure. Compared to slow-gelling systems and shear-thinning hydrogels, macroporous hydrogels offer researchers the ability to design hydrogels with predetermined geometries and volumes for controlled drug delivery. The pores in their structure also allow for cell infiltration. However, the interconnected porous structure can result in faster drug release due to the shortened diffusion path and a reduced polymer volume fraction, which in turn leads to a lower drug loading capacity. Wang and colleagues developed a macroporous hydrogel composed of gelatin, oxidized sodium alginate, and adipic acid dihydrazide using a high-speed shearing method. They successfully created a macroporous structure for drug delivery without the use of porogens or molds [12].

11.1.1.2 Microgels and Nanogels

Microgels and nanogels have certain advantages compared to macrogels. The small size of micro and nanogels, which is less than the diameter of the inner part of many needles (~1 mm), allows them to be easily injectable. Additionally, their small size offers a substantial surface area for bioconjugation and allows for easy passage through tissue barriers. Microgels smaller than 5 μm are suitable for oral and pulmonary drug delivery but due to their rapid clearance from circulation, they are not suitable for intravenous use.

Nanogels, ranging in size from 10 to 100 nm, can pass through endothelial gaps in small blood vessels, making them suitable for systemic applications. Hydrogels smaller than 10 nm can be filtered out by the kidneys, while hydrogels sized between 0.5 and 10 nm can be taken up by macrophages through phagocytosis. Nanogels are particularly useful in gene therapy for delivering nucleotide-based drugs, such as plasmid DNA [13]. Gene therapy is a treatment approach used for conditions like cancer, hemophilia, and viral infections. The release of DNA from nanogels not only facilitates cellular uptake but also prolongs its circulation time. The permeable vascular structure in the tumor region allows the nanogel to accumulate in that area and the drug to be effective there. Cationic nanogels containing polyethylene oxide (PEO) and polyethyleneimine (PEI) have been reported to transport oligonucleotides across gastrointestinal epithelial tissues and even the blood–brain barrier, reaching the target tissue [14, 15]. Additionally, nanogels can be modified to create protein–polymer conjugates, which can extend the plasma half-life of proteins and enhance their stability [16]. The application areas and methods of hydrogels vary according to their sizes. Hydrogel size (micro or nano) can be controlled by varying production parameters like needle diameter and flow rate or by modifying gelation conditions like polymer and surfactant concentrations. Microfluidic systems and micromolding techniques are used to create microgels, whereas emulsion and nanomolding techniques are used to create nanogels [17, 18].

Size distribution, deformability, shape, and surface chemistry are all crucial design considerations for controlled drug delivery systems, in addition to the hydrogels' dimensions. It has been shown that nanogels with a positive zeta potential or a high aspect ratio are taken up by cells more quickly, especially in cellular uptake [19]. Moreover, experiments conducted on mice have demonstrated that the size and deformability of microgels determine their biodistribution and circulation time in the body [20].

11.1.2 Bioadhesion

Bioadhesion property is another important characteristic that influences the method and application of hydrogels. The ability of the hydrogel to adhere to the tissue where it is applied is crucial for ensuring long-term and effective drug release at the targeted site, thereby achieving the desired therapeutic effect. This is especially important in nasal, oral, and intestinal applications, where the epithelial and mucosal structures are wet and slippery, making it difficult for the hydrogel to adhere. An approach to developing bioadhesive hydrogels for drug delivery in these tissues is to use polymers like chitosan or poly(acrylic acid), which can form hydrogen bonds with the mucosal structure [21, 22]. In these systems, the positive-charged polymers also interact electrostatically with negatively charged cell membranes and tissues, ensuring a long-term effect. Another approach to producing bioadhesive hydrogels involves the use of the L-3,4-dihydroxyphenylalanine (DOPA) molecule, derived from aquatic organisms like mussels, which is known for its strong adhesive properties [23].

11.2 Drug Release Mechanisms

Drug release from hydrogels used as controlled drug delivery systems can be regulated through various mechanisms, including diffusion, degradation, swelling, mechanical deformation, drug–polymer interactions, or environmental stimuli (smart systems).

11.2.1 Controlled Drug Release Via Diffusion

In diffusion-controlled drug release, reservoir or matrix systems are used to allow the drug to diffuse from the hydrogel's network structure or pores (Fig. 11.3). The drug is contained in a central core of the reservoir system, which is covered by the hydrogel membrane. These systems can be produced in capsule, sphere, or plate forms, and the drug located at the center of the system is released at a constant rate. In the matrix system, drug release occurs through the macromolecular pores or network structures within the hydrogel. While drug release in the reservoir system occurs at a constant rate, independent of time, in the matrix system, the release is time-dependent, with the initial release rate being directly proportional to the square

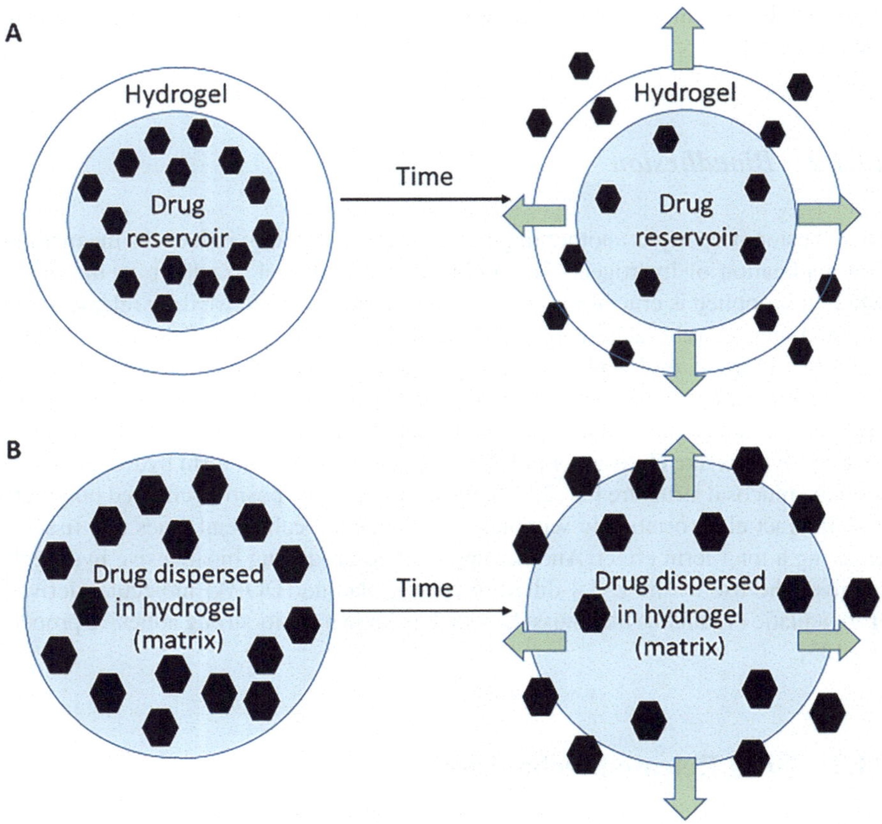

Fig. 11.3 Diffusion controlled drug release systems. (**a**) The hydrogel membrane in the reservoir system is positioned on a drug-containing structure. These systems allow the drug to be released at a constant rate. (**b**) The drug is released in the matrix system via network structures or macromolecular pores in the hydrogel

root of time. In diffusion-based drug release, the pore size in the network is the most important factor affecting the release.

The typical pore size in the network refers to the space between polymer chains and generally ranges from 5 to 100 nm [24]. The pore size in hydrogels can be directly determined using techniques such as electron microscopy, atomic force microscopy, confocal microscopy, small-angle X-ray scattering, or small-angle neutron scattering. Additionally, the pore size in the network can be indirectly determined through mechanical tests such as rheology, indentation, compression, and tensile testing. The most important factors affecting the pore size in the network are polymer concentration, cross-linker concentration, and especially stimuli such as temperature or pH in smart systems. The heterogeneity of the network structure and the polymer chains (polydispersity) can cause the pore size to vary across a wide range. A homogeneous distribution can only be achieved by using symmetrical tetrahedral macromers (such as four-armed PEG) of the same size. The pore size in the network controls the drug's diffusion through the hydrogel by influencing steric interactions between the drug and the polymer network. When the pore size of the network is larger than the size of the drug molecule ($r_{pore}/r_{drug} > 1$), the release process progresses based on diffusion. Diffusion takes place regardless of the size of the pores, and smaller drug molecules are able to move throughout the network.

In such cases, the diffusion coefficient (D) of the drug can be determined according to the Stokes–Einstein equation, depending on the size of the drug molecule (r_{drug}), the viscosity of the solvent (η), the temperature of the environment (T), and the gas constant (R):

$$D = \frac{RT}{6\pi\eta r_{drug}}$$

The size of the drug molecule (r_{drug}) is generally scaled with its molecular weight. Macromolecular drugs (as hormones or proteins) are larger, which results in slower release rates. The length of the diffusion path, which is defined by the hydrogel's thickness, determines the drug's diffusion time in the hydrogel (t_{diff}). The drug's diffusion time can be estimated by H^2/D.

Steric hindrance has a stronger effect on diffusion when the network's pore size is near that of the drug molecule ($r_{pore}/r_{drug} \sim 1$). Factors that reduce the pore size include increasing the concentration of polymers or cross-linkers. In such cases, the polymer chains exert significant drag force on the drug molecule, increasing the length of the diffusion path and resulting in slower and prolonged release.

In cases where the pore size is much smaller than the drug molecule ($r_{pore}/r_{drug} < 1$), strong steric hindrance immobilizes the drug. The drug becomes physically trapped within the network and can only be released if the network degrades, swells, or undergoes mechanical deformation.

11.2.2 Controlled Drug Release Via Degradation

Controlling the degradation of the network structure in hydrogels is an approach used to regulate drug release. As the polymer chains break down and the network degrades, the size of the pores within the network increases, which facilitates drug release. Degradation can occur via hydrolysis or enzymatic reactions, either in the main polymer chain or at the cross-linking sites. The reduction in polymer mass resulting from degradation is called erosion, which can be classified into two types: surface erosion or bulk erosion, depending on where it occurs. If the diffusion rate of water or enzymes entering the hydrogel is higher than the bond degradation rate, the hydrogel degrades via bulk erosion.

Hydrogels composed of oxidizable polysaccharides, like chitosan and alginate, break down by bulk erosion and the rate of oxidation regulates the rate of degradation. In contrast, surface erosion occurs when the bond-breaking rate exceeds the diffusion rate of water or enzymes. Hydrogels that use hydrophobic interactions for cross-linking inhibit water entry, resulting in surface erosion. In this case, weight loss progresses linearly over time, and the drug release follows zero-order kinetics, meaning a constant drug release rate over time. The hydrogel's erosion mechanism and degradation reaction can be changed to produce the required release kinetics. In this way, a long-term release lasting for months or even years can be achieved. A very important factor to consider here is that the degradation products must either be nontoxic or small enough to be excreted from the body.

11.2.3 Controlled Drug Release Via Swelling

In swelling-controlled drug release, the drug is dispersed within a solid polymer, and when the dry hydrogel comes into contact with body fluids, it begins to swell. As the polymer chains relax during the swelling process, the network size increases, allowing drug release to occur. The degree of swelling of the hydrogel is determined by the balance between the interaction forces that prevent the deformation of the network structure and the osmosis that allows water absorption. Drug release that occurs in this manner is independent of time and takes place at a constant rate. The drug diffuses from the region of higher concentration to the region of lower concentration because of the concentration gradient between the drug dispersed in the hydrogel and the surrounding environment. Therefore, both diffusion and swelling are involved in such systems. The swelling of the hydrogel can be influenced by various external factors such as temperature, pH, light, ionic strength, or an electric field. In macroscopic hydrogels controlled by swelling, water diffusion is slow, which also slows down the release rate. The diffusion path can be shortened by decreasing the hydrogel's thickness, or macropores may be created inside the hydrogel to accomplish faster release.

11.2.4 Controlled Drug Release Via Mechanical Deformation

Another approach to ensure drug release from hydrogels involves mechanically deforming the network structure to increase or decrease its size, allowing drug release through convective flow. In this method, ultrasound and magnetic fields are typically used to mechanically deform the network structure, resulting in pulsatile release. The disadvantage of this approach is that cumulative deformation can mechanically weaken the hydrogel. However, this issue can be addressed by using self-healing hydrogels. Alginate hydrogels, for instance, can undergo reversible cross-linking with divalent cations and, following ultrasound application, revert to their initial physiological state.

11.2.5 Controlled Drug Release Via Drug Polymer Interaction

Another method for controlling drug release from hydrogels involves chemically bonding the drug molecules to the chains of polymers. The release decreases with increasing drug–polymer chain interaction in strength. This method is particularly used when the drug size is much smaller than the adjustable network size; otherwise, the drug may exit the network structure in a very short time. High-sensitivity interactions are achieved through chemical or physical means, using covalent bonding, electrostatic interactions, and hydrophobic interactions.

Covalent bonding immobilizes the drug on the polymer chains, and this bonding can be either stable or degradable. When a drug molecule is bound using highly stable covalent bonds, its release can only occur once the network structure degrades. In contrast, if the drug is attached via degradable bonds, release occurs over time or in response to a specific stimulus that causes the bonds to break. Examples of highly stable covalent bonds include those formed via amide bonds [25], thiolene bonds [26], and metal-free click reactions [27]. Cleavable bonds can be macromolecular, like peptide sequences, or small, like ester bonds [28], disulfide bonds [29], and β-elimination bonds [30].

Electrostatic interactions create a nonspecific but strong attraction between the drug and the polymer, with drug release occurring as the hydrogel degrades. This approach is used when both the drug and the polymer carry charges [31]. An additional agent can be added to the system to accomplish the desired interaction if the drug and polymer's electrostatic interaction is not strong enough to regulate release. In hydrogel drug delivery systems designed for the release of positively charged agents, alginate is frequently used due to its negative charge. However, as mentioned earlier, stronger electrostatic interactions can be achieved by functionalizing alginate with sulfonic acid groups, thereby extending the release time of protein-based drugs [32].

Hydrophobic interactions are used for the release of hydrophobic drugs that cannot be encapsulated within hydrogels due to their hydrophilic nature. In this approach, hydrophobic binding sites are created for hydrophobic drugs within hydrogels that include hydrophobic polymers or components. One method to create hydrophobic regions within a hydrogel is through the copolymerization of the hydrogel with hydrophobic monomers such as cyclodextrin or cholesterol [33, 34]. However, creating hydrophobic regions within the hydrogel can reduce the amount of water it can absorb, potentially altering the biochemical or physical properties of the hydrogel. In this regard, the external hydrophilic and internal hydrophobic nature of macromolecular oligosaccharides, like cyclodextrin, allows for the preparation of a drug release system without compromising the hydrogel structure. The separation of the soluble drug between the hydrogel and the release medium, regardless of the hydrogel's network size, is the main factor governing drug release in such systems.

While there are numerous suitable regions on polymer chains to create interactions between the hydrogel and drug molecules, it is important to note that controlling drug release via chemical interactions may affect the chemical modification of the drug or the hydrogel's biocompatibility. Modifying the active site of a drug, or in the case of protein-based drugs, altering the protein's conformation, could reduce the activity of the active ingredient. For instance, lysine amino acids can be used to modify a protein-based bioactive agent with polyethylene glycol, which can result in the loss of bioactivity [35].

11.3 Release Kinetics in Hydrogel-Based Systems

It is challenging to predict drug release profiles theoretically because of the numerous chemical, physical, and biological interactions that occur during drug release from hydrogels. In studies conducted on this topic, systems have been developed through trial and error, and basic information about drug release mechanisms has been obtained based on the experimental data collected. In release kinetics, experimental data are generally represented as the amount of drug released over time. Most commonly, the drug release profile is defined by the experimental Ritger–Peppas equation [36, 37]:

$$\frac{M_t}{M_\infty} = kt^n$$

In the equation, M_t represents the amount of drug released at a given time t, M_∞ represents the initial amount of drug, k is a kinetic constant, and n is the diffusion exponent. The release mechanism, the hydrogel's geometry, and the polymer's polydispersity all affect the diffusion exponent (n). If the drug is released by Fickian diffusion, it is expressed as $n = 0.5$, while if surface erosion is effective in the release, it is expressed as $n = 1.0$. If a release in a system is controlled by multiple mechanisms, n can take values ranging from 0.5 to 1.0.

11.4 Transition of Hydrogels to Clinical Use in Drug Delivery

Although many hydrogel systems have been established in the field of drug delivery, the number of those that have progressed to clinical application is quite limited (Table 11.1). Researchers must consider factors such as production, storage, regulations, and cost when moving from laboratory to clinic. A significant challenge in hydrogels is their high water content, which complicates sterilization of the final product. Thus, sterility must be validated across all material sources and production processes. Additionally, to prevent degradation of the hydrogel or the hydrolysis of bonds between the drug and the hydrogel during storage, the hydrogel may require to be dried before storage. However, this drying process should not change the

Table 11.1 Hydrogel drug delivery systems used in the clinic

Product	Hydrogel	Drug	Therapeutic application
INFUSE	Collagen	Recombinant human bone morphogenetic protein 2 (BMP2)	Bone fractures, oral and facial reconstruction, spinal fusion
MASTERGRAFT	Calcium phosphate and collagen	BMP2	Spinal fusion
OP 1, OP 1 Putty	Collagen	BMP2	Long bone fractures, spinal fusion
VANTAS	Poly(2-hydroxyethyl methacrylate), poly(2-hydroxypropyl methacrylate)	Histrelin acetate	Subcutaneous implant for prostate cancer treatment
SUPPRELIN LA	Poly(2-hydroxyethyl methacrylate)	Histrelin acetate	Subcutaneous implant for treating central precocious puberty in children
AzaSite	Poly(acrylic acid)	Azithromycin	Bacterial conjunctivitis
Besivance	Poly(acrylic acid)	Besifloxacin	Bacterial conjunctivitis
Cervidil	PEG or urethane polymer	Dinoprostone (prostaglandin E2)	Vaginal suppository for cervical dilation during childbirth
Differin	Carbomer 940	Adapalene	Topical treatment for acne vulgaris
AndroGel	Carbomer 940	Testosterone	Topical gel for hypogonadism treatment
Calamine-zinc gelatin	Gelatin	Calamine zinc oxide	Wound dressing to reduce pain and itching
ALGICELL Ag, Suprasorb A + Ag	Alginate	Silver	Antimicrobial wound dressing containing silver
Prontosan	Hydroxyethyl cellulose	Polyhexanide	Antiseptic wound dressing
ReGel	Paclitaxel	PLGA-PEG	Cancer treatment

structure of the hydrogel or the bioactivity of the drug. On the other hand, if the hydrogel is stored in a liquid environment, it should be kept in a sealed storage container to minimize water evaporation and drug loss, and transported through a temperature-controlled supply chain. Storage conditions such as temperature, humidity, and time must be carefully considered. Regulatory and cost-related issues are also significant challenges in developing hydrogel release systems. In regulations, all systems that release drugs or cells, including hydrogels, are classified as combination products, and obtaining regulatory approval can take 7–10 years. Considering the long approval times and the limited protection periods of patents, these types of systems face difficulties in transitioning to commercial clinical use. Additionally, the cost of developing a hydrogel release system is quite high, making the commercialization of the produced system difficult. Among the commercial products listed in Table 11.1, the INFUSE product, which is a collagen gel releasing BMP2 for bone regeneration, has a commercial revenue of approximately 750 million dollars, which is quite high. Other systems that have entered clinical application hold promise for the coming years.

11.5 Conclusion

Studies on hydrogel-based drug delivery systems are gaining importance with each passing year. The development of numerous hydrogel-based drug delivery systems currently in clinical use clearly demonstrates the significance of hydrogels in this field. The method of application and the ability to control drug release can be tailored through strategies such as adjusting the hydrogel size, porosity, and interaction with the drug. By using these strategies together in different combinations, it is possible to create various designs with different release kinetics for different applications. Considering factors such as maintaining drug efficacy, storage conditions, and cost, it is expected that the number of injectable macro, micro, and nano hydrogel drug delivery systems used in clinical practice will continue to grow in the coming years.

References

1. Wichterle, O., Lim, D.: Hydrophilic gels for biological use. Nature. **185**(4706), 117–118 (1960)
2. Olthof, M.G.L., Kempen, D.H.R., Liu, X., Dadsetan, M., Tryfonidou, M.A., Yaszemski, M.J., Dhert, W.J.A., Lu, L.: Bone morphogenetic protein-2 release profile modulates bone formation in phosphorylated hydrogel. J. Tissue Eng. Regen. Med. **12**(6), 1339–1351 (2018)
3. Ahsan, A., Tian, W.X., Farooq, M.A., Khan, D.H.: An overview of hydrogels and their role in transdermal drug delivery. Int. J. Polym. Mater. Polym. Biomater. **70**(8), 574–584 (2021)
4. Hassan, M.M., Heins, K., Zheng, H.: Wound dressing based on silver nanoparticle embedded wool keratin electrospun nanofibers deposited on cotton fabric: preparation, characterization, antimicrobial activity, and cytocompatibility. ACS Appl. Bio Mater. **7**(4), 2164–2174 (2024)
5. Kalaf, E.A.G., Flores, R., Bledsoe, J.G., Sell, S.A.: Characterization of slow-gelling alginate hydrogels for intervertebral disc tissue-engineering applications. Mater. Sci. Eng. C. **63**, 198–210 (2016)
6. Wang, X., Abernathy, H.G., Kemp, L.K., Morgan, S.E.: Sprayable adhesive glycopolymer hydrogels with rapid in-situ gelation. Polym. Chem. **15**(31), 3139–3148 (2024)
7. Sahoo, G.P., Rai, V.K., Pradhan, D., Halder, J., Rajwar, T.K., Mahanty, R., Saha, I., Mishra, A., Dash, P., Al-Tamimi, J., Manoharadas, S., Kar, B., Gosh, G., Rath, G.: A doxorubicin loaded chitosan–poloxamer in situ implant for the treatment of breast cancer. RSC Adv. **14**(46), 33952–33967 (2024)
8. Cai, Y., Xiang, Y., Dong, H., Huang, W., Liu, Y., Zhao, C., Yuan, D., Li, Y., Shi, J.: Injectable self-assembling peptide hydrogel as a promising vitreous substitute. J. Control. Release. **376**, 402–412 (2024)
9. Shuai, F., Zhang, Y., Yin, Y., Zhao, H., Han, X.: Fabrication of an injectable iron (III) crosslinked alginate-hyaluronic acid hydrogel with shear-thinning and antimicrobial activities. Carbohydr. Polym. **260**, 117777 (2021)

10. Mahmoudi, C., Tahraoui Douma, N., Mahmoudi, H., Iurciuc, C.E., Popa, M.: Hydrogels based on proteins cross-linked with carbonyl derivatives of polysaccharides, with biomedical applications. Int. J. Mol. Sci. **25**(14), 7839 (2024)
11. Sharma, P.K., Taneja, S., Singh, Y.: Hydrazone-linkage-based self-healing and injectable xanthan–poly (ethylene glycol) hydrogels for controlled drug release and 3D cell culture. ACS Appl. Mater. Interfaces. **10**(37), 30936–30945 (2018)
12. Wang, L., Deng, F., Wang, W., Li, A., Lu, C., Chen, H., Wu, G., Nan, K., Li, L.: Construction of injectable self-healing macroporous hydrogels via a template-free method for tissue engineering and drug delivery. ACS Appl. Mater. Interfaces. **10**(43), 36721–36732 (2018)
13. Itzhakov, R., Hak, H., Sadhasivam, S., Belausov, E., Fallik, E., Spiegelman, Z., Sionov, E., Poverenov, E.: Nanogel particles based on modified nucleosides and oligosaccharides as advanced delivery system. ACS Nano. **17**(22), 23020–23031 (2023)
14. Vinogradov, S.V., Bronich, T.K., Kabanov, A.V.: Nanosized cationic hydrogels for drug delivery: preparation, properties and interactions with cells. Adv. Drug Deliv. Rev. **54**(1), 135–147 (2002)
15. Liu, J., Li, M., Huang, Y., Zhang, L., Li, W., Cao, P., Min, W., Li, J., Jing, W.: A nanogel with effective blood-brain barrier penetration ability through passive and active dual-targeting function. J. Nanomater. **2021**, 6623031 (2021)
16. Liu, X., Jin, L.L., Zhao, L.L., Wang, Y.C., Zhang, L., Huang, Z.Z., Jin, H.Q., Liu, J.Y., Liang, Z.Y., Liu, X., Tan, H., Ren, L.J.: In-vivo thrombolytic efficacy of RGD modified protein-polymer conjugated urokinase nanogels. Polym. Test. **104**, 107392 (2021)
17. Feng, Q., Li, D., Li, Q., Cao, X., Dong, H.: Microgel assembly: fabrication, characteristics and application in tissue engineering and regenerative medicine. Bioact. Mater. **9**, 105–119 (2022)
18. Kaur, M., Sudhakar, K., Mishra, V.: Fabrication and biomedical potential of nanogels: an overview, international journal of polymeric materials and polymeric. Biomaterials. **68**(6), 287–296 (2019)
19. Guo, H., Li, F., Xu, W., Chen, J., Hou, Y., Wang, C., Ding, J., Chen, X.: Mucoadhesive cationic polypeptide nanogel with enhanced penetration for efficient intravesical chemotherapy of bladder cancer. Adv. Sci. **5**(6), 1800004 (2018)
20. Merkel, T.J., Chen, K., Jones, S.W., Pandya, A.A., Tian, S., Napier, M.E., Zamboni, W.E., DeSimone, J.M.: The effect of particle size on the biodistribution of low-modulus hydrogel PRINT particles. J. Control. Release. **162**(1), 37–44 (2012)
21. Hamedi, H., Moradi, S., Hudson, S.M., Tonelli, A.E., King, M.W.: Chitosan based bioadhesives for biomedical applications: a review. Carbohydr. Polym. **282**, 119100 (2022)
22. Conejo-Cuevas, G., Lopes, A.C., Badillo, I., Del Campo, F.J., Ruiz-Rubio, L., Pérez-Álvarez, L.: Self-healing, piezoresistive and temperature responsive behaviour of chitosan/polyacrylic acid dynamic hydrogels. J. Colloid Interface Sci. **678**, 320–333 (2025)
23. Vale, A.C., Pereira, P.R., Alves, N.M.: Polymeric biomaterials inspired by marine mussel adhesive proteins. React. Funct. Polym. **159**, 104802 (2021)
24. Hasirci, V., Hasirci, N.: controlled release systems. In: Fundamentals of Biomaterials, 2nd edn. Springer Nature, Berlin (2024). https://doi.org/10.1007/978-3-031-54046-2_18
25. Khorsandi, H., Karimi, A.R., Azadikhah, F.: Preparation and evaluation of a Sulfadimethoxine-conjugated hydrogel based on N-isopropylacrylamide as a sustained release drug delivery system. ChemistrySelect. **7**(37), e202202428 (2022)
26. Lee, Y., Lim, S., Kim, J.A., Chun, Y.H., Lee, H.J.: Development of Thiol–Ene reaction-based HA hydrogel with sustained release of EGF for enhanced skin wound healing. Biomacromolecules. **24**(11), 5342–5352 (2023)
27. Wei, H., Geng, L., Zhu, H., Li, J.: Facile fabrication of thermosensitive hydrogel microspheres based on a combination of metal-free click chemistry and spray drying. Chem. Eng. Process. Process. Intensification. **136**, 116–122 (2019)
28. Mauri, E., Rossetti, A., Mozetic, P., Schiavon, C., Sacchetti, A., Rainer, A., Rossi, A.F.: Ester coupling of ibuprofen in hydrogel matrix: a facile one-step strategy for controlled anti-inflammatory drug release. Eur. J. Pharm. Biopharm. **146**, 143–149 (2020)

29. Jiao, C., Obst, F., Geisler, M., Che, Y., Richter, A., Appelhans, D., Gaitzsch, J., Voit, B.: Reversible protein capture and release by redox-responsive hydrogel in microfluidics. Polymers. **14**(2), 267 (2022)
30. Schneider, E.L., Henise, J., Reid, R., Ashley, G.W., Santi, D.V.: Hydrogel drug delivery system using self-cleaving covalent linkers for once-a-week administration of exenatide. Bioconjug. Chem. **27**(5), 1210–1215 (2016)
31. Shim, J., Kang, J., Yun, S.I.: Chitosan–dipeptide hydrogels as potential anticancer drug delivery systems. Int. J. Biol. Macromol. **187**, 399–408 (2021)
32. Freeman, I., Kedem, A., Cohen, S.: The effect of sulfation of alginate hydrogels on the specific binding and controlled release of heparin-binding proteins. Biomaterials. **29**(22), 3260–3268 (2008)
33. Choi, J.H., Park, A., Lee, W., Youn, J., Rim, M.A., Kim, W., Kim, N., Song, J.E., Khang, G.: Preparation and characterization of an injectable dexamethasone-cyclodextrin complexes-loaded gellan gum hydrogel for cartilage tissue engineering. J. Control. Release. **327**, 747–765 (2020)
34. Wang, Y., Deng, M., Wu, Y., Hu, C., Zhang, B., Guo, C., Song, H., Kong, Q., Wang, Y.: Sustained gene delivery from inflammation-responsive anti-inflammatory hydrogels promotes extracellular matrix metabolism balance in degenerative nucleus pulposus. Compos. Pt. B. **236**, 109806 (2022)
35. Lee, S., Greenwald, R.B., McGuire, J., Yang, K., Shi, C.: Drug delivery systems employing 1, 6-elimination: releasable poly (ethylene glycol) conjugates of proteins. Bioconjug. Chem. **12**(2), 163–169 (2001)
36. Ritger, P.L., Peppas, N.A.: A simple equation for description of solute release I, Fickian and non-fickian release from non-swellable devices in the form of slabs, spheres, cylinders or discs. J. Control. Release. **5**(1), 23–36 (1987)
37. Ritger, P.L., Peppas, N.A.: A simple equation for description of solute release II. Fickian and anomalous release from swellable devices. J. Control. Release. **5**(1), 37–42 (1987)

Chapter 12
Hydrogels Used in Nerve Therapy

Damla Arslantunalı Şahin

Abstract The nervous system represents the most intricate system within the human body, characterized by processes that are distinct from those governing other systems. In 1 min, millions of information pieces pass through neurons and the nervous system processes the information to produce the required responses, and thus voluntary or involuntary body functions are performed. Any impairment in the nervous system can disrupt the communication between the brain and peripheral organs, leading to consequences that significantly impact daily life. Mechanical, thermal, chemical, or ischemic damage that may occur in the nervous system is mostly caused by traumatic accidents or some degenerative disorders and may cause a decrease or complete loss of neural functions. Among the various strategies employed to address peripheral nerve injuries, the use of hydrogels has gained attention. The success rates of the therapy depend on the extent of the damage. This section discusses the structure of the nervous system, treatment methods used after injuries that may occur in the nervous system, and the properties of the hydrogels applied in the nerve treatments.

Keywords Nervous system · Nerve repair · Nerve guide · Spinal cord · Peripheral nervous system · Brain injuries · Peptide-based hydrogels · Polysaccharide-based hydrogels

---

D. A. Şahin (✉)
BIOMATEN Center of Excellence in Biomaterials and Tissue Engineering, Middle East Technical University, Ankara, Türkiye
e-mail: adamla@metu.edu.tr

N. Hasirci, V. Hasirci (eds.), *Hydrogels and Bioinks in Tissue Engineering*, Biological and Medical Physics, Biomedical Engineering,
https://doi.org/10.1007/978-3-031-90966-5_12

12.1 Use of Hydrogels in Nerve Repair

Hydrogels are gaining increasing recognition for their role in the treatment of injuries to the peripheral nerves, spinal cord, and brain. These biomaterials can be engineered to exhibit essential properties such as biocompatibility, porosity, and flexibility, which are crucial for constructing effective nerve guide channels. By providing a suitable environment for cellular activities, hydrogels facilitate the regeneration of damaged tissues. Moreover, the ability to incorporate therapeutic molecules that either promote healing or inhibit detrimental factors allows for controlled release mechanisms, making hydrogels particularly appealing for nerve regeneration studies. In the treatment of peripheral nerve injuries, both natural and synthetic hydrogels have demonstrated efficacy in facilitating tissue repair and functional recovery. Polymers like alginate, hyaluronic acid, and polyethylene glycol (PEG) have proven effective in promoting cell adhesion, enhancing cell survival, and supporting axonal growth. The versatility of these materials enables their use in various forms, such as scaffolds or drug delivery systems, thus improving overall treatment efficacy. For spinal cord injuries, hydrogels have demonstrated the ability to bridge gaps and facilitate axonal regeneration. By combining specific growth factors with hydrogels, researchers have been able to enhance functional recovery in animal models. This synergy not only promotes cell proliferation and differentiation but also modulates inflammatory responses, creating an optimal healing environment.

The structural properties of hydrogels, such as their pore size and mechanical attributes, can be customized to address the specific requirements of injured spinal cord tissues, thereby enhancing their potential for clinical applications. Injectable hydrogels hold particular promise in the realm of brain injuries, especially in situations where extensive surgical interventions are not viable. These hydrogels can be delivered through minimally invasive techniques, allowing for precise placement within the injured brain tissue. Their ability to gel in situ provides immediate support for damaged areas, facilitating cellular infiltration and recovery. As research and technology go forward, the potential of hydrogels to restructure neural tissue engineering and regenerative medicine is becoming increasingly important. Studies yielding promising results could greatly enhance the quality of life of the patients.

12.2 Anatomy of the Nervous System

The central nervous system (CNS) and the peripheral nervous system (PNS) are the two most important structural components of the nervous system. The PNS is made of cranial nerves, sensory nerve bundles, and nerves that emerge from the spinal cord, while the CNS consists of the brain and the spinal cord. Neurons, the basic building blocks of the nervous system, enable people to perceive stimuli from the outside world and regulate their movements accordingly. The heart of this system is

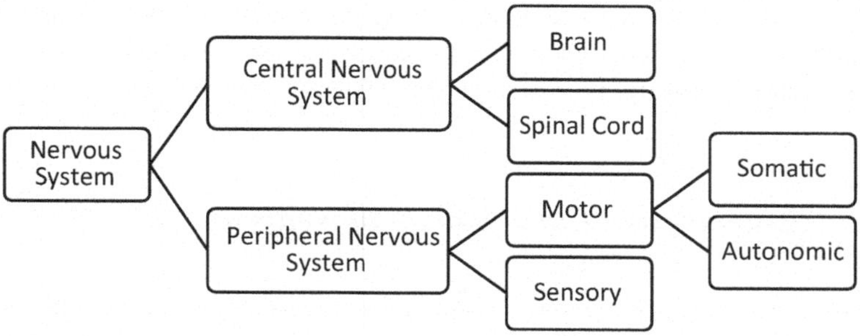

Fig. 12.1 Classification of the nervous system

the brain, which receives signals from the peripheral systems, interprets them, and transmits them to the corresponding surrounding tissues [1].

12.2.1 Central Nervous System (CNS)

The brain and spinal cord make up the Central Nervous System (CNS), which is the body's command center (Fig. 12.1). Three layers of membranes called meninges shield both. The brain is housed inside the skull's hard bones, while the spine's bony vertebrae shield the spinal cord. The cerebrospinal fluid, which acts as a cushion to reduce the impact between the brain and skull or between the spinal cord and vertebrae, is a third line of defense. The CNS is separated into gray matter and white matter based on tissue makeup. The "butterfly"-shaped gray matter, which is mostly located in the outer layers of the brain, is made up of capillaries, glial cells, and neuron cell bodies and their dendrites. Sensory, autonomic, and motor control are the three primary roles of the spinal cord. In order to control these processes, spinal nerves that descend from the brain receive afferent signals from the periphery and transmit them to the brain, as well as efferent commands that travel from the brain to the periphery. The white matter of the spinal cord consists of pathways that transmit sensory information (from receptors in the skin, skeletal muscles, tendons, joints, and various internal organs) to the brain, and motor information from the brain to peripheral tissues.

12.2.2 Peripheral Nervous System (PNS)

The second structural component of the nervous system is the peripheral nervous system (PNS) [2]. It links the central nervous system to the outside world. Since peripheral nerves transmit and receive information to and from the spinal cord and peripheral organs, they might be referred to as "spinal." Twelve cranial nerves, in addition to spinal nerves, transmit and receive information between the brain and surrounding organs. Thus, the autonomic nervous system, which governs involuntary activity, and the somatic nervous system, which governs voluntary activities, make up the PNS (Fig. 12.1).

12.2.3 Structure and Classification of Neurons

The main function of the nervous system is the electrical-chemical conductivity of signals produced both inside the nervous system and from the surroundings. The basic building block of the nervous system, the neuron, is in charge of producing and sending incoming signals [1]. Signals are received, integrated, processed, and transmitted to other neurons by neurons. Despite differences in size, shape, number of projections, number of receiving fibers (dendrites), and chemical messengers (neurotransmitters), all neurons function according to the same fundamental principle.

The cell body (also known as the soma or perikaryon), dendrites, axon, and axon terminal are the four anatomical components that neurons typically have (Fig. 12.2). Electrical insulation is provided by the myelin sheath that envelops the axon, and electrical transmission is accelerated by the spaces between the sheath (nodes of Ranvier) [3]. Additionally, neuronal branching is specialized to support electrical communication. Dendritic projections originating from the cell body are the main source of this branching. The neuron's capacity to receive incoming signals is determined by the presence of dendrites and their projections. The axon terminal, a branch of the neuronal cell dedicated to signal transmission, receives the messages. Depending on the size of the body and the type of neuron, the axon may be several 100 μm long. The axon's distinct cytoskeleton, which preserves its functional integrity, is another characteristic. Action potentials and neurotransmitters are conveyed

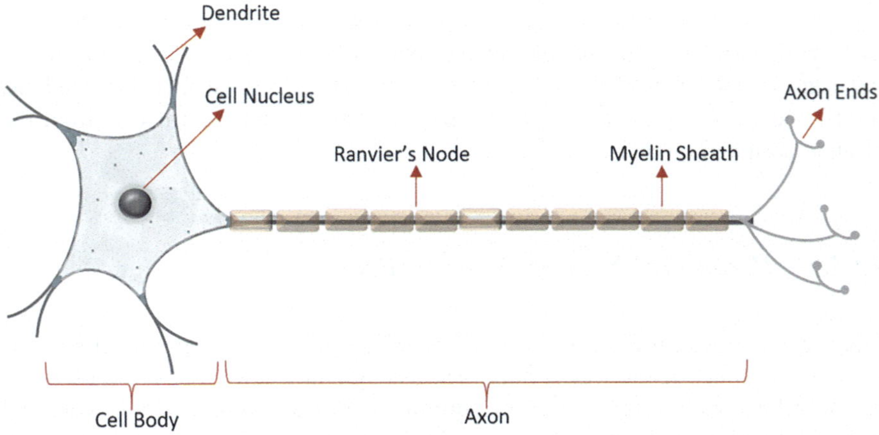

Fig. 12.2 Neuron structure

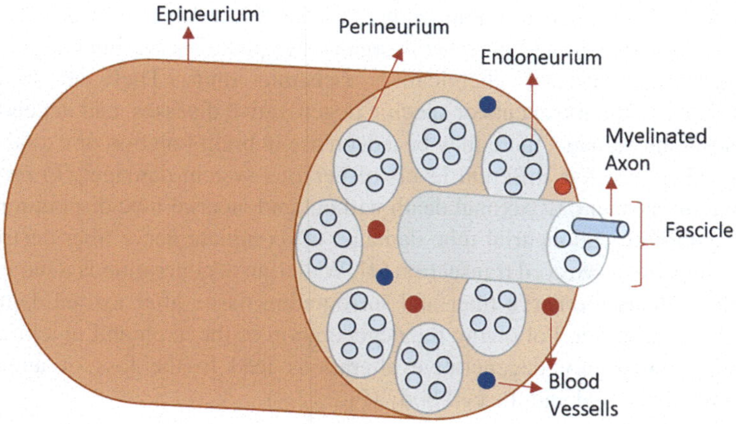

Fig. 12.3 Structure of peripheral nerve fiber

and released at the axon terminal. Myelin, a fatty substance produced by Schwann cells, covers some axons in the peripheral nervous system (Fig. 12.2). As electrical signals move along the axon, the insulation provided by the myelin coating keeps them from dissipating. An axon from a human motor neuron can reach a length of 1 m, from the base of the spine to the toes, making this insulation essential.

In CNS, some neurons in the human brain may have axons that are a few millimeters long, while others may have no axons at all. The 31 pairs of sensory and motor nerves in the spinal cord exit from both sides and spread to the periphery. The cell bodies of these nerve cells are located in the gray matter within the spinal cord, and the nerve fibers extend outward from the gray matter through the surrounding white matter. The structure of neurons in the PNS is different from that in the CNS because they are found in nerve cords. In the peripheral system, nerve fibers are held together with connective tissue (Fig. 12.3). The endoneurium, a loose connective tissue, envelops each nerve fiber and the Schwann cells that maintain it. Collagen fibrils in the connective tissue hold a bundle of nerve fibers together, forming fascicles that are encircled by a dense layer of connective tissue known as the perineurium. The epineurium, the outermost layer of the connective tissue sheaths, is a dense and uneven layer of connective tissue that completely covers all of the nerve fascicles within a nerve trunk [3].

12.3 Peripheral Nerve Injuries

The patient's quality of life is significantly impacted when peripheral nerves are injured. As a result, nerve damage is a major clinical issue on a global scale. A reduction in motor and sensory functions brought on by injury results in painful neuropathies, which significantly affect patients' everyday activities. Injury also causes a loss of neuronal communication via sensory and motor nerves between the central nervous system and peripheral organs. Peripheral nerve injuries can result from mechanical, thermal, chemical, or ischemic injury. These are frequently brought on by traumatic events or specific degenerative diseases, and depending on the extent of the damage, there may be a decrease in brain function or a total loss of function. There are five different levels of nervous system damage: (1) reversible functional impairment, (2) axonal death without endoneurial tube discontinuity, (3) axonal death with endoneurial tube damage, (4) complete nerve fiber destruction, and (5) complete nerve cord transection [4]. Wallerian degeneration is a sequence of events that affects the nerve fiber and surrounding tissue after axonal damage or total nerve transection. Following the degeneration of the axon and myelin at both ends, these post-injury degenerative alterations lead to the loss of attachment between the distal end and the cell body (Fig. 12.4).

12.3.1 Approaches Used in Nerve Injuries

Peripheral nerve injuries are treated using a variety of techniques, and the degree of damage determines the success rate. The end-to-end anastomosis technique, which involves surgically joining two ends, is initially recommended for injuries in which nerve terminals are not separated. Grafting techniques, however, are used in greater injuries where the nerve ends are split. To cover the injured area for recovery, segments of nerve tissue taken from the patient (autografts), another person (allografts), or an animal (xenografts) might be used. These techniques can have drawbacks, though, including tissue rejection, contamination, sensory loss at the donor location, and donor recruitment challenges. In such applications, nerve guide conduits made of biomaterials can be preferred. The reasons for using nerve guide conduits include ease of production, biocompatibility, and lack of contamination risk [4].

The brain and spinal cord make up the central nervous system. The majority of spinal cord injuries are caused by vertebral column fractures or dislocations from auto accidents, sports, and work-related incidents. These injuries are divided into major and secondary categories based on their severity, and the available therapeutic options—such as end-to-end anastomosis, grafting, and nerve guide conduits—are comparable to those for peripheral nerve injuries. The creation of scar tissue is the main obstacle to treating the central nervous system; hence, bioactive substances like growth factors are needed to stop this [5].

Nerve regeneration is impacted by a number of factors, including the patient's age and condition, the duration of the nerve gap following injury, and the interval between injury and treatment. End-to-end anastomosis is the gold standard when the lesion is minor and there is no segment loss, and regeneration may happen on its own [6]. This technique involves using fascicular and/or epineural sutures to connect the nerve ends at the site of injury.

In the cases, where there is a gap due to segmental loss at the injury site, performing end-to-end suturing is not possible due to sensory loss related to nerve tension.

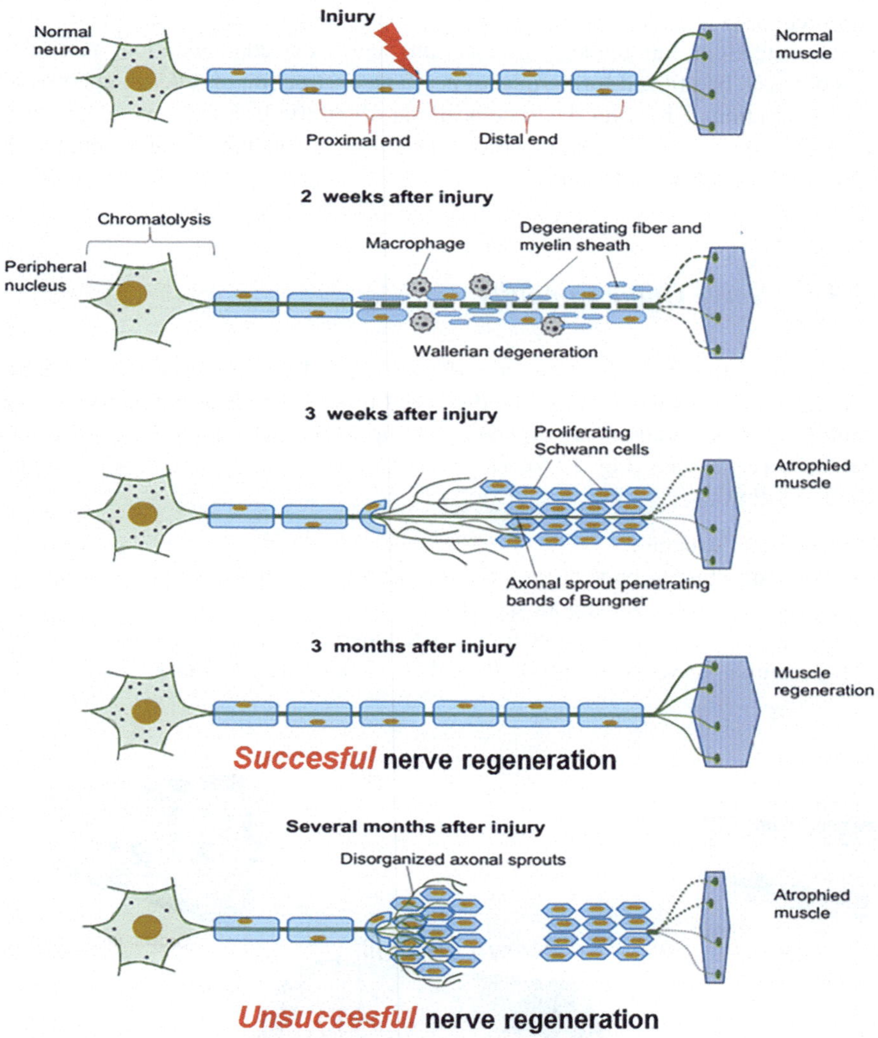

Fig. 12.4 Peripheral nerve injury and Wallerian degeneration

In such cases, alternative methods such as transplantation or application of grafts or nerve guide conduits are utilized. The damaged axons' distal and proximal ends are connected with materials such as conduits or grafts. Autografts are regarded as the gold standard when end-to-end surgery is not feasible. The nerve graft is usually transplanted to the location of injury after being removed from another part of the patient's body. Because the patient's own Schwann cells and undamaged endoneurial tubes are used, this treatment does not elicit an immunogenic reaction. The development of neuromas (nerve tumors), sensory loss at the donor location, and restricted graft availability are some of its drawbacks. The drawbacks of xenografts (from animals) and allografts (from other people) include supply constraints, secondary infection, immunological rejection, and cross-contamination. Acellular allografts can be utilized for peripheral nerve reconstruction in order to get around these restrictions. However, because acellular allografts lack Schwann cells, they are not as good as autografts. As a result, nerve guide conduits are now being used in place of transplant techniques.

12.3.2 Nerve Guide Conduits

In addition to the methods of end-to-end suturing and grafting used in nerve injuries, nerve conduits that connect the distal and proximal ends at the injury site are employed. These conduits are typically cylindrical and have polymeric structures that aid nerve healing (Fig. 12.5). The essential features of a nerve conduit can be listed as follows [7]:

- Cylindrical structure.
- Porous structure to prevent infiltration of fibrous scar tissue while allowing diffusion of nutrients (ideal pore size: 10–20 μm).
- Mechanical properties suitable for nerve regeneration (maximum tensile strength for acellular nerve tissue: 1400 kPa and elastic modulus: 576 kPa).

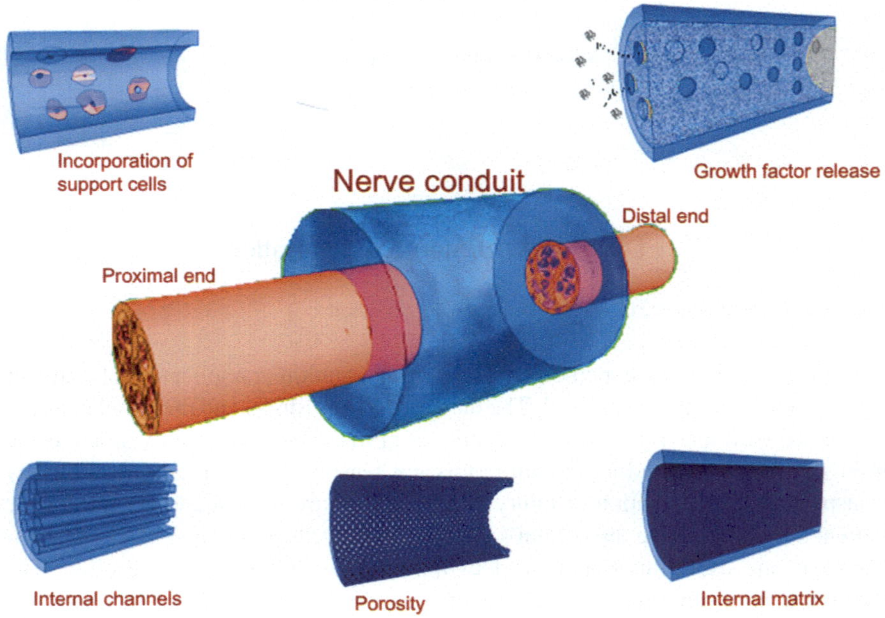

Fig. 12.5 Structural properties of nerve guide conduits

- Compatibility with biological tissues, capable of creating a suitable microenvironment for cell adhesion, proliferation, and regeneration.
- Architecture containing supportive structures, such as inner channels, inner matrix, and growth factor release regions for nerve guidance.

As an alternative to autografts, a number of conduit materials have been approved by the FDA (Table 12.1). These conduits, which are usually 3–10-mm-long hollow tubes, are used to treat nerve damage. In these systems, the objective of nerve regeneration is to encourage the creation of a fibrin cable and use the conduit as a scaffold at the site of injury. Extracellular matrix proteins like collagen and laminin are deposited at the site of injury by both neuronal and non-neuronal cells. This regulated setting promotes nerve tissue regeneration, decreases scarring, and boosts growth factor synthesis by nerve endings.

Nerve guide conduits are made from both natural biopolymers and synthetic polymers. Although natural biopolymers are frequently chosen for designs because they work well with other materials, their application might be complicated by their poor mechanical qualities. Conversely, the biodegradability of synthetic polymers

Table 12.1 FDA-approved nerve conduits available on the market

Product name	Material	Structural properties	Company
Neurogen®	Collagen Type 1	Semipermeable, fibrillar collagen	Integra LifeSciences Co, Plainsboro NJ, USA
Revolnerv®	Collagen Type 1&2	Semipermeable, tubular collagen	Orthomed SA-06640 Saint-Jeannet, France
NeuroFlex™	Collagen Type 1	Flexible, semipermeable tubular collagen matrix	Collagen Matrix, Inc., Franklin, USA
Neuromatrix™	Collagen Type 1	Semipermeable, tubular collagen	Collagen Matrix, Inc., USA
NeuroWrap™	Collagen Type 1	Tubular structure formed by longitudinally aligned collagen fibers	Integra LifeSciences Co., USA
NeuroMend™	Collagen Type 1	A self-opening and self-coiling collagen wrap	Collagen Matrix, Inc., USA
Neurotube®	Polyglycolic Acis	An absorbable woven PGA (polyglycolic acid) mesh tube	Synovis Micro Companies Alliance, Birmingham, AL, USA
Axoguard™	Polymeric material produced from the submucosa of porcine small intestine	A soft, flexible, and porous hydrogel	Cook Biotech Incorporated, Indiana, USA
Neurolac™	Poly(D,L-lactide-co-caprolactone)	A synthetic and transparent tube	Polyganics BV, Groningen, Netherlands
Salutunnel™	Polyvinyl alcohol (PVA)	A non-biodegradable tube	Salumedica LCC, Atlanta, GA, USA

restricts their use, despite their simplicity of manufacturing. Because they have appropriate mechanical qualities, ease of production, and tunable deterioration rates, hybrid conduits are therefore seen as a good substitute for nerve guiding channels.

12.4 Spinal Cord Injuries

The three primary roles of the spinal cord, which is a component of the central nervous system (CNS), are motor, autonomic, and sensory control. In order to control these processes, spinal neurons that descend from the vertebral column receive efferent commands from the brain, integrate them, and transmit them to the periphery. At the same time, the periphery sends back to the brain afferent signals. Pathways that carry motor information (to skeletal muscles, smooth muscles, cardiac muscles, and glands) and sensory information (from the skin, skeletal muscles, tendons, joints, and other visceral receptors) make up the spinal cord's white matter. Each of the 31 spinal nerve pairs has a motor ventral root and a sensory dorsal root. Due to the spinal cord's incapacity to carry information from the brain or peripheral, any damage to these nerves may cause loss of function. The majority of spinal cord injuries result from vertebral column fractures or dislocations brought on by auto accidents and work-related dangers. Peripheral nerve injuries are thought to impact about 20 million people in the USA; the market is expected to grow from $407.85 million in 2023 to $827.39 million in 2033 [8].

Any damage to the spinal cord is followed by a series of cellular and molecular events. Spinal cord injuries can be categorized into two main types based on the severity of the damage: primary and secondary injuries. Any mechanical trauma resulting in focal damage to neural tissue is termed a primary injury. In the area of primary injury, a region of hemorrhagic necrosis is observed due to disrupted neuronal and endothelial cell membranes. The necrosis zone is primarily found in the more vascularized gray matter. As a result, axons near the gray matter are severely damaged, while axons in the white matter may survive. In other words, primary injury involves no damage beyond hemorrhage in the spinal cord.

The main issue in repairing axons after spinal cord injury is that secondary pathological events begin after the initial physical trauma, spreading damage to otherwise intact axons. This is known as secondary injury [9]. Following hemorrhagic necrosis, both gray and white matter begin to swell, and integrity is compromised due to edema. At the microscopic level, secondary injury is characterized by various symptoms including necrosis, apoptosis, infarction, axonal disruption, and glial and fibrotic scar formation. Necrosis begins at the moment of primary injury and spreads to adjacent spinal segments, and almost 8 h after the injury, this damage becomes irreversible. Secondary injury primarily arises from apoptosis occurring 6 h post-injury. Both apoptosis and necrosis, along with infarction, vasospasm, thrombosis, and neurogenic shock, impact neurons. Additionally, axons swell, the myelin sheath surrounding the axon deteriorates, and Wallerian degeneration occurs. Scar

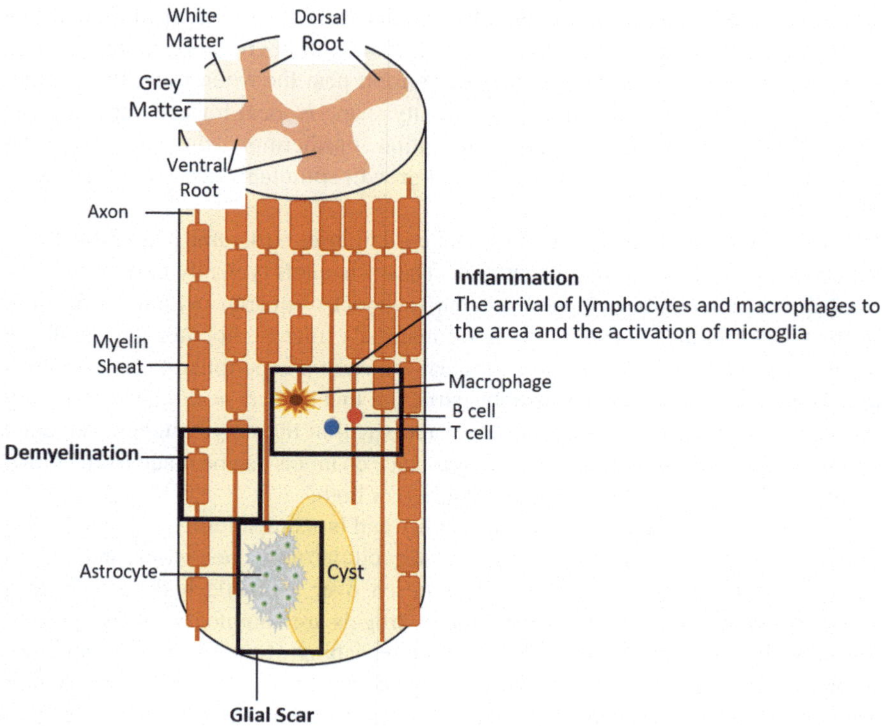

Fig. 12.6 Scar formation after spinal cord injury

formation in the lesion area after spinal cord injury obstructs the repair process. This scar formation includes glial or fibrotic components. Residual debris from damaged myelin sheaths is removed by microglial cells. In addition to microglia, macrophages, astrocytes, and oligodendrocytes are recruited to the injury site. Previous studies have shown that axonal degeneration leads to astrocyte activation, consequently resulting in glial scar formation (Fig. 12.6).

12.5 Use of Hydrogels in Peripheral Nervous System and Spinal Cord Injuries

Because hydrogels have mechanical characteristics similar to those of nerve tissue, they are favored for use in the construction of cylindrical nerve guidance conduits intended to connect the proximal and distal ends of the damage site in peripheral nerve injuries. The conduit's structural material, filler material, and/or medication release supports that promote nerve repair are all made of hydrogels [10]. Hydrogels provide a suitable environment for neuron regeneration because of their three-dimensional structure and physical characteristics, which are similar to those of real nerve tissue. When hydrogels come into contact with other biological fluids, they absorb water, swell, and have low interfacial tension. In in vivo applications, hydrogels' soft and elastic properties reduce discomfort near the wound site. Particularly in tissue engineering applications, its porosity shape is ideal for cellular functions like cell adhesion, migration, and proliferation. Furthermore, the cells inside the hydrogel may readily receive nutrition and growth stimulants because of its three-dimensional structure.

Nerve guidance channels used in spinal cord injuries structurally resemble those utilized for peripheral nerve regeneration. These channels typically have a cylindrical shape (which can be designed as a half-cylinder or quarter-cylinder depending on the type of injury) and use polymeric materials (matrix, fiber, or gel) as filling materials tailored to the needs of healing. However, unlike peripheral nerve injuries, spinal cord injuries feature a dystrophic structure that hinders healing, necessitating the inclusion of various enzyme–drug systems within the nerve guidance channel structure. The outer wall of these nerve guidance channels can be made from hydrogel materials, as can the inner filling material, or both.

Because of their superior cytocompatibility and biocompatibility, suitable biodegradability, minimal immunogenicity, specific cellular/tissue responses, and antigenicity, natural hydrogels are frequently chosen over synthetic hydrogels from a polymer standpoint. Furthermore, natural hydrogels are commonly used in peripheral nerve injuries because they facilitate the reshaping of the local hydrogel environment during nerve extension growth and provide sufficient cell adhesion on their surfaces. However, natural hydrogels also have disadvantages, including low mechanical properties, variability in features from batch to batch, challenges in processability (for both polysaccharides, such as cellulose and chitosan, and polypeptides, such as silk), resource limitations, and sometimes high production costs (e.g., hyaluronic acid). They might also show high rates of microbial breakdown (particularly for polypeptides) and biodegradation and catabolism (like gelation). In order to make hybrid hydrogels, it may be necessary to chemically modify natural polymers in addition to using other natural or synthetic polymers that are biodegradable and biocompatible (such as poly(vinyl alcohol) (PVA), poly(acrylamide) (PAAm), etc.). The benefits of both synthetic polymers with adjustable physical properties and naturally functional polymers are combined in these natural/synthetic hybrid hydrogels [11]. After being injected as a liquid into the lesion site, certain hydrogels can be made to solidify there. Others that are sensitive to heat may solidify at body temperature. Temperature changes or the combining of two reactive chemicals can cause solidification. For spinal cord injuries, injectable hydrogels may be more beneficial than conventional nerve guiding channels, particularly in cases when the lesion sites are small and complicated because of scar formation. These hydrogels offer optimal interaction between the biomaterial and the severed nerve ends while accommodating the uneven shape of the lesion. For nerve repair in spinal cord injuries, a variety of nerve guiding channels composed of various biomaterials are employed. Significant studies have been done in this field, although their use is not as common as in peripheral nerve injuries. In addition to preventing misdirected regeneration toward the peripheral nerve system, cylindrical nerve guidance tubes isolate the lesion from the surrounding environment. Uneven growth can still happen inside the tube, though, because hollow tube channels do not resemble the internal perineurium and endoneurium [12]. This could be the reason that in big gap nerve injuries, hollow tube nerve channels frequently work less well than nerve autografts. Thus, regenerating axons can stick to the structure and align along the holes within a nerve channel filled with hydrogel, which can be made with a regular topology or a porous structure. Hydrogels used in nerve treatment can be peptide-based, polysaccharide-based, synthetic, or hybrid in nature.

12.5.1 Peptide-Based Hydrogels

These hydrogels have protein structures having peptide bonds in their chemical formula. The most commonly used ones are collagen, gelatin, fibrin, silk, and self-assembled peptides.

12.5.1.1 Collagen

Biomaterials based on collagen are being researched for PNS and CNS regeneration. Because of their accessibility, acceptable biodegradability, biocompatibility, low immunological response, and bodily absorbability, these materials hold great promise for neural tissue engineering. They do, however, have certain drawbacks, including limited elasticity, poor mechanical qualities, and inadequate thermal and dimensional stability. Several crosslinking methods are used to increase their robustness. The FDA has approved the use of type I collagen in peripheral nerve injuries, and it has been utilized in the structure of commercially produced nerve guidance channels (such as NeuraGen®, NeuroMatrixTM, NeuraWrapTM, and NeuroMendTM) [13].

For PNS regeneration, collagen-based hydrogels—mainly Type I collagen—are frequently employed to promote axonal development and cell culture in three-dimensional settings. Peripheral nerve regeneration can be improved by using collagen hydrogels or matrices to fill channels, tubes, and vascular grafts. In rat dorsal root ganglion neurons (DRG), Bozkurt et al. created a highly aligned three-dimensional collagen scaffold that stimulates neurite and axonal development [14]. Their research demonstrated that Schwann cells co-cultured with DRGs migrate into the collagen scaffold and align themselves appropriately within the structure, while DRG cells take on the morphologies of natural nerve tissue within the collagen scaffold. Similarly, hydrogels based on aligned collagen promote directed axon migration, which supports peripheral nerve healing in vitro and in vivo by increasing the quantity and size of myelinated fibers [15]. Cell adhesion molecules can also be added to hydrogels based on collagen. According to some research, Type I collagen that has been altered with laminin or peptides generated from laminin improves neural cell survival, encourages directed neurite formation on scaffolds, and aids mesenchymal stem cell neuronal differentiation [16]. Rein et al. (2024) more recently implanted collagen or chitosan nerve conduits in two males and two females (ages 17–57) after nerve grafting and neuroma removal, or in cases of traumatic nerve injuries. Due to ongoing neuropathic discomfort, revision surgery was required to remove the conduits, and all removed conduits underwent histological examinations. One patient experienced a mechanical problem following nerve conduit implantation over finger joints, while four patients had scar neuromas. There were very few indications of nerve conduit deterioration during surgery. With a vascularized fibrous membrane coating that contained deposits from macrophages that were probably caused by dissolved chitosan, chitosan conduits retained their structure and form. Collagen conduits, on the other hand, showed typical histopathologic deterioration, disintegrating into tiny, round strands. None of the nerve conduits that were removed showed signs of a foreign body reaction [17].

Peripheral nerve regeneration has reportedly been accelerated by the inclusion of certain nanoparticles in collagen-based hydrogels. In three-dimensional collagen/single-walled carbon nanotube hydrogels, for example, the function and interaction of Schwann cells (SCs) and PC12 cells have been assessed for neural regeneration. The combined effects of electrical stimulation and topography have been demonstrated to promote neuronal development in the resulting electroconductive hydrogels [18]. For pharmaceutical applications, collagen is used in a variety of physical forms, such as microspheres, nanoparticles, films, coatings, fibers, and sponges. At physiological temperatures, it can also gel in situ as an injectable hydrogel. As a result, it can be utilized as an injectable hydrogel for spinal cord injury or to build nerve guiding channels. Over the course of 6 weeks, Cholas et al. investigated how adult rat neural stem cells (NSC) repaired and healed in response to laminin-loaded collagen and dehydrothermally crosslinked collagen implants as well as pGDNF (plasmid encoding for glial cell line-derived neurotrophic factor). Following the damage for 6 weeks, they found that collagen, either by itself or as a carrier for NSCs or pGDNF, decreased the development of cysts in the defect area [19].

The ability of acetic acid-cured electrospun Type I collagen nanofibers to lessen protein denaturation in astrocytes and dorsal root ganglia was assessed by Liu et al. They discovered that neurite development from dorsal root ganglia (DRG) explants was supported by aligned collagen fibers [20]. In contrast to astrocytes implanted into random collagen fibers and glass slides, those implanted into aligned collagen fibers displayed longer morphologies, greater elongation factors, and cell shape indices. In a rat injury model, cell penetration into the scaffold structures was detected 10 days after implantation. Following PRISMA criteria, a systematic review and meta-analysis were carried out with the conduit type—either porcine SIS (small intestine submucosa) or bovine collagen—as the predictor variable [21]. The statistical significance of variations in overall time-to-functional sensory recovery (FSR) and at the 6- and 12-month points was evaluated using a log-rank test. Within 6 months, 42% of patients with bovine collagen conduits and 64% of patients with porcine SIS conduits achieved functional sensory recovery (FSR). Additionally, 94% of patients with bovine collagen conduits and 82% of those with porcine SIS conduits achieved FSR within 12 months.

12.5.1.2 Gelatin

Because of its low cost, non-immunogenicity, biodegradability, and biocompatibility, gelatin is a biopolymer that is frequently employed in biomedical research and applications. Gelatin is typically made by physically or chemically hydrolyzing collagen; there are two varieties: "Type A," which is treated with acid, and "Type B," which is produced by alkaline hydrolysis. In tissue engineering, both kinds are used, especially in research on nerve regeneration.

In order to differentiate multipotent stem cells generated from the human placenta, Li et al. studied chemically crosslinked gelatin Type A and Type B hydrogels [22]. They confirmed the development of dental pulp stem cells (DPSCs) into neuron-like cells by demonstrating that both hydrogels are appropriate for cell adhesion with an attachment rate of about 80%. Furthermore, they found that, in comparison to other polymeric scaffolds under the same circumstances, the time needed for DPSCs cultivated on gelatin hydrogels to differentiate was noticeably shorter (around 2 h). They came to the conclusion that neural cell culture, adhesion, and differentiation can be accomplished with both Type A and Type B gelatin hydrogels.

In a different study, genipin was added as a crosslinking agent to an injectable hydrogel based on agar/gelatin for use as a filler in nerve guidance channels [23]. The outcomes demonstrated that the generated hydrogel was biocompatible with Schwann cells and fibroblasts, and 2D and 3D migration analyses demonstrated that the cells could move across the hydrogel's surface and enter its internal structure. They proposed using the hydrogel for medication delivery and nerve development applications. In order to treat peripheral nerve injury, this group has created a hydrogel that releases vascular endothelial growth factor (VEGF) continuously for up to 65 days. According to the study, neurite sprouting from dorsal root ganglions (DRGs) cultivated on VEGF-containing hydrogels demonstrated that the straightforward and appropriate production method of genipin-crosslinked agar/gelatin was adequate for maintaining the activity of sensitive bioactive molecules like VEGF.

Dursun-Usal et al. developed a nerve channel made of methacrylated gelatin and poly(2-hydroxyethyl methacrylate) (GelMA-pHEMA) [24]. Channels with varying amounts of pHEMA were studied for porosity, pore size, mechanical properties, and swelling capacity, and their suitability for nerve regeneration was tested. The results indicated that the channel resembled natural nerve tissue, providing a suitable environment for the adhesion and proliferation of Schwann cells, making it appropriate for use in peripheral nerve injuries. In the same study, directed fibers were added to the produced channels, and it was shown that PC12 cells migrated directionally along these fibers within the channel.

More recently, complex hydrogel structures were created utilizing GelMA and chitosan (CS) and digital light processing (DLP) printing. To build conducting routes inside the hydrogel, poly(3,4-ethylenedioxythiophene) (PEDOT) nanoparticles were included using interfacial polymerization [25]. During in vitro culture, this GelMA/CS-PEDOT hydrogel encouraged axon outgrowth and supported cell proliferation in PC12 and Schwann cells. Additionally, axon extension in PC12 cells on the conductive substrate was improved by in vitro direct current electrical stimulation. In vivo experiments showed that the GelMA/CS-PEDOT scaffold was successful in using a conductive nerve conduit to repair a 10-mm sciatic nerve lesion in rats.

12.5.1.3 Fibrin

When thrombin catalyzes blood clotting, fibrin—a fibrous biopolymer—is created from fibrinogen. Fibrin has been demonstrated to be useful in tissue engineering and wound healing in addition to its functions in blood coagulation and hemostasis. Schwann cells (SCs) were shown to stay alive within the hydrogels for up to 7 days when Hodde et al. placed highly aligned electrospun polycaprolactone (PCL) nanofibers in three-dimensional fibrin hydrogel scaffolds [26]. The quantity of embedded SCs in the hydrogel was constant over the testing period, despite the fact that SCs multiplied on the 2D fibrin substrate. The branching of SCs was arranged by the addition of aligned PCL nanofibers, indicating that fibrin-based hydrogels with aligned PCL nanofibers may be a viable scaffold for peripheral nerve regeneration.

Du et al. electrospun a polymer solution of fibrinogen treated with thrombin and polyethylene oxide (PEO) to create a 3D hierarchically aligned nanofiber fibrin hydrogel (AFG). They found that the elasticity and hierarchical alignment of AFG were comparable to those of the extracellular matrix of nerves (~1.5 kPa) [27]. Dorsal root ganglion (DRG) cells and SCs migrated and attached orientedly along the nanofibers, according to in vitro experiments. In vivo tests on rats with peripheral nerve injuries that were 10 mm long revealed that the hydrogel that was created had morphological, electrophysiological, and histological characteristics that were comparable to those of autologous nerve grafts.

In addition to being used as a filling material or wall component in nerve channels, fibrin has also been utilized as an adhesive [28]. Fibrin adhesive aids in preventing the separation of nerve endings. Although the natural nerve tissue has a lower surface area compared to the channels, making them potentially inadequate to sustain the tensile forces of daily movements, the efficacy of fibrin adhesive has been demonstrated to help provide additional strength. Isaacs et al. conducted a comparative analysis of autologous and commercial fibrin adhesives used in sutured injuries of lower extremity nerves [29]. This study revealed that two of the main formulations (Tisseel™ and Evicel™) performed better than other nerve adhesives while providing similar results. The authors noted that when the adhesive was applied around the repair site, it could form a semi-nervous tube, serving as an effective barrier to scar tissue and a guide for regenerating axons.

Another study investigated the effects of Wnt5a protein-loaded fibrin hydrogels as a release system placed within nerve channels on a 10-mm rat sciatic nerve injury [30]. This study observed positive effects on the proliferation of Schwann cells and concluded that the developed nerve channel supported greater axonal regeneration compared to tubular fibrin structures in in vivo studies.

12.5.1.4 Silk

The strength and flexibility of silk fibroin (SF), a naturally occurring amphiphilic block copolymer, are derived from hydrophilic (more complex, less structured) and hydrophobic (highly organized) segments. Silk fibroin's mechanical qualities, biocompatibility, predictable rate of breakdown, low molecular weight, and tolerable immunogenicity make it a strong contender for applications involving peripheral nerve regeneration. As a result, numerous investigations have confirmed the efficacy of silk-based biomaterials in nerve tissue engineering.

In vitro and in vivo tests of a hydrogel based on silk fibroin peptide (SF16) as a scaffold for peripheral nerve tissue engineering were conducted by Wei et al. [31]. The outcomes demonstrated that PC12 cells' survival, proliferation, and morphology could all be preserved by the hydrogel. Functional recovery was improved by the application of the generated hydrogel, according to in vivo electrophysiological tests. In addition, the hydrogel produced a greater average axon diameter, thicker myelin, and higher axon density than the negative control group (which received physiological saline treatment).

In another study, a pure SF hydrogel with high mechanical properties and aligned micro-grooved topography was prepared, and its biocompatibility was tested with Schwann cells and dorsal root ganglion (DRG) cells [32]. It was observed that Schwann cells adhered significantly to the surface and exhibited a spindle morphology after 24 h of incubation. The Schwann cells on the grooved hydrogels aligned with a small orientation angle ($7.2 \pm 2.6°$), while those on the non-grooved hydrogels showed random distribution. Similarly, DRG cells produced neuronal extensions in an aligned manner on the grooved hydrogel, whereas a scattered extension occurred on the non-grooved surface. As a result, it was observed that silk hydrogels provided significant support for nerve regeneration, with axonal extensions being longer compared to other nerve channels. This suggests that the use of silk hydrogel is effective in long-distance peripheral nerve injuries.

A biomimetic multichannel silk nerve conduit with multicellular spatiotemporal distributions for effective spinal cord injury (SCI) repair was presented by Yuan et al. [33]. The biomimetic silk nerve conduit was crafted using a modified directional freeze-casting strategy, featuring hierarchical channels and aligned pore structures. These hierarchical structures created optimal environments for mesenchymal stem cells (MSCs) and Schwann cells (SCs) to reside in designated channels, facilitating effective repair mechanisms. In vitro results demonstrated enhanced SC migration and MSC differentiation within the conduit, which further promotes tube formation, endothelial cell migration, and M2 polarization of macrophages. Additionally, the conduit effectively facilitated tissue repair and functional recovery by reducing glial scar formation and promoting neuron regeneration and myelin sheath reconstruction. In an 8-mm sciatic nerve deficit model in rats, another group examined the application of macroscopic perforations in the tubular walls of silk-based tubular nerve guides (tNGCs) and contrasted the properties of these enhanced silk+tNGCs with those of standard tNGCs and autologous nerve transplants [34]. It was shown that silk+tNGCs encouraged the creation of blood vessels from nearby tissues into the intraluminal area, resulting in neovascular formation, using a combination of micro-computed tomography and histological investigations. When evaluated 7 weeks after surgery, the silk-added group showed enhanced axon regeneration at the distal coaptation point and a much greater number of blood vessels than both the autologous nerve transplants and the tNGCs without silk.

12.5.1.5 Self-Assembled Peptides

A family of self-assembling nanofiber hydrogels known as self-assembled peptides (SAPs) is made from natural amino acids that alternate between hydrophilic and hydrophobic amino acid groups. In addition to having distinct qualities, they are made to be used in the treatment of peripheral nerve damage by encouraging cell division or continuous drug release. Following nerve damage, polymeric biomaterials are typically employed as carriers that offer a three-dimensional spatial structure for the release of growth factors, bioactive medications, and implanted nerve cells. In this manner, they facilitate the repair of injured neural networks, direct axonal regeneration in a particular direction, foster functional rehabilitation, and establish a milieu for cell differentiation and proliferation. SAP hydrogels are great matrices for nerve cell seeding since they can contain more than 99% water. Furthermore, it is simple to design and alter the peptide sequence at the molecular level. SAPs are made using extremely sophisticated synthesis techniques, and as they are made of natural amino acids, the breakdown products are small molecular amino acids that don't harm the body. As a result, the drawbacks of typical animal-derived tissue materials that may be linked to cytotoxicity and immunological rejection in humans are eliminated, as are those associated with harmful chemical residues and foreign elements. Additionally, SAP hydrogels help injured peripheral nerve axons myelinate [35]. IKVAV (isoleucine–lysine–valine–alanine–valine), RGD (arginine–glycine–aspartic acid), YIGSR (tyrosine–isoleucine–glycine–serine–arginine), and RADA (arginine–alanine–aspartic acid–alanine) are peptide sequences that are frequently employed for nerve healing. During the peripheral nerve healing process, IKVAV inhibits the adhesion and differentiation of glial cells, stimulates the adhesion, differentiation, and axonal growth of neurons in the injured peripheral nerve region, and replicates the ideal milieu required for nerve cell growth. RGD and YIGSR promote cell adhesion. In an in vivo rat model of sciatic nerve damage, bioactive RGD worked in conjunction with topographic cues from nanofibers as guide channel fillers to complete nerve regeneration in 4 weeks, resulting in total recovery [36]. RADA is a complementary hydrophilic and lipophilic peptide made up of successive residues of arginine, alanine, and aspartic acid. By altering two functional sequences, IKVAV and RGD, Zhang et al. were able to successfully develop the RADA16-I self-assembled peptide structure. By joining the chains, they aided in nerve injury regeneration and repair [37].

12.5.2 Polysaccharide-Based Hydrogels

These hydrogels have saccharide structures and the most commonly used ones are chitosan, alginate, hyaluronic acid, and agarose.

12.5.2.1 Chitosan

A deacetylated form of chitin, the second most prevalent polymer after cellulose, is chitosan. A variety of shapes and structures, such as liquid gels, films, tubes, microbeads, nano/microfibers, and membranes, can be created from chitosan hydrogels. These hydrogels' blood compatibility, cytocompatibility, environmental responsiveness, mechanical properties that can be adjusted, ability to combine with other materials to achieve desired properties, and capacity to be produced in a variety of well-designed porous structures make them suitable for use in nerve healing. Studies recommending the use of chitosan channels in vitro and in vivo have demonstrated encouraging outcomes for the repair of peripheral nerves. For the restoration of peripheral nerves, chitosan is frequently utilized in hybrid forms with other biomaterials. Alginate and chitosan complexes were used by Pfister et al. to create nerve channel hydrogels [38]. These hydrogels' mechanical characteristics and excellent permeability were comparable to those of normal neural tissue due to their porous microstructure. Chitosan tube hydrogels enhanced with hyaluronic acid and/or collagen components have been created with consideration for channels that resemble the epineurium required for peripheral nerve regeneration [39]. These channels promoted the adhesion and growth of mouse hippocampus cells and showed no signs of cytotoxicity. Since it is biocompatible and promotes axonal regeneration, calcium ion-enriched chitosan-based hydrogel has also been utilized as a nerve guiding conduit.

To support neural cell adhesion, cell affinity, proliferation, and neurite growth, chitosan is also used in combination with laminin, laminin peptides (YIGSR (Tyr-Ile-Gly-Ser-Arg) and IKVAV (Ile-Lys-Val-Ala-Val) sequences) (REF), collagen and its derivatives (REF), fibronectin (REF), and combinations of poly(L-lysine) (REF). In one study, a double-layer composite hydrogel nerve conduit was created by injecting a conductive, self-healing, injectable double-crosslinked chitosan hydrogel into a non-conductive chitosan tube [40]. When examined for degradability, biocompatibility, antioxidant activity, and Schwann cell proliferation activity, this hydrogel channel containing 7,8-dihydroxyflavone (DHF) produced results that were appropriate for peripheral nerve regeneration. Compared to a hollow chitosan channel, this double-layer composite hydrogel nerve conduit considerably aided sciatic nerve regeneration in a rat sciatic nerve damage model, and the results of the nerve healing were comparable to those of autologous transplantation.

12.5.2.2 Alginate

Alpha-L-guluronic acid and beta-D-mannuronic acid combine to form alginate, a polymer based on polysaccharides. The source of this natural polymer is brown seaweed. Numerous animal models have been used to investigate the regenerative capacity of alginate in peripheral nerve damage. For regeneration in peripheral nerve injury (PNI), a study team has shown that the canal does not have to have a specific shape; a sponge-like shape can show regenerative effects even without suture fixation [41]. The use of sutures to connect the two nerve endings in canals not only requires a strict microsurgical procedure but also causes significant damage to the nerve endings.

Rod-shaped neural conduits with aligned nerve fibers that can be used to reconnect severed nerves were created by Kato-Negishi et al. [42]. They used a coaxial laminar flow microfluidic system that creates a three-layer laminar flow to achieve this core–shell configuration. Cortical cells, sodium alginate, and calcium chloride solutions with type I collagen formed the core, shell, and mantle, respectively. The cells in the alginate hydrogel were aligned by the laminar flow of the microfluidic device. The resulting structure allowed the rod-shaped neural units to connect at both ends, separated the sodium alginate hydrogel shell from the surrounding tissue at the site of nerve injury, and supported the alignment of nerve fiber production. The authors suggested that the developed rod-shaped neural units could be suitable not only for the clinical treatment of peripheral nerve injuries but also for modeling and studying neural cell interactions.

A covalent crosslinking technique was used in a different investigation to increase the rate at which alginate-based hydrogels degraded [43]. Ethylenediamine was utilized to crosslink alginate, and the resulting hydrogels were applied to a cat model to treat PNS injuries. Both hydrogel and tubular forms of alginate assisted PNS healing, according to in vivo investigations. This study suggested that hydrogels might have therapeutic value for clinical usage, despite the fact that commercially available nerve guiding conduits are tubular.

Alginate is selected for research on spinal diseases and is also classified as a "pro-regenerative" and "anti-degenerative" agent. Alginate can be utilized as a hydrogel to promote nerve regeneration, create a milieu for axonal adhesion, and fill gaps in injured spinal cords. In one work, regenerated axons after alginate hydrogel implantation filled a 2 mm gap between the two severed ends of a rat spinal cord [44]. Without the help of growth factors that were implanted, the axons developed in the direction of the implant, and there were fewer glial cells and glial scarring.

12.5.2.3 Hyaluronic Acid

A member of the family of naturally occurring polysaccharide-based polymers, hyaluronic acid (HA) is produced by practically all living things. N-acetylglucosamine and glucuronic acid residues repeat as disaccharide units to form HA, a linear, highly hydrated, negatively charged glycosaminoglycan. With its hydrated, interconnected pores that promote axonal regeneration, angiogenesis, and neuronal sprouting, it has been shown to be a promising biomaterial for treating damage to peripheral nerves [45].

Rajaram et al. produced functional 3D porous alginate/HA hydrogels using a bioplotting technique, demonstrating the long-term survival of Schwann cells [46]. Additionally, the researchers filled the hollow lumen of nerve conduits with hydrogels based on HA. By employing coagulation factor XIII (FXIIIa) as a crosslinking agent for HA hydrogels, Broguiere et al. were able to do away with the necessity for chemical crosslinkers [47]. They claimed that the resulting hydrogel was biocompatible, long-lasting, stable in solution, injectable, and cell-friendly. The findings demonstrated proper neuron polarization, strong synaptic connections, fast and sustained electrical activity, and rapid neurite growth. In order to improve spinal cord regeneration by creating an ideal microenvironment, Huang et al. have developed a platform that combines natural silk nanofibrils, hyaluronic acid, and basic fibroblast growth factor [48]. Because of its unique nanostructure and polysaccharide composition, this bioactive scaffold exhibits notable porosity and hydrophilicity. The concentration of hyaluronic acid can be changed to change the pore size from 90 μm to 120 μm. Human umbilical vein endothelial cells used in in vitro experiments showed enhanced migration, proliferation, and viability of the cells on the scaffold. Rats used in in vivo trials showed that the scaffold improved spinal cord regeneration and aided in the restoration of motor functions. This outcome may be linked to the scaffold's ability to reduce glial scar formation, as shown by lower levels of growth-associated protein-43, microtubule-associated protein 2, and neurofilament-200.

12.5.2.4 Agarose

Red algae's cell walls contain repeating units of β-D-galactopyranose and anhydro-α-L-galactopyranose, which combine to form agarose, a naturally occurring polymer. Agarose's stiffness, porosity, and biocompatibility allow it to be tailored for a range of nerve tissue engineering uses [49]. Agarose is biocompatible but non-immunogenic, non-biodegradable, and may generate non-cytotoxic hydrogels that can be employed as cell-encapsulated structures for nervous system regeneration, drug/gene carriers, and nerve conduits. Cellular fibrin-agarose hydrogels have also been demonstrated in studies to greatly improve peripheral nerve healing. The research investigated the implantation of nano-structured fibrin-agarose bioartificial nerve conduits (NFABNS) either alone or in combination with NeuraGen® collagen type I conduits to close a 10-mm sciatic nerve gap in rats. NFABNS successfully closed the gap by 14 weeks, significantly improving functional recovery and peripheral nerve regeneration [50].

In vitro, agarose can also be employed to control neurite development and cell adhesion. Immobilizing different laminin oligopeptides improves the biocompatibility of agarose scaffolds and boosts the gels' ability to encourage neurite outgrowth in nerve tissues. Stokols and colleagues supplemented agarose scaffolds with freeze-dried brain-derived neurotrophic factors (BDNF) to encourage linear axonal development along the scaffold. In order to promote spinal cord healing when neurites passed through the scaffold, agarose hydrogels containing BDNF-releasing microtubules were inserted into a damaged region of a rat spinal cord [51].

In another study, by varying quantities, special agarose/gelatin/polypyrrole (Aga/Jel/PPy, AGP3) hydrogels were created that had mechanical characteristics and conductivity similar to spinal cord tissue [52]. Noncovalent interactions caused the gelation, and because of their physically cross-linked properties, AGP3 hydrogels are injectable. These hydrogels' great biocompatibility was shown in vitro, allowing neural stem cells that adhered to the structures to extend their neurites. According to in vivo research, the hydrogel helped in neurogenesis and fully covered the tissue damage area. Furthermore, it was shown that AGP3 promotes nerve repair by inhibiting the production of glial fibrous tissue.

12.5.3 Synthetic Hydrogels

These hydrogels are generally synthesized in the labs. The most commonly used ones are methacrylate-based hydrogels, polyethylene glycol, and polyvinyl alcohol.

12.5.3.1 Methacrylate-Based Hydrogels

For many years, poly(hydroxyethyl methacrylate) (pHEMA) has been recognized as one of the most significant and extensively utilized hydrogel biomaterials. Since 1955, it has been altered using a variety of natural and synthetic materials in a variety of ways and used for tissue engineering, drug administration, and the creation of contact lenses and dressings [53]. The manufacturing process, polymer content, degree of cross-linking, temperature, and final application environment all affect pHEMA's characteristics. The hydrogel's usage in the creation of neural guide conduits is made easier by its capacity to adjust its structural characteristics to suit nerve tissue.

Like other hydrogels, pHEMA can be utilized as a filler inside the guide channel as well as the cylindrical structural material of the channel. Multi-walled carbon nanotubes (mwCNTs) were added to pHEMA in one study to offer conductivity for nerve guide channel construction [54]. The usefulness of pHEMA hydrogels for nerve regeneration was evaluated in vitro using SHSY5Y human neuroblastoma cells. These hydrogels were created with a porous structure to help transfer vital substances for cell development and feeding while eliminating waste. SHSY5Y neuroblastoma cells planted in mwCNT-loaded pHEMA channels remained viable and developed a neural network topology when an electrical current was applied. However, no neuronal network organization was seen in channels composed entirely of pHEMA, even if the cells were still viable. Furthermore, the same group created an exterior conduit called poly(2-hydroxyethylmethacrylate) (pHEMA) that contained an interpenetrating network (IPN) comprising methacrylated gelatin (GelMA) and methacrylated hyaluronic acid (HaMA) [55]. To support the SH-SY5Y cell model, Schwann cells (SCs) were added to the pHEMA, and netrin-1 was injected into the tube's distal end. The presence of netrin-1 caused SHY5Y cells to move along the conduit. NeuN expression increased after 2 weeks, suggesting increased neuronal activity.

In another work, 10-mm surgically generated holes in rat sciatic nerves were filled with porous, 12-mm-long poly(2-hydroxyethyl methacrylate-co-methyl methacrylate) (pHEMA-MMA) tubes, which had an outer diameter of 1.8 mm and an inner diameter of 1.3 mm [56]. Collagen matrix alone or collagen matrix enhanced with neurotrophin-3, brain-derived neurotrophic factor, or acidic fibroblast growth factor (FGF-1) was used to fill the inner lumen of the tubes. It was discovered that the tubes were biocompatible and biologically stable, supporting nerve regeneration in more than 90% of cases.

For the repair of spinal cord injury lesions, poly(hydroxypropyl methacrylate) (pHPMA) exhibits promise as a more biocompatible hydrogel than pHEMA. Adhesive peptides and other compounds are frequently combined with pHPMA hydrogels. According to one study, 3 months after a severe injury, inserting pHPMA into a hole made in the rat spinal cord not only decreased the void's volume but also promoted axonal development along the grafts, which aided in functional recovery [57]. Furthermore, via facilitating binding to integrin receptors, adhesion-promoting peptides (like RGD) added to pHPMA hydrogels are known to promote axonal development and decrease the expression of stress proteins.

In one study, a series of 1-, 4-, and 7-channel nerve growth conduits (NGCs) were made from poly(caprolactone) methacrylate (PCLMA) using sophisticated projection-based 3D printing [58]. The morphological, mechanical, and biological characteristics of these NGCs were assessed using in vitro Schwann cell culture and in vivo rat nerve healing procedures. The findings indicated that multi-channel NGCs are more successful in directing nerve repair since 4- and 7-channel NGCs considerably exceeded 1-channel NGCs in fostering nerve regeneration.

12.5.3.2 Polyethylene Glycol (PEG)

PEG-based hydrogels are widely used in nerve tissue engineering because of their great biocompatibility, non-toxicity to adjacent tissues, and water solubility. PEG applied directly to damaged sites restores motor function by repairing cell membranes, lowering oxidative stress, and encouraging axonal regeneration. Furthermore, PEG can be covalently or noncovalently coupled with proteins, peptides, and nanoparticles to decrease its immunogenicity and restrict its clearance by the reticuloendothelial system for use in spinal cord injury [59]. In order to promote brain tissue regeneration, cross-linked PEG can also operate as a release mechanism for cells, such as bone marrow stromal cells, and bioactive substances that can alter growth factors and inflammatory reactions. Additionally, PEG hydrogels can be used as injectable formulations that, when cross-linked, gel in situ at the site of injury [60]. It is well known that Ca^{2+} enters neurons as a result of initial damage in spinal cord accidents. Axon degeneration can result from the production of free oxidative radicals by this buildup of Ca^{2+}, which can trigger apoptosis and activate proteases. PEG functions as a surface-active agent, encouraging membrane fusion and fluidity because of its hydrophilic nature. PEG can stop nerve fiber degradation and foster an environment that is conducive to nerve fiber regeneration in spinal cord injury. Applying an electric field and sealing cell membranes with PEG may work together to prevent the entry of Ca^{2+} ions, facilitating functionality restoration. PEG also stimulates angiogenesis, isolates or reduces local glial scar invasion, and thus facilitates and directs axonal regeneration while helping to re-establish synaptic connections with target tissues [61].

To produce injectable platforms that can carry cells or other medications straight to the site of harm, several PEG formulations have been developed. In the early phases of injury in rats, for example, a study evaluating the effects of PEG-600, matrigel, and alginate revealed that PEG produced the maximum vascularization and axonal density in the lesion area [62]. In contrast, no axonal growth was observed with matrigel, and no differences were found between the alginate and resection-only groups. Thus, PEG was interpreted as the hydrogel that provided the greatest vascularization within the lesion area.

12.5.3.3 Polyvinyl Alcohol (PVA)

Hydrogels based on polyvinyl alcohol (PVA) are notable for their superior mechanical qualities and capacity to sustain a persistently damp environment. One of the only FDA-approved non-biodegradable synthetic nerve conduits is PVA, which is sold under the brand name SaluTunnelTM (Table 12.1) [63]. SaluTunnel™ features a longitudinal slit to facilitate surgical placement, with an inner diameter of 10 mm and a standard length of 6.35 cm. This nerve protective wrap is used in the treatment of peripheral nerve injuries without loss of nerve tissue. It is presented sterile and hydrated in saline at the surgical site and is designed for single use only. However, the non-absorbable nature of this material introduces disadvantages, such as nerve compression and tension-related issues at the suture sites after nerve regeneration is complete. This indicates that PVA-based nerve conduits should be carefully evaluated in clinical applications.

12.6 Injectable Hydrogels Used in Brain Injuries

The anatomy and function of the brain are extremely complicated in comparison to other organs. Damage to the brain can occur in deep areas and affect functional tissues. In order to treat brain damage, tissue engineering and the choice of biomaterials are crucial. The implanted material should fill different topological features, interact well with brain tissue, avoid irreparable harm to tissue function, and not put undue pressure on the patient's brain tissue. A specific level of cell infiltration, proliferation, and differentiation should be facilitated by the substance [64].

The necessary geometric configurations can be provided by a variety of injectable hydrogels. However, because of shear stress and high injection pressure, injecting liquid hydrogels into the cranial cavity using small-diameter needles or catheters may result in further tissue injury. Furthermore, hydrogels injected in liquid form might have different solidification times; delayed solidification can hinder the dispersion of implanted cells and medications, while quick solidification can clog needles and catheters [60].

Cell movement and migration depend on the stiffness of the hydrogel. Neurons, astrocytes, and endothelial cells tend to survive better in softer gels, while astrocytes perform better in stiffer gels. For instance, neurons exhibit the best survival in hydrogels with stiffness less than 1 kPa, whereas astrocytes show improved survival in gels up to 10 kPa [65]. This indicates that the mechanical properties of the gel must match the characteristics of brain tissue, which can vary with age, gender, and disease of the patient.

The most widely used natural hydrogel materials for brain injuries are derived from extracellular matrix (ECM) constituents such as collagen, cellulose, gelatin, and hyaluronic acid [66]. Despite drawbacks including production variability and immunological reactions, these natural hydrogels are favored for their high biocompatibility, suitable breakdown rates, and efficiency in promoting cellular biological processes. In contrast, synthetic polymer hydrogels have gained significant interest in recent years. Synthetic materials like polyethylene glycol (PEG), polyhydroxyethyl methacrylate (PHEMA), and poly-N-hydroxyethyl acrylamide (PHPMA) offer better controllability compared to natural materials, allowing for improved compatibility with the intracranial injury environment. Therefore, synthetic hydrogels, produced with the desired properties, potentially provide more effective solutions for brain injuries.

12.7 Conclusion

Hydrogels are increasingly being used in the treatment of injuries to the brain, spinal cord, and peripheral nerves due to their ability to mimic the extracellular matrix and provide an optimal environment for cellular processes such as adhesion, proliferation, and differentiation. To develop nerve guiding conduits—which are crucial for promoting nerve regeneration—key characteristics such as biocompatibility, porosity, flexibility, and ease of manufacture are required. These properties can be effectively achieved by hydrogels. Moreover, the incorporation of therapeutic molecules leads to targeted therapy, enhances the regenerative potential of the surrounding tissues, and allows nerve regeneration at the injury site. In the context of brain injuries, injectable hydrogels have particular importance. Their minimally invasive nature allows for easy application, reducing the risk of additional trauma during the delivery process. Injectable hydrogels can adapt to the complex architecture of brain tissue, filling voids and creating a supportive environment for neuronal recovery. As the understanding of hydrogel properties and their interactions with neural tissue continues to advance, the scope of their applications in clinical settings is expected to expand further. Future research will likely focus on optimizing hydrogel formulations to match the mechanical properties and biochemical signals of specific tissue types, ensuring effective integration and functionality within the host environment. Additionally, studies may explore the long-term effects of hydrogel applications on tissue regeneration and functional recovery. Ultimately, the ongoing exploration of hydrogels represents a promising frontier in the treatment of nerve injuries and brain trauma, with the potential to revolutionize patient care and recovery processes in neurology and rehabilitation.

References

1. Marinho, L.S.R., Ikebara, J.M., Higa, G.S.V., de Lima Vasconcellos, T.H., Móvio, M.I., Takada, S.H., Kihara, A.H.: Overview of the structure and function of the nervous system. In: Handbook of Neural Engineering, pp. 17–48. Academic Press, New York (2025)
2. Ashaber, M., Tomina, Y., Kassraian, P., Bushong, E.A., Kristan, W.B., Ellisman, M.H., Wagenaar, D.A.: Anatomy and activity patterns in a multifunctional motor neuron and its surrounding circuits. Elife. **10**, e61881 (2021)
3. Zhang, T., Zeng, Y., Zhang, Y., Zhang, X., Shi, M., Tang, L., et al.: Neuron type classification in rat brain based on integrative convolutional and tree-based recurrent neural networks. Sci. Rep. **11**(1), 7291 (2021)
4. Lopes, B., Sousa, P., Alvites, R., Branquinho, M., Sousa, A.C., Mendonça, C., et al.: Peripheral nerve injury treatments and advances: one health perspective. Int. J. Mol. Sci. **23**(2), 918 (2022)
5. Duan, L., Yu, X.: Fibroblasts: new players in the central nervous system? Fundam. Res. **4**(2), 262–266 (2024)
6. Samadian, H., Maleki, H., Fathollahi, A., Salehi, M., Gholizadeh, S., Derakhshankhah, H., et al.: Naturally occurring biological macromolecules-based hydrogels: potential biomaterials for peripheral nerve regeneration. Int. J. Biol. Macromol. **154**, 795–817 (2020)
7. Irisarri, C.: History of peripheral nerve injuries. J. Hand Surg. **49**(6), 812–823 (2024)
8. Nova One Advisor. U.S. peripheral nerve injuries market size—companies. Report Code: 8121. https://www.novaoneadvisor.com/report/us-peripheral-nerve-injuries-market (2024)
9. Shen, H., Fan, C., You, Z., Xiao, Z., Zhao, Y., Dai, J.: Advances in biomaterial-based spinal cord injury repair. Adv. Funct. Mater. **32**(13), 2110628 (2022)
10. Kaplan, B., Levenberg, S.: The role of biomaterials in peripheral nerve and spinal cord injury: a review. Int. J. Mol. Sci. **23**(3), 1244 (2022)
11. Kong, F., Mehwish, N., Niu, X., Lin, M., Rong, X., Hu, F., Lee, B.H.: Personalized hydrogels for individual health care: preparation, features, and applications in tissue engineering. Mater. Today Chem. **22**, 100612 (2021)
12. Ramesh, P.A., Sethuraman, S., Subramanian, A.: Multichannel conduits with fascicular complementation: significance in long segmental peripheral nerve injury. ACS Biomater Sci. Eng. **10**(4), 2001–2021 (2024)
13. Krauss, E.M., Weber, R.V., Mackinnon, S.E.: Nerve injury, repair, and reconstruction. In: Plastic Surgery-Principles and Practice, pp. 803–825. Elsevier, Amsterdam (2022)
14. Bozkurt, A., Brook, G.A., Moellers, S., Lassner, F., Sellhaus, B., Weis, J., et al.: In vitro assessment of axonal growth using dorsal root ganglia explants in a novel three-dimensional collagen matrix. Tissue Eng. **13**(12), 2971–2979 (2007)

15. Gonzalez-Perez, F., Cobianchi, S., Heimann, C., Phillips, J.B., Udina, E., Navarro, X.: Stabilization, rolling, and addition of other extracellular matrix proteins to collagen hydrogels improve regeneration in chitosan guides for long peripheral nerve gaps in rats. Neurosurgery. **80**(3), 465–474 (2017)
16. Hiraoka, M., Kato, K., Nakaji-Hirabayashi, T., Iwata, H.: Enhanced survival of neural cells embedded in hydrogels composed of collagen and laminin-derived cell adhesive peptide. Bioconjug. Chem. **20**(5), 976–983 (2009)
17. Rein, S., Schober, R., Poetschke, J., Kremer, T.: Non degradation of chitosan and initial degradation of collagen nerve conduits used for protection of nerve coaptations. Microsurgery. **44**(1), e31093 (2024)
18. Behan, B.L., DeWitt, D.G., Bogdanowicz, D.R., Koppes, A.N., Bale, S.S., Thompson, D.M.: Single-walled carbon nanotubes alter Schwann cell behavior differentially within 2D and 3D environments. J. Biomed. Mater. Res. A. **96**(1), 46–57 (2011)
19. Cholas, R., Hsu, H.P., Spector, M.: Collagen scaffolds incorporating select therapeutic agents to facilitate a reparative response in a standardized hemiresection defect in the rat spinal cord. Tissue Eng. A. **18**(19–20), 2158–2172 (2012)
20. Liu, T., Houle, J.D., Xu, J., Chan, B.P., Chew, S.Y.: Nanofibrous collagen nerve conduits for spinal cord repair. Tissue Eng. A. **18**(9–10), 1057–1066 (2012)
21. Jacobs, T., Patil, D., Ziccardi, V.B.: Both type I bovine collagen conduits and porcine small intestine submucosa conduits result in functional sensory recovery following peripheral nerve microsurgery: a systematic review and meta-analysis. J. Oral Maxillofac. Surg. **82**(12), 1559–1568 (2024)
22. Contessi Negrini, N., Angelova Volponi, A., Sharpe, P.T., Celiz, A.D.: Tunable cross-linking and adhesion of gelatin hydrogels via bioorthogonal click chemistry. ACS Biomater Sci. Eng. **7**(9), 4330–4346 (2021)
23. Tonda-Turo, C., Gnavi, S., Ruini, F., Gambarotta, G., Gioffredi, E., Chiono, V., et al.: Development and characterization of novel agar and gelatin injectable hydrogel as filler for peripheral nerve guidance channels. J. Tissue Eng. Regen. Med. **11**(1), 197–208 (2017)
24. Usal, T.D., Yucel, D., Hasirci, V.: A novel GelMA-pHEMA hydrogel nerve guide for the treatment of peripheral nerve damages. Int. J. Biol. Macromol. **121**, 699–706 (2019)
25. Han, Y., Sun, M., Lu, X., Xu, K., Yu, M., Yang, H., Yin, J.: A 3D printable gelatin methacryloyl/chitosan hydrogel assembled with conductive PEDOT for neural tissue engineering. Compos. Part B. **273**, 111241 (2024)
26. Hodde, D., Gerardo-Nava, J., Wöhlk, V., Weinandy, S., Jockenhövel, S., Kriebel, A., et al.: Characterisation of cell–substrate interactions between Schwann cells and three-dimensional fibrin hydrogels containing orientated nanofibre topographical cues. Eur. J. Neurosci. **43**(3), 376–387 (2016)
27. Du, J., Liu, J., Yao, S., Mao, H., Peng, J., Sun, X., et al.: Prompt peripheral nerve regeneration induced by a hierarchically aligned fibrin nanofiber hydrogel. Acta Biomater. **55**, 296–309 (2017)
28. Chow, N., Miears, H., Cox, C., MacKay, B.: Fibrin glue and its alternatives in peripheral nerve repair. Ann. Plast. Surg. **86**(1), 103–108 (2021)
29. Isaacs, J.E., McDaniel, C.O., Owen, J.R., Wayne, J.S.: Comparative analysis of biomechanical performance of available "nerve glues". J. Hand Surg. Am. **33**(6), 893–899 (2008)
30. Liu, Y.J., Chen, X.F., Zhou, L.P., Rao, F., Zhang, D.Y., Wang, Y.H.: A nerve conduit filled with Wnt5a-loaded fibrin hydrogels promotes peripheral nerve regeneration. CNS Neurosci. Ther. **28**(1), 145–157 (2022)
31. Wei, G.J., Yao, M., Wang, Y.S., Zhou, C.W., Wan, D.Y., Lei, P.Z., et al.: Promotion of peripheral nerve regeneration of a peptide compound hydrogel scaffold. Int. J. Nanomedicine. **8**, 3217 (2013)
32. Yuan, T., Li, W., Zhou, M., Wang, X., Wang, B., Zhao, Y.: Biomimetic multichannel silk nerve conduits with multicellular spatiotemporal distributions for spinal cord injury repair. Adv. Mater. **36**(44), 2411628 (2024)

33. Hromada, C., Heimel, P., Kerbl, M., Gál, L., Nürnberger, S., Schaedl, B., et al.: Silk-based nerve guidance conduits with macroscopic holes modulate the vascularization of regenerating rat sciatic nerve. Neural Regen. Res. **20**(6), 1789–1800 (2025)
34. Sun, W., Taylor, C.S., Gao, Z., Gregory, D.A., Haycock, J.W., Zhao, X.: Co-assembling bioactive short peptide nanofibers coated silk scaffolds induce neurite outgrowth of PC12 cells. Int. J. Biol. Macromol. **278**, 134774 (2024)
35. Xie, C., Chen, Y., Wang, L., Liao, K., Xue, B., Han, Y., et al.: Recent research of peptide-based hydrogel in nervous regeneration. Bioact. Mater. **40**, 503–523 (2024)
36. Cavanaugh, M., Silantyeva, E., Pylypiv Koh, G., Malekzadeh, E., Lanzinger, W.D., Willits, R.K., Becker, M.L.: RGD-modified nanofibers enhance outcomes in rats after sciatic nerve injury. J. Funct. Biomater. **10**(2), 24 (2019)
37. Zhang, N., Luo, Y., He, L., Zhou, L., Wu, W.: A self-assembly peptide nanofibrous scaffold reduces inflammatory response and promotes functional recovery in a mouse model of intracerebral hemorrhage. Nanomedicine. **12**(5), 1205–1217 (2016)
38. Pfister, L.A., Papaloïzos, M., Merkle, H.P., Gander, B.: Hydrogel nerve conduits produced from alginate/chitosan complexes. J. Biomed. Mater. Res. A. **80**(4), 932–937 (2007)
39. Nawrotek, K., Tylman, M., Rudnicka, K., Gatkowska, J., Wieczorek, M.: Epineurium-mimicking chitosan conduits for peripheral nervous tissue engineering. Carbohydr. Polym. **152**, 119–128 (2016)
40. Deng, P., Chen, F., Zhang, H., Chen, Y., Zhou, J.: Multifunctional double-layer composite hydrogel conduit based on chitosan for peripheral nerve repairing. Adv. Healthc. Mater. **11**(13), 2200115 (2022)
41. Wu, S., Suzuki, Y., Tanihara, M., Ohnishi, K., Endo, K., Nishimura, Y.: Repair of facial nerve with alginate sponge without suturing: an experimental study in cats. Scand. J. Plast. Reconstr. Surg. Hand Surg. **36**(3), 135–140 (2002)
42. Kato-Negishi, M., Onoe, H., Ito, A., Takeuchi, S.: Rod-shaped neural units for aligned 3D neural network connection. Adv. Healthc. Mater. **6**(15), 1700143 (2017)
43. Abdelbasset, W.K., Jasim, S.A., Sharma, S.K., Margiana, R., Bokov, D.O., Obaid, M.A., et al.: Alginate-based hydrogels and tubes, as biological macromolecule-based platforms for peripheral nerve tissue engineering: a review. Ann. Biomed. Eng. **50**(6), 628–653 (2022)
44. Jahandideh, A., Noori, H., Rahimi, B., Hamblin, M.R., Behroozi, Z., Ramezani, M., Ramezani, F.: Alginate scaffolds improve functional recovery after spinal cord injury. Eur. J. Trauma Emerg. Surg. **48**(3), 1711–1721 (2022)
45. Nelson, D.W., Gilbert, R.J.: Extracellular matrix-mimetic hydrogels for treating neural tissue injury: a focus on fibrin, hyaluronic acid, and elastin-like polypeptide hydrogels. Adv. Healthc. Mater. **10**(22), 2101329 (2021)
46. Rajaram, A., Schreyer, D., Chen, D.: Bioplotting alginate/hyaluronic acid hydrogel scaffolds with structural integrity and preserved schwann cell viability. 3D Print. Addit. Manuf. **1**(4), 194–203 (2014)
47. Broguiere, N., Cavalli, E., Salzmann, G.M., Applegate, L.A., Zenobi-Wong, M.: Factor XIII cross-linked hyaluronan hydrogels for cartilage tissue engineering. ACS Biomater Sci. Eng. **2**(12), 2176–2184 (2016)
48. Huang, A., Huang, Y., Yang, W., Wang, L., You, R., Wang, J., et al.: Fabrication of multifunctional silk nanofibril/hyaluronic acid scaffold for spinal cord repair. Int. J. Biol. Macromol. **263**, 130287 (2024)
49. Shinde, S.S., Sarkate, A.P., Kumar, P., Surwade, S.P., Salunkhe, T.A., Rathod, S.S., et al.: Smart sustainable biomaterials and their recent advancement for targeting the nervous system. In: Advances in Sustainable Biomaterials, pp. 117–140
50. Chato-Astrain, J., Campos, F., Roda, O., Miralles, E., Durand-Herrera, D., Sáez-Moreno, J.A., et al.: In vivo evaluation of nanostructured fibrin-agarose hydrogels with mesenchymal stem cells for peripheral nerve repair. Front. Cell. Neurosci. **12**, 501 (2018)

51. Stokols, S., Tuszynski, M.H.: Freeze-dried agarose scaffolds with uniaxial channels stimulate and guide linear axonal growth following spinal cord injury. Biomaterials. **27**(3), 443–451 (2006)
52. Jahandideh, A., Noori, H., Rahimi, B., Hamblin, M.R., Behroozi, Z., Ramezani, M., Ramezani, F.: Alginate scaffolds improve functional recovery after spinal cord injury. Eur. J. Trauma Emerg. Surg. **48**(3), 1711–1721 (2021)
53. Omidian, H., Chowdhury, S.D., Cubeddu, L.X.: Hydrogels for neural regeneration: exploring new horizons. Materials. **17**(14), 3472 (2024)
54. Arslantunali, D., Budak, G., Hasirci, V.: Multiwalled CNT-pHEMA composite conduit for peripheral nerve repair. J. Biomed. Mater. Res. A. **102**(3), 828–841 (2014)
55. Arslantunalı-Sahin, D., Son, C.D., Hasirci, V.: Fabrication and characterization of pHEMA hydrogel conduit containing GelMA-HaMA IPN for peripheral nerve regeneration. Explor. BioMat-X. **1**(1), 34–57 (2024)
56. Midha, R., Munro, C.A., Dalton, P.D., Tator, C.H., Shoichet, M.S.: Growth factor enhancement of peripheral nerve regeneration through a novel synthetic hydrogel tube. J. Neurosurg. **99**(3), 555–565 (2003)
57. Woerly, S., Pinet, E., De Robertis, L., Van Diep, D., Bousmina, M.: Spinal cord repair with PHPMA hydrogel containing RGD peptides (NeuroGel™). Biomaterials. **22**(10), 1095–1111 (2001)
58. Li, H., Yao, K., Chen, Y., Ye, W., Shu, Q.: Projection-based 3D printing of multichannel poly (caprolactone) methacrylate nerve guidance conduit for peripheral nerve regeneration. Mater. Adv. **5**(11), 4753–4763 (2024)
59. Yu, L., Bennett, C.J., Lin, C.H., Yan, S., Yang, J.: Scaffold design considerations for peripheral nerve regeneration. J. Neural Eng. **21**(4), 041001 (2024)
60. Gao, Y., Zhang, T.L., Zhang, H.J., Gao, J., Yang, P.F.: A promising application of injectable hydrogels in nerve repair and regeneration for ischemic stroke. Int. J. Nanomedicine. **19**, 327–345 (2024)
61. Lu, X., Perera, T.H., Aria, A.B., Callahan, L.A.S.: Polyethylene glycol in spinal cord injury repair: a critical review. J. Exp. Pharmacol. **10**, 37–49 (2018)
62. Estrada, V., Brazda, N., Schmitz, C., Heller, S., Blazyca, H., Martini, R., Müller, H.W.: Long-lasting significant functional improvement in chronic severe spinal cord injury following scar resection and polyethylene glycol implantation. Neurobiol. Dis. **67**, 165–179 (2014)
63. Taylor, C.S., Haycock, J.W.: Biomaterials and scaffolds for repair of the peripheral nervous system. In: Peripheral Nerve Tissue Engineering and Regeneration, pp. 245–279 (2022)
64. Podell, J.E., Morris, N.A.: Traumatic brain injury and traumatic spinal cord injury. Continuum. **30**(3), 721–756 (2024)
65. Lin, P.H., Dong, Q., Chew, S.Y.: Injectable hydrogels in stroke and spinal cord injury treatment: a review on hydrogel materials, cell–matrix interactions and glial involvement. Mater. Adv. **2**(8), 2561–2583 (2021)
66. Sun, H., Zhang, L., Cheng, W., Hao, F., Zhou, L., Li, Q.: Injectable hydrogels in repairing central nervous system injuries. Adv. Mater. Sci. Eng. **2021**(1), 7381980 (2021)

Chapter 13
Hydrogels Used in Cartilage Treatment

Nisa İrem Büyük, Didem Aksu, Fatma Neşe Kök, and Gamze Torun Köse

Abstract Cartilage tissue is a viscoelastic connective tissue that is mainly responsible for load bearing, load transfer, and enabling movement of joints. Cartilage injuries occur due to trauma, aging, sports activities, and pathological conditions. The avascular and aneural nature of cartilage tissue limits its reparative ability, requiring clinical intervention for treatment. The challenges encountered with traditional surgical methods have led to a rise in interest in the advancement of methods based on cartilage tissue engineering in recent years. The unique advantages of hydrogels make them a popular choice in the tissue engineering approach. These biomaterials are biocompatible and biodegradable, possess high water retention capacity, and can replicate native tissue, resulting in an appropriate scaffold for the treatment of cartilage damage. Natural polymers, including alginate, agarose, fibrin, hyaluronic acid, collagen, gelatin, chitosan, chondroitin sulfate, cellulose, gellan gum, and dextran, alongside synthetic polymers such as polyethylene glycol, polylactic acid, polycaprolactone, and polylactic-co-glycolic acid, are utilized in hydrogel production, commonly employed in cartilage tissue engineering approaches. Besides the utilization of natural or synthetic polymers individually in hydrogel fabrication, the potential of composite hydrogels in cartilage treatment is significantly promising.

Keywords Cartilage · Cartilage defects · Cartilage regeneration · Tissue engineering · Natural polymers · Synthetic polymers · Hydrogels

---

N. İ. Büyük · D. Aksu · G. Torun Köse (✉)
Department of Genetics and Bioengineering, Faculty of Engineering, Yeditepe University, Istanbul, Türkiye
e-mail: gamzekose@yeditepe.edu.tr

F. N. Kök
Department of Molecular Biology and Genetics, Faculty of Science and Letters, Istanbul Technical University, Istanbul, Türkiye

N. Hasirci, V. Hasirci (eds.), *Hydrogels and Bioinks in Tissue Engineering*, Biological and Medical Physics, Biomedical Engineering,
https://doi.org/10.1007/978-3-031-90966-5_13

13.1 Cartilage Tissue and Structural Components

Cartilage is connective tissue characterized by a viscoelastic nature that occurs prior to the formation of bone tissue during embryogenesis. In mammals, this tissue serves as a connective interface that reduces friction between bones, improves load transfer, and enables movement of joints. Cartilage types can be classified by their distinct biochemical, structural, and mechanical features [1–3] and are found in the human body in three different forms: fibrous, elastic, and hyaline (articular) cartilage.

Fibrous cartilage, the most resilient cartilage tissue in the human body, has both collagen type I and collagen type II. This cartilage is predominantly located in regions requiring substantial pressure and tensile strength, such as intervertebral discs and the endpoints of tendons and ligaments. Elastic cartilage is a type of cartilage tissue that is distinguished by its elasticity owing to the presence of elastin fibers, and it lacks nerves, blood vessels, and lymphatic vessels. Due to its ability to bend and stretch, this variety of cartilage is present in parts of the body, including the trachea and ears. The primary locations for hyaline cartilage, commonly referred to as articular cartilage, are the shoulder, elbow, hip, and knee joints. The outstanding mechanical properties of hyaline cartilage tissue originate from its structural composition, composed of proteins (such as collagens and proteoglycans), macromolecules, and a water content of 60–80% (Fig. 13.1) [1].

Hyaline cartilage comprises four fundamental layers: superficial tangential, middle, deep, and calcified zones. The thickness of the multiple layers of this complex cartilage tissue changes based on its structure, chondrocyte situation, and extracellular matrix (ECM) composition [2]. Chondrocytes, which constitute 5% of the cartilage volume, are responsible for the generation of ECM in cartilage tissue and the reduction of the coefficient of friction in the cartilage joint area. The composition of ECM in cartilage tissue is mainly composed of proteoglycans, comprising sulfated glycosaminoglycan (sGAG) groups attached to a hyaluronic backbone [1, 2]. Hyaline cartilage has a load-bearing surface at joint locations, capable of supporting approximately 3.5 times body weight at the knee and 1.5 times at the shoulder [1].

13.2 Cartilage Defects

The treatment of articular cartilage injury represents an important orthopedic issue in clinical practice. It is usually caused by aging, trauma, osteonecrosis, and osteochondral disorders. Avascular, hypocellular, and annular cartilage possesses a restricted capacity for self-repair following injury. The limited self-repair capacity of cartilage may result in joint deterioration, discomfort, functional impairment, and degenerative arthritis due to localized lesions [2, 3].

Osteoarthritis, a prevalent form of degenerative arthritis, is defined by a reduction in cartilage and structural alterations within the joint. The alterations

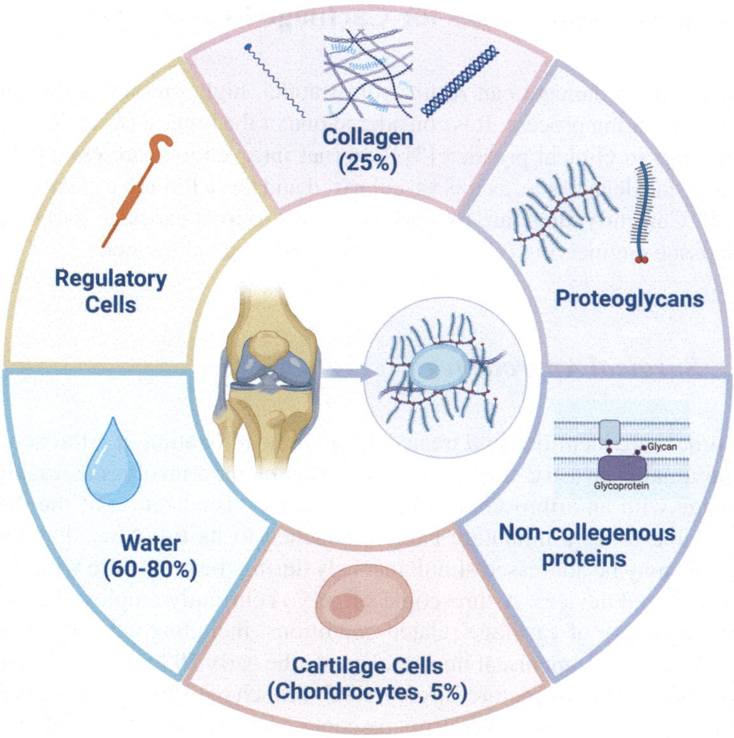

Fig. 13.1 Structural components of cartilage tissue [1, 2]. (Created in https://BioRender.com)

encompass a reduction in collagen and proteoglycan levels, degradation of ECM composition, fibrillation, meniscal degeneration, and hypertrophy of the subchondral bone. symptoms of osteoarthritis include pain in the joints, joint noises, stiffness, and inflammation following periods of immobilization. Arthritis can be classified as primary or secondary, with primary arthritis having undetermined causes and secondary arthritis resulting from prior damage. The primary risk factors for osteoarthritis include gender, age, genetic susceptibility, obesity, and previous joint injury. Unlike osteoarthritis, rheumatoid arthritis is an immune-mediated inflammatory disorder [1, 2]. Additionally, microfractures are encountered in the clinic, in addition to osteochondral and chondral injuries, which are prominent cartilage injuries. Microfractures may transform into macroscopic chondral fissures due to stress or deterioration. Chondral fissures can develop into osteochondral fissures caused by insufficient metabolic activity following injuries. Osteochondral injuries can penetrate the cartilage to the osteochondral bone, resulting in bleeding. Osteoarthritis may arise as a result of these neglected advanced cartilage diseases [1].

13.3 Current Approaches for Cartilage Regeneration

Articular cartilage damage can result from trauma, high stresses during physical exercise, or the aging process. It is considered one of the biggest obstacles for orthopedic surgeons in clinical practice [3]. External interventions are essential for the treatment of cartilage tissue, as it is vascularized and has a limited capacity to regenerate itself. Currently, there are two primary ways to treat cartilage tissue damage: cartilage tissue engineering (CTE) approaches and surgical methods.

13.3.1 Surgical Approaches

Joint disorders are identified and treated through the utilization of arthroscopic surgery, a minimally invasive surgical method. This method involves assessing cartilage damage with an arthroscope, which allows for visualization of the damage's size and shape using a miniature camera attached to its tip. After diagnosis, the affected area may be addressed simultaneously during the procedure with mechanical and motorized devices. Arthroscopic surgery is currently employed extensively to address a variety of cartilage-related conditions, including osteoarthritis of the knee and degenerative meniscal injuries. During the early 1980s, Steadman and colleagues developed microfracture surgery, a sub-branch of arthroscopic surgery. This technique requires drilling several microscopic apertures into the joint surface to activate the body's intrinsic healing mechanisms. For cartilage injury, the microfracture surgical technique is regarded as an effective arthroscopic treatment. An alternative surgical method is mosaicplasty surgery. The procedure involves transplanting tiny, cylindrical bone and cartilage tissues from non-load-bearing regions of the body to the injured area in a mosaic configuration [1].

13.3.2 Cartilage Tissue Engineering (CTE)

Different methods and applications are available for the therapy of cartilage damage [4]. CTE applications may be performed using cells only, scaffolds only, or through the integration of both components. Several cell types have been differentiated in vitro for cartilage regeneration purposes. The primary consideration is to prevent the immune system's rejection of the implanted cells. These methods use high-density cells to enhance cell-to-cell contacts and replicate embryonic development via the natural ECM generated by these cells, resulting in tissue regeneration. The stem cells, which are commonly employed, possess the capability to differentiate into bone, cartilage, adipose tissue, muscle, tendon/ligament, dermis, and various other connective tissue types [5].

Autologous chondrocytes, bone marrow-derived mesenchymal stem cells (BMSCs), adipose tissue-derived stem cells, and synovial-derived stem cells are the four cell types that can be used for CTE [5]. Despite achieving successful clinical outcomes with this technology, its application has been restricted by requirements for a further procedure, significant dermal injury, potential biological responses, and the limited capacity of recently differentiated cells. At the same time, it has been observed that usage of this differentiated cell type is ineffective for the treatment of osteochondral injuries. On the other hand, the advantages of mesenchymal stem cells (MSCs) have contributed to their increasing significance. Currently, autologous MSC therapies are frequently used for cartilage tissue degeneration. Although it is shown that differentially generated MSCs possess analogous characteristics, they do not exhibit the same phenotypes or differentiation potentials. BMSCs are the most substantially specified and commonly used for skeletal tissue regeneration. Adipose-derived stem cells are commonly used due to their minimally invasive nature, ease of collection, ability for cartilage differentiation, and high proliferation potential. Mesenchymal stem cells derived from synovial tissue are obtained from human synovial fluid and the synovium of the knee and hip and are utilized in the damaged tissue [5].

In addition to the stem cell approach, growth factors play a significant role in cartilage regeneration [5]. Platelet-rich plasma (PRP), fibroblast growth factor (FGF), insulin-like growth factor-I (IGF-I), and the transforming growth factor beta family (TGF-β) are all examples of growth factors that have been used for this purpose.

A distinct approach includes the targeted delivery of bioactive molecules to the specific site. Bioactive molecules utilized in CTE are categorized into two types: (1) small molecules and certain natural biomaterials, and (2) high molecular weight bioactive agents, which include synthetic polymers. A few examples of small molecules include simvastatin and cartogenin. Another approach requires the use of tissue scaffolds in CTE. Tissue scaffolds used for CTE may be classified as either natural or synthetic. Natural materials frequently utilized in cartilage repair involve agarose, alginate, chitosan, collagen, fibrin, and hyaluronan [5–7].

13.4 Application of Hydrogels in Cartilage Treatment

Hydrogels are a prominent component of tissue engineering strategies for the treatment of cartilage injuries. Hydrogels provide multipurpose biomaterials in tissue engineering and cell therapy, characterized by their important high-water content, biodegradability, and biocompatibility. Hydrogels can be developed from natural and synthetic polymers or their composites, with their mechanical strength enhanced through various cross-linking mechanisms, depending on their unique characteristics and specific functional groups. Natural and synthetic polymers utilized in hydrogel formation can effectively reproduce the unique physicochemical characteristics of the extracellular matrix.

13.4.1 Natural Polymer-Based Hydrogels

Natural hydrogels commonly used for CTE applications include alginate, agarose, fibrin, hyaluronic acid, collagen, gelatin, chitosan, chondroitin sulfate, cellulose, and gellan gum.

13.4.1.1 Alginate

Alginate is a natural polysaccharide that is anionic and water-soluble, normally produced from brown algae. It is composed of two copolymers: mannuronic acid and guluronic acid. Because of its high viscosity and gelation properties, it is widely used in CTE applications. Rapid gelation can be completed by employing cations with a valence of two or higher [3, 8]. The most implemented cation for this purpose is Ca^{+2}. Alginate can be administered into the damaged area without invasiveness once it has been converted to gel form. Alginate hydrogels are typically employed in conjunction with chondrocytes in CTE applications. For surface adhesion, cells can be placed into wet/dry hydrogels or encapsulated in gel form. Cells can be integrated into an alginate hydrogel and eventually printed with a 3D bioprinter [8]. While alginate is a natural material with certain advantages, it demonstrates limitations regarding surface proteins that affect cell interaction.

Alginate is utilized in applications as a composite material, generally formed with several types of materials.

By cross-linking hyaluronic acid (HA) and alginate (Alg) at various ratios, Thanh et al. [9] evaluated the biodegradability and 3D printability of the hydrogel. Additionally, they investigated the chondrogenic differentiation of the hydrogel by culturing MSCs. Consequently, it was found that the mechanical and degradability properties of HA-Alg hydrogel not only contribute to the development of cartilage tissue but also to the formation of other surrounding tissue. In a study performed by Wang et al. [10], a hybrid hydrogel was generated using sodium alginate and polyglutamic acid. The degradation time was determined to be extended by up to 125%, and the swelling rate decreased to 47% by including cellulose in a microcrystalline structure in the system.

Alginate hydrogel has shown the ability to improve regulatory gene expression in cartilage tissue. Multiple studies have been conducted to create gel microspheres by combining cells and/or growth factors with alginate solution and cross-linking with $CaCl_2$. Within the gel microspheres, the cells have a uniform distribution, as indicated by the results. Research showed that alginate spheres comprising cells can function as tissue scaffolds to promote chondrogenesis for cartilage regeneration [11]. The chondrogenic differentiation of adipose-derived mesenchymal stem cells in agarose, alginate, and gelatin tissue scaffolds has been researched by Awad et al. [12]. There was a substantial rise in protein and proteoglycan biosynthesis in alginate and gelatin hydrogels compared to agarose hydrogels, and all three varieties of hydrogels revealed significant results in cartilage differentiation when compared to control groups.

13.4.1.2 Agarose

Agarose is a linear natural polymer extracted from red algae, and its solid form can be dissolved in boiling water. Upon cooling this solution below 45 °C, extensive hydrogen bonding occurs between the chains. These bonds change the solution into a gel. The gelation temperature is affected by the monomer composition and the concentration of the solution. Agarose is a polymer that is especially appropriate for performing cell encapsulation experiments, particularly with chondrocytes. Adipose tissue-derived stem cells encapsulated in agarose have shown the ability to induce glycosaminoglycan production and differentiate into chondrocytes [12]. Agarose hydrogels have been reported to be employed in combination with stem cell therapy. The hydrogels were provided with several kinds of signaling molecules, and their influence on the maintenance of cartilage homeostasis was investigated [13]. The 3D magnetic biomaterial was developed by Bonhome-Espinoza et al. [14] through the integration of magnetic nanoparticles into fibrin-agarose hydrogels to produce cartilage-like tissue. The research showed that magnetic nanoparticles do not induce toxic effects within the hydrogel matrix; however, the scaffold demonstrates enhanced and more stable mechanical strength. Growth factors were released, and the proliferation rate of chondrocytes improved when platelet-enriched plasma (PRP) and chondrocyte cells were placed into agarose hydrogels [15].

To preserve the advantages of agarose and prevent its drawbacks, Singh et al. [16] developed an agarose/silk hydrogel. They proved that this composite structure induced the formation of GAG and collagen and was compatible with the immune system, particularly by reducing the secretion of tumor necrosis factor (TNF-α) by mouse macrophages.

13.4.1.3 Fibrin

Fibrin is a natural ECM component that is generated by the interaction between fibrinogen and thrombin, which plays a role in blood coagulation. The structure comprises glycoproteins that can be composed of several pairs of polypeptide chains. It is frequently used in ECM due to its biocompatible, biodegradable, and nanofibrous structural characteristics [17]. Studies have demonstrated that fibrin hydrogel is an ideal gel for implanting chondrocytes to create a new cartilage matrix. The newly formed tissue can establish a mechanical bond between cartilage disks and encourage healing. For cartilage tissue engineering, the US Food and Drug Administration (US-FDA) has approved human fibrin hydrogels. The comprehensive study conducted by Peretti et al. [18] on mice and piglets demonstrated that fibrin hydrogels are a sufficient cell-based therapy for articular cartilage lesions. Despite the suitability of fibrin polymer as a hydrogel for cartilage development, it was found to experience a substantial volume reduction of up to 50% post-implantation.

Due to its biocompatible structure and appropriate surface properties for cell adhesion, fibrin is a suitable material for 3D applications. Fibrin has been noted to have structural weaknesses and stiffness, making it incapable of maintaining its shape during 3D printing. This issue has been solved by integrating fibrin with different polymers. The mechanical properties of fibrin were improved, its structural density was raised, and a more biomechanically stable material was generated by cross-linking it with genipin [19]. The studies conducted by Almeida et al. [20] intended to achieve cartilage regeneration by utilizing fibrin hydrogel with cartilage ECM microparticles and transforming growth factor (TGF)-β3. In vitro experiments revealed that sGAG and type II collagen deposition increased in comparison to control groups, an outcome that was confirmed by in vivo studies on rodents.

13.4.1.4 Hyaluronic Acid

Hyaluronic acid (HA, or hyaluronan) is a nonsulfated glycosaminoglycan that serves as an essential component of cartilage ECM. HA is a highly suitable material for CTE due to its physicochemical properties such as viscosity, high water retention capacity, and elasticity. Due to the existence of receptors, it contributes to cell proliferation, differentiation, and interaction [21]. The bioorthogonal chemical reaction cross-linking of HA by Han et al. [22] demonstrated its potential for in situ clinical studies. The hydrogel generated by this investigation was shown to enhance cell survival and promote cell migration to cartilage tissue both in vitro and in vivo. Due to its specific characteristics, HA has potential as a 3D bioink material. Nevertheless, its low stability could result in difficulties when it is used alone. To solve these issues, HA can be cross-linked or combined with many different types of materials [17]. Davachi et al. [23] enzymatically cross-linked HA with chitosan to generate biocompatible and sustainable hydrogels demonstrating ideal mechanical properties. This approach facilitated cell encapsulation by enabling efficient operation in the liquid phase, and the usage of natural materials helped cartilage regeneration without any detrimental impacts. In another investigation [24], the gelation and mechanical properties of HA were considerably enhanced by enzymatically cross-linking it with silk fibroin, a natural material. Furthermore, the expression of cartilage matrix proteins that increase ECM secretion by chondrocytes was raised.

13.4.1.5 Collagen

Collagen, the most abundant protein in animal structure, is known for providing mechanical strength to tissues such as skin, bone, and cartilage. Due to its presence in cartilage ECM and its biocompatible and safe certification and clinical approval by the FDA, type I collagen is widely used as a tissue scaffold in CTE applications.

Collagen hydrogels trigger ECM synthesis, reduce the risk of immune rejection, and promote cartilage regeneration [25]. Type II collagen gel in injectable form has been used to treat articular cartilage defects. Although type II collagen is more abundant in cartilage, type I collagen-based hydrogels have been used for cartilage repair due to their applicability and homogeneous structure [3]. Vázquez-Portalatín [26] used a mixture of type I and II collagen in his study. With this mixture, it was aimed to utilize the biological activity of type II collagen and the mechanical properties of type I. It was reported that stable hydrogels with superior mechanical properties were formed and suitable for CTE applications.

Collagen may be utilized independently for hydrogel production; however, its properties can be enhanced through hybridization with various natural or synthetic components. Saghati et al. [27] obtained hydrogels by mixing collagen with phenolated alginate, and it was observed that the levels of cartilage-specific proteins such as SOX9 and COL2A1 increased, and the swelling rate and degradation rate of collagen slowed down. In another study, chitosan, a natural polymer, was combined with nano-sized hydroxyapatite and collagen to enhance its mechanical capabilities [28]. Likewise, in a study [29], in which the rapid degradation and poor mechanical properties of collagen were tried to be prevented by mixing type II collagen with polyvinyl alcohol, a synthetic material, the elastic modulus value was strengthened twofold with hybrid hydrogel.

13.4.1.6 Gelatin

Gelatin is a protein derived from the hydrolysis of collagen. It is a material that is noteworthy for CTE regarding its biocompatibility, biodegradability, thermal stability, and low melting point [1, 3, 17]. However, to utilize it for 3D printer technology, several modifications or materials are needed to overcome its weak mechanical qualities [3]. Gelatin-based hydrogels are frequently incorporated with alginate composites. The aim here is to increase biocompatibility, strength, and flexibility. Kreller et al. [30] modified gelatin by physicochemical methods to increase its 3D printability and formed a composite hydrogel with alginate. This study demonstrated the printability of hydrogels with the desired shape and stability without the use of extra additives or toxic chemicals, such as cross-linkers. In one study, alginate and chitosan were added to the medium in addition to gelatin to form a composite [31]. Additionally, in another study, it was shown that silicon nanoparticles were added to the gelatin-alginate duo, and the mechanical properties were strengthened up to five times. It has also been suggested that nanoparticles can provide long-term support at the healing site by prolonging the degradation time of the material [32]. Gelatin has been used not only with alginate but also in composites with various natural and/or synthetic materials. These include carboxymethyl cellulose and collagen [2]. The material known as methacrylated gelatin (GELMA) is a modified form of gelatin. Due to the methacrylate groups in its structure, it instantly cross-links that mimic ECM and creates an environment that is conducive to cell proliferation. GELMA is the preferred material for 3D bioprinting tissue scaffolding due to its ability to incorporate cells [17].

13.4.1.7 Chitosan

Chitosan is obtained by deacetylation of chitin, which is particularly present in the shell of shellfish, and is the second most abundant biopolymer after cellulose. This biopolymer is a linear aminopolysaccharide composed of β (1,4)-linked glucosamine and N-acetyl glucosamine subunits. Chitosan's properties, such as biocompatibility, biodegradability, favorable cell adhesion, low immunological reaction, antibacterial activity, and low cost, support its use in tissue engineering and regenerative medicine. At the same time, the molecular structure of chitosan resembles some ECM components in cartilage, such as glycosaminoglycans (GAGs), which increases the biomimetic properties of the biopolymer and encourages its use in CTE [3, 5]. Chitosan's low solubility at neutral pH and high solubility in acidic media limit its use in tissue engineering and regenerative medicine. Different strategies have been developed to overcome this problem that negatively affects cell viability and proliferation. Many chemical modifications or hybridizations of chitosan have been used to create new biological activities and increase its mechanical properties. Chitosan hydrogels prepared by the enzymatic cross-linking method support the proliferation of cartilage cells and MSCs and help to maintain cartilage morphology [5]. In a study using this method, collagen type II and chondroitin sulfate were added to chitosan hydrogels and compared with unmodified chitosan hydrogels. As a result of the study, it was observed that the modified chitosan hydrogels had a positive effect on cell proliferation and chondrogenesis [33]. He et al. [34] created a 3D bioprinting bioink using carboxymethylated chitosan. Cross-linking was performed between the carboxyl groups in the carboxymethyl chitosan structure and calcium, and thus the mechanical properties of the chitosan-based bioink were improved. In addition, it was shown to increase cell viability, proliferation, and chondrogenic expression in in vitro and in vivo experiments. A hybrid hydrogel composed of chitosan and carrageenan was considered a promising material for cartilage regeneration in another study due to its ability to induce chondrogenic differentiation [35].

13.4.1.8 Chondroitin Sulfate

Chondroitin sulfate (CS) is a linear polysaccharide composed of repeating sulfated disaccharide units with 1–3 D-glucuronic acid and N-acetylgalactosamine linkages. It is the main component of cartilage tissue and is found in connective tissue in all parts of the body. CS, a polyanionic saccharide with hydroxyl groups in its structure, exhibits an adhesive property. The hydroxyl groups facilitate the formation of hydrogen bonds, making it easier to interact with the cell membrane. Due to this feature, CS facilitates cell adhesion and has a positive effect on hydrogel formation as in other polyanionic molecules [2, 3]. It has been reported that implanted CS-containing hydrogel-based tissue scaffolds show high performance in providing cartilage deposition by supporting chondrocyte metabolism in CTE [36]. At the same time, research [37] using MSCs has shown that CS-containing scaffolds help cartilage ECM form and guide mesenchymal cells to become chondrogenic.

CS, a biocompatible polysaccharide, has limited mechanical properties and a rapid degradation profile. Its high water solubility is another factor limiting its use in hydrogel formation. In order to avoid these limitations, there are many studies in the literature in which composites are made with different biomaterials (natural fibers, nanostructures, etc.) or the structure of CS is modified to strengthen the hydrogel structure instead of using CS alone [2, 3]. Catechol chemistry, which changes the surface chemistry of any material, forms covalent cross-links on the material's surface. The surface of CS hydrogels has been modified using this method, and it has been demonstrated that the modified CS hydrogel generates an environment that is comparable to cartilage tissue and enhances the expression of chondrogenic markers specific to cartilage (e.g., SOX9, collagen type II) [36]. In a study conducted to form CS hydrogel, CS was synthesized to form a composite with pullulan and tyramine. The study demonstrated that the composite CS hydrogel, which was able to recreate the cartilage microenvironment and increase the protein and gene expression of aggrecan, collagen type I, and collagen type II, was capable of doing so. The optimum ratios of the hydrogel were determined [38].

13.4.1.9 Cellulose

D-glucose forms the linear-chain polymer of polysaccharide, known as cellulose. Numerous biological sources can provide cellulose for isolation, as it is the most prevalent organic macromolecule in nature. The use of cellulose in tissue engineering applications has limitations through its insufficient physical stability, low adsorption capacity, and limited solubility in water. To avoid limitations in the use of cellulose (solubility, stability, etc.) and improve its properties, structural modification is required. Following alteration, cellulose's ability to retain water can be enhanced, which will favorably impact the production of hydrogel. Modification methods such as carboxymethylation, methylation, and acetylation are particularly preferred for structural modification of cellulose. The most common modified cellulose types used in tissue engineering include carboxymethyl cellulose, methyl cellulose, and hydroxypropyl cellulose [3, 17].

Generally, CTE uses cellulose-based hydrogels by either modifying the cellulose or forming a composite with another material. Boyer et al. [39] formed a composite hydrogel containing silylated hydroxypropyl cellulose and nano-sized clay particles known as laponite for use in CTE applications, and as a result of the study, it was observed that the mechanical strength of the hydrogel increased. Another study combined hydroxypropyl cellulose with an active exopolysaccharide to form hydrogels. The study demonstrated that the cellulose-based hydrogel induced the production of cartilage-like ECM [40].

Due to the limited applications of plant-derived cellulose derivatives, bacterial cellulose (BC), which is a fine network of nanoscale fibers produced by bacteria such as *Acetobacter*, has a wide range of applications in CTE. BC has more advanced properties than other cellulose types with its porous structure, high water retention capacity, high purity, and biocompatibility. Biomedical applications widely prefer BC due to its many usage advantages, particularly in the production of tissue scaffolds for tissue engineering [3]. Martinez Avila et al. [41] showed that a double-layer hydrogel with bacterial nanocellulose makes a good environment for chondrocytes to make cartilage because it has the same structure as collagen fibrils. Phatchayawat et al. [31] created a biocompatible, biodegradable, and cell adhesion-promoting composite hydrogel containing bacterial cellulose. Research demonstrated that the BC-based hydrogel stimulated the differentiation of MSCs into chondrocytes and boosted their proliferation.

13.4.1.10 Gellan Gum

Gellan gum (GG) is an anionic polysaccharide produced by the bacterium *Sphingomonas elodea* and composed of repeating quaternary saccharide units (D-glucose, L-rhamnose, and D-glucuronic acid). Gelation of gellan gum occurs when a divalent cation is added to its solution, followed by cooling the medium to physiological temperature. Furthermore, studies have demonstrated that adding one or two valent cations to GG solutions influences their viscoelastic properties. GG can be used in a variety of applications, particularly in food and tissue engineering, due to its easy acquisition, translucent gel form, and resistance to heat and acidic solutions. GG exists in two forms: acetylated and deacetylated. Deacetylated GG possesses diverse applications and is accessible for commercial use.

The research on CTE applications indicates that the use of GG in isolation is inadequate, requiring its integration with other natural materials to improve its qualities. Lee et al. [42] formed a hydrogel containing GG and enriched with glycol chitosan, which can be used in CTE applications, in a structure similar to cartilage tissue ECM. The study revealed that the addition of glycol chitosan enhanced the hydrogel's mechanical properties, positively impacted cell proliferation, and stimulated the expression of genes specific to cartilage tissue. A study for CTE utilized GG in hydrogel creation with silk fiber due to its inadequate mechanical qualities and restricted cell adhesion. This composite hydrogel exhibited superior mechanical durability and increased proliferation of cartilage cells [43].

13.4.1.11 Dextran

Dextran is a bacterial polysaccharide composed of α-1,6-linked D-glucopyranosyl units and α-1,3-linked side chains. It is suitable for many biomedical applications, including CTE. It is a polymer with high potential for use, especially in hydrogel formation. The structure of hydrogels derived from this polymer exhibits biocompatibility and can undergo biological and enzymatic degradation. This renders it an excellent polymer for application in bone and cartilage tissue engineering.

Hydrogels containing dextran show promise in therapeutic approaches for cartilage damage. Fan et al. [44] encapsulated rabbit cartilage cells, rabbit adipose-derived stem cells, and human adipose-derived stem cells in dextran-based hydrogels and investigated the therapeutic potential of cell-containing dextran-based hydrogels following osteochondral damage in rabbits. The treatment demonstrated the biocompatibility of dextran-based hydrogels containing various cell types, their positive impact on cell viability and proliferation, and their ability to support chondrogenesis both in vitro and in vivo. In a different study, researchers mixed cellulose nanocrystal/dextran-containing hydrogel with magnetic nanoparticles attached to the small molecule of cartogenin. Dextran-based hydrogels containing cartogenin magnetic nanoparticles have demonstrated the ability to control the release profile of the small molecule, facilitate the differentiation of bone marrow-derived MSCs into cartilage cells, and demonstrate significant potential for cartilage regeneration in in vivo experiments [45].

13.4.2 Synthetic Polymer-Based Hydrogels

The most commonly used synthetic hydrogels for CTE purposes are classified as polyethylene glycol, polylactic acid, polycaprolactone, and poly(lactic-co-glycolic acid).

13.4.2.1 Polyethylene Glycol

Polyethylene glycol (PEG) is a hydrophilic linear synthetic ethylene oxide polymer. PEG hydrogel has garnered significant interest in tissue engineering applications due to its biocompatibility, high solubility, and ease of modification [17]. Researchers have noted that cross-linking PEG, a polymer with partial inert characteristics, with hydrogels can stimulate chondrogenesis and enhance cartilage tissue development [1]. The FDA has approved PEG for medical applications and considers it safe for oral consumption, making it a popular synthetic polymer candidate in clinical applications [3]. Research has demonstrated that hydrogels containing PEG possess a mechanically stable structure and are resilient materials with high tensile strength [46]. Due to its structural durability, CTE uses PEG to reinforce natural materials in composite form, in addition to its standalone applications. Research indicates that PEG-based hydrogels promote the adhesion and proliferation of chondrocytes, MSCs, and embryonic stem cells [5]. For these reasons, many research groups have chosen PEG, one of the fundamental materials, to design hydrogels for use in cartilage regeneration.

PEG exhibits insufficient structural biodegradability. Researchers have conducted various studies to enhance the biodegradability of PEG. Neumann et al. [47]

created a unique PEG hydrogel capable of photopolymerization and hydrolytic degradation. During the hydrogel synthesis, caprolactone was added to the environment to form ester bonds, and these bonds facilitated hydrolytic degradation. After encapsulating chondrocytes in the formed hydrogel and conducting a culture study, an increased accumulation of sulfated glycosaminoglycans and collagen in the hydrogels was observed by the end of day 21. However, studies have also demonstrated an increase in the material's modulus of elasticity.

PEG hydrogels are materials that are sensitive to environmental stimuli. Thus, there are applications in the field of controlled release of model drugs. Chemically linked PEG hydrogels are a unique biomaterial capable of releasing peptide drugs, growth factors, and live cells gradually. One of the cost-effective growth factors, platelet lysate, is frequently used in CTE. In the study conducted by Tang [48], platelet lysate was not directly added to the hydrogel as in other studies; instead, heparin and nanoparticles were used to equalize the particle distribution of platelet lysate, and the developed poly-d,l-lactic acid–PEG–poly-d,l-lactic acid (PDLLA-PEG-PDLLA) hydrogel was loaded to preserve the gelation capacity. The in vivo application reveals that this system enhances early cartilage degeneration and facilitates cartilage repair during the late stages of osteoarthritis. This study highlights the significance of achieving successful outcomes with acellular hydrogels for cardiac tissue engineering.

13.4.2.2 Polylactic Acid

Polylactic acid (PLA) is a semicrystalline aliphatic polyester derived from lactic acid, characterized by its ability to undergo biological degradation and hydrolysis. The predominant tissue scaffolds are composed of poly(α-hydroxy esters). These tissue scaffolds have been used for cartilage regeneration since the early 1990s and contain PLA, poly(glycolic acid) (PGA), and their copolymers. PLA has a hydrophobic structure and a slow degradation rate. The copolymers of PLA and PGA are materials that can be optimized for mechanical and degradation properties [1].

Tamai et al. [49] conducted a study on rabbits loading BMP-2 onto PLA-PEG hydrogels and showed that the subchondral damage began to heal after 1 week and was completely repaired after 6 weeks. Narayanan [50] has developed a fibrous hydrogel bioink composed of PLA, alginate, and human adipose-derived stem cells. In vitro results have shown that human adipose-derived stem cell viability and proliferation increased over 8 weeks. In vivo experiments revealed higher levels of cell proliferation, collagen, and proteoglycan, along with chondrogenic differentiation and cartilage repair, compared to the control group.

13.4.2.3 Polycaprolactone

Polycaprolactone (PCL) is an aliphatic polyester approved by the FDA in terms of biocompatibility and safety of use. PCL is a polymer synthesized by the ring-opening polymerization of ε-caprolactone monomers. PCL is suitable for biomedical applications due to its biodegradability, hydrophilic character, superior mechanical properties, ability to adjust surface roughness, and modifiability through various modification techniques. The degradation profile has a tendency to decompose over a period of approximately 2–3 years longer compared to polylactides. Implants, drug delivery systems, and tissue engineering applications, among others, have a wide range of applications in the medical field [51].

Many studies involving hydrogels in CTE applications have used PCL as a composite. In one study, researchers formed a photo-cross-linked hydrogel of PCL-PEG-PCL using PCL and PEG polymers. The created composite PCL hydrogel not only provides a suitable environment for the chondrogenesis of bone marrow stem cells but also contributes to the production of cartilage tissue matrix [52]. In another study using composite PCL hydrogel, PCL was added to the hydrogel containing alginate and GELMA to enhance its mechanical properties. The synthesis of the PCL-containing hydrogel demonstrates mechanical properties similar to those of healthy articular cartilage [53].

13.4.2.4 Poly(Lactic-Co-Glycolic Acid)

Poly (lactic-co-glycolic acid) (PLGA) is a copolymer formed by the ring-opening polymerization technique of lactic acid and glycolic acid monomers. PLGA has been approved by the FDA for use in clinical studies due to its biocompatibility and biodegradability. This polymer has a wide range of applications in drug delivery systems, implants, and tissue engineering. The degradation rate of PLGA can be adjusted by changing the ratio of glycolic acid units to lactic acid units. PLGA, in addition to its advantages, also has properties that limit its use in biomedical applications. PLGA's suboptimal mechanical properties and low osteoconductivity make its use alone in biomedical applications preferable when combined with other materials [17].

Many studies have shown that using PLGA as a composite enhances cartilage regeneration in CTE applications. The limited mechanical properties of hydrogels for cartilage applications and the insufficient biological activity of synthetic polymers restrict their standalone use. For this reason, Tang et al. [54] have created a hybrid tissue scaffold with PLGA tissue scaffolds and hydrogels enriched with plasma proteins containing MSCs. Studies have demonstrated that the PLGA in this hybrid tissue scaffold enhances its mechanical durability and stimulates the differentiation of MSCs into cartilage cells.

13.5 Conclusion

Tissue engineering is an interdisciplinary field that spans from materials engineering to life sciences. This field's goals include developing potential approaches to healing tissue damage and imitating the relevant tissue. Researchers are exploring various cells, biomaterials, and growth factors for these purposes. Among biomaterials, hydrogels stand out with their ability to mimic cartilage structure. The internal porous structure of hydrogels provides a suitable environment for cells to grow and multiply. Therefore, in this section, natural and synthetic hydrogels used in cartilage tissue engineering have been extensively searched from a broad perspective, and the advantages and disadvantages of the materials in their individual or composite forms have been provided. In particular, researchers' approaches and orientations in recent years have been examined. In the future, it is thought that the properties of hydrogel-based tissue scaffolds can be developed using 3D printing techniques for the combination and production of natural or synthetic hydrogels. To facilitate cartilage tissue engineering applications, hydrogels with biocompatibility, mechanical durability, and biodegradability properties should be studied in more detail. Therefore, understanding medical needs while simultaneously reducing the difficulty of hydrogel production should be the focus of future research in this field. In order for the studies to translate into products and reach patients, their practicality and functionality must be improved, and their use in tissue damage treatment should be prioritized.

References

1. Leone, G.: Cartilage replacement implants using hydrogels. In: Biomedical Hydrogels, pp. 149–183 (2011)
2. Hafezi, M., Nouri Khorasani, S., Zare, M., Esmaeely Neisiany, R., Davoodi, P.: Advanced hydrogels for cartilage tissue engineering: recent progress and future directions. Polymers (Basel). **13**(23), 4199 (2021)
3. Wei, W., Ma, Y., Yao, X., Zhou, W., Wang, X., Li, C., et al.: Advanced hydrogels for the repair of cartilage defects and regeneration. Bioact. Mater. **6**(4), 998–1011 (2021)
4. Nordberg, R.C., Bielajew, B.J., Takahashi, T., Dai, S., Hu, J.C., Athanasiou, K.A.: Recent advancements in cartilage tissue engineering innovation and translation. Nat. Rev. Rheumatol. **20**(6), 323–346 (2024)
5. Maia, F.R., Carvalho, M.R., Oliveira, J.M., Reis, R.L.: Tissue engineering strategies for osteochondral repair. Adv. Exp. Med. Biol. **1059**, 353–371 (2018)
6. Bordbar, S., Li, Z., Lotfibakhshaiesh, N., Ai, J., Tavassoli, A., Beheshtizadeh, N., et al.: Cartilage tissue engineering using decellularized biomatrix hydrogel containing TGF-β-loaded alginate microspheres in mechanically loaded bioreactor. Sci. Rep. **14**(1), 11991 (2024)
7. Dixit, A., Mahajan, A., Saxena, R., Chakraborty, S., Katti, D.S.: Engineering sulfated polysaccharides and silk fibroin based injectable IPN hydrogels with stiffening and growth factor presentation abilities for cartilage tissue engineering. Biomater. Sci. **12**(8), 2067–2085 (2024)
8. Liu, W., Madry, H., Cucchiarini, M.: Application of alginate hydrogels for next-generation articular cartilage regeneration. Int. J. Mol. Sci. **23**(3), 1147 (2022)

9. Thanh, T.N., Laowattanatham, N., Ratanavaraporn, J., Sereemaspun, A., Yodmuang, S.: Hyaluronic acid crosslinked with alginate hydrogel: a versatile and biocompatible bioink platform for tissue engineering. Eur. Polym. J. **166**, 111027 (2022)
10. Wang, P., Pu, Y., Ren, Y., Yang, R., Zhang, W., Tan, X., et al.: Dynamic regulable sodium alginate/poly(γ-glutamic acid) hybrid hydrogels promoted chondrogenic differentiation of stem cells. Carbohydr. Polym. **275**, 118692 (2022)
11. Pleumeekers, M., Nimeskern, L., Koevoet, W., Kops, N., Poublon, R., Stok, K., et al.: The in vitro and in vivo capacity of culture-expanded human cells from several sources encapsulated in alginate to form cartilage. Eur. Cell. Mater. **27**, 264–280 (2014)
12. Awad, H.A., Wickham, M.Q., Leddy, H.A., Gimble, J.M., Guilak, F.: Chondrogenic differentiation of adipose-derived adult stem cells in agarose, alginate, and gelatin scaffolds. Biomaterials. **25**, 3211–3222 (2004)
13. Salati, M.A., Khazai, J., Tahmuri, A.M., Samadi, A., Taghizadeh, A., Taghizadeh, M., et al.: Agarose-based biomaterials: opportunities and challenges in cartilage tissue engineering. Polymers (Basel). **12**(5), 1150 (2020)
14. Bonhome-Espinosa, A.B., Campos, F., Durand-Herrera, D., Sánchez-López, J.D., Schaub, S., Durán, J.D.G., et al.: In vitro characterization of a novel magnetic fibrin-agarose hydrogel for cartilage tissue engineering. J. Mech. Behav. Biomed. Mater. **104**, 103619 (2020)
15. Yin, Z., Yang, X., Jiang, Y., Xing, L., Xu, Y., Lu, Y., et al.: Platelet-rich plasma combined with agarose as a bioactive scaffold to enhance cartilage repair: an in vitro study. J. Biomater. Appl. **28**(7), 1039–1050 (2014)
16. Singh, Y.P., Bhardwaj, N., Mandal, B.B.: Potential of agarose/silk fibroin blended hydrogel for in vitro cartilage tissue engineering. ACS Appl. Mater. Interfaces. **8**(33), 21236–21249 (2016)
17. Han, X., Chang, S., Zhang, M., Bian, X., Li, C., Li, D.: Advances of hydrogel-based bioprinting for cartilage tissue engineering. Front. Bioeng. Biotechnol. **9**, 816 (2021)
18. Peretti, G.M., Xu, J.W., Bonassar, L.J., Kirchhoff, C.H., Yaremchuk, M.J., Randolph, M.A.: Review of injectable cartilage engineering using fibrin gel in mice and swine models. Tissue Eng. **12**(5), 1151–1168 (2006)
19. Gupta, N., Cruz, M.A., Nasser, P., Rosenberg, J.D., Iatridis, J.C.: Fibrin-Genipin hydrogel for cartilage tissue engineering in nasal reconstruction. Ann. Otol. Rhinol. Laryngol. **128**(7), 640–646 (2019)
20. Almeida, H.V., Eswaramoorthy, R., Cunniffe, G.M., Buckley, C.T., O'Brien, F.J., Kelly, D.J.: Fibrin hydrogels functionalized with cartilage extracellular matrix and incorporating freshly isolated stromal cells as an injectable for cartilage regeneration. Acta Biomater. **36**, 55–62 (2016)
21. Asari, A.: Medical application of Hyaluronan. Chem. Biol. Hyaluronan., 457–473 (2004)
22. Han, S.-S., Yoon, H.Y., Yhee, J.Y., Cho, M.O., Shim, H.-E., Jeong, J.-E., et al.: In situ cross-linkable hyaluronic acid hydrogels using copper free click chemistry for cartilage tissue engineering. Polym. Chem. **9**(1), 20–27 (2018)
23. Davachi, S.M., Haramshahi, S.M.A., Akhavirad, S.A., Bahrami, N., Hassanzadeh, S., Ezzatpour, S., et al.: Development of chitosan/hyaluronic acid hydrogel scaffolds via enzymatic reaction for cartilage tissue engineering. Mater. Today Commun. **30**, 103230 (2022)
24. Ziadlou, R., Rotman, S., Teuschl, A., Salzer, E., Barbero, A., Martin, I., et al.: Optimization of hyaluronic acid-tyramine/silk-fibroin composite hydrogels for cartilage tissue engineering and delivery of anti-inflammatory and anabolic drugs. Mater. Sci. Eng. C. **120**, 111701 (2021)
25. Yuan, T., Zhang, L., Li, K., Fan, H., Fan, Y., Liang, J., et al.: Collagen hydrogel as an immunomodulatory scaffold in cartilage tissue engineering. J. Biomed. Mater. Res. Pt. B Appl. Biomater. **102**(2), 337–344 (2014)
26. Vázquez-Portalatı, N.N., Kilmer, C.E., Panitch, A., Liu, J.C.: Characterization of collagen type I and II blended hydrogels for articular cartilage tissue engineering. Biomacromolecules. **17**(10), 3145–3152 (2016)

27. Saghati, S., Rahbarghazi, R., Baradar Khoshfetrat, A., Moharamzadeh, K., Tayefi Nasrabadi, H., Roshangar, L.: Phenolated alginate-collagen hydrogel induced chondrogenic capacity of human amniotic mesenchymal stem cells. J. Biomater. Appl. **36**(5), 789–802 (2021)
28. Kaviani, A., Zebarjad, S.M., Javadpour, S., Ayatollahi, M., Bazargan-Lari, R.: Fabrication and characterization of low-cost freeze-gelated chitosan/collagen/hydroxyapatite hydrogel nanocomposite scaffold. Int. J. Polym. Anal. Charact. **24**(3), 191–203 (2019)
29. Lan, W., Xu, M., Zhang, X., Zhao, L., Huang, D., Wei, X., et al.: Biomimetic polyvinyl alcohol/type II collagen hydrogels for cartilage tissue engineering. J. Biomater. Sci. Polym. Ed. **31**(9), 1179–1198 (2020)
30. Kreller, T., Distler, T., Heid, S., Gerth, S., Detsch, R., Boccaccini, A.R.: Physico-chemical modification of gelatine for the improvement of 3D printability of oxidized alginate-gelatine hydrogels towards cartilage tissue engineering. Mater. Des. **208**, 109877 (2021)
31. Phatchayawat, P.P., Khamkeaw, A., Yodmuang, S., Phisalaphong, M.: 3D bacterial cellulose-chitosan-alginate-gelatin hydrogel scaffold for cartilage tissue engineering. Biochem. Eng. J. **184**, 108476 (2022)
32. Ghanbari, M., Salavati-Niasari, M., Mohandes, F., Firouzi, Z.: Modified silicon carbide NPs reinforced nanocomposite hydrogels based on alginate-gelatin by with high mechanical properties for tissue engineering. Arab. J. Chem. **15**(1), 103520 (2022)
33. Choi, B., Kim, S., Lin, B., Wu, B.M., Lee, M.: Cartilaginous extracellular matrix-modified chitosan hydrogels for cartilage tissue engineering. ACS Appl. Mater. Interfaces. **6**(22), 20110–20121 (2014)
34. He, Y., Derakhshanfar, S., Zhong, W., Li, B., Lu, F., Xing, M., et al.: Characterization and application of carboxymethyl chitosan-based bioink in cartilage tissue engineering. J. Nanomater. **2020**, 1–11 (2020)
35. Liang, X., Wang, X., Xu, Q., Lu, Y., Zhang, Y., Xia, H., et al.: Rubbery chitosan/carrageenan hydrogels constructed through an electroneutrality system and their potential application as cartilage scaffolds. Biomacromolecules. **19**(2), 340–352 (2018)
36. Shin, J., Kang, E.H., Choi, S., Jeon, E.J., Cho, J.H., Kang, D., et al.: Tissue-adhesive chondroitin sulfate hydrogel for cartilage reconstruction. ACS Biomater Sci. Eng. **7**(9), 4230–4243 (2021)
37. Wang, T., Yang, F.: A comparative study of chondroitin sulfate and heparan sulfate for directing three-dimensional chondrogenesis of mesenchymal stem cells. Stem Cell Res. Ther. **8**(1), 284 (2017)
38. Chen, F., Yu, S., Liu, B., Ni, Y., Yu, C., Su, Y., et al.: An injectable enzymatically crosslinked carboxymethylated pullulan/chondroitin sulfate hydrogel for cartilage tissue engineering. Sci. Rep. **6**(1), 20014 (2016)
39. Boyer, C., Figueiredo, L., Pace, R., Lesoeur, J., Rouillon, T., Le, V.C., et al.: Laponite nanoparticle-associated silated hydroxypropylmethyl cellulose as an injectable reinforced interpenetrating network hydrogel for cartilage tissue engineering. Acta Biomater. **65**, 112–122 (2018)
40. Rederstorff, E., Rethore, G., Weiss, P., Sourice, S., Beck-Cormier, S., Mathieu, E., Maillasson, M., Jacques, Y., Colliec-Jouault, S., Fellah, B.H., Guicheux, J., Vinatier, C.: Enriching a cellulose hydrogel with a biologically active marine exopolysaccharide for cell-based cartilage engineering. J. Tissue Eng. Regen. Med. **11**(4), 1152–1164 (2017)
41. Martínez Ávila, H., Feldmann, E.-M., Pleumeekers, M.M., Nimeskern, L., Kuo, W., de Jong, W.C., et al.: Novel bilayer bacterial nanocellulose scaffold supports neocartilage formation in vitro and in vivo. Biomaterials. **44**, 122–133 (2015)
42. Lee, S., Choi, J.H., Park, A., Rim, M., Youn, J., Lee, W., et al.: Advanced gellan gum-based glycol chitosan hydrogel for cartilage tissue engineering biomaterial. Int. J. Biol. Macromol. **158**, 452–460 (2020)
43. Kim, W., Choi, J.H., Kim, P., Youn, J., Song, J.E., Motta, A., et al.: Preparation and evaluation of gellan gum hydrogel reinforced with silk fibers with enhanced mechanical and biological properties for cartilage tissue engineering. J. Tissue Eng. Regen. Med. **15**(11), 936–947 (2021)

44. Fan, L., Lin, C., Zhao, P., Wen, X., Li, G.: An Injectable Bioorthogonal Dextran Hydrogel for Enhanced Chondrogenesis of Primary Stem Cells, vol. 24, p. 504 (2018)
45. Yang, W., Zhu, P., Huang, H., Zheng, Y., Liu, J., Feng, L., et al.: Functionalization of novel theranostic hydrogels with Kartogenin-grafted USPIO nanoparticles to enhance cartilage regeneration. ACS Appl. Mater. Interfaces. **11**, 34744–34754 (2019)
46. Ebhodaghe, S.O.: Hydrogel—based biopolymers for regenerative medicine applications: a critical review. Int. J. Polym. Mater. Polym. Biomater. **71**(3), 155–172 (2022)
47. Neumann, A.J., Quinn, T., Bryant, S.J.: Nondestructive evaluation of a new hydrolytically degradable and photo-clickable PEG hydrogel for cartilage tissue engineering. Acta Biomater. **39**, 1–11 (2016)
48. Tang, Q., Lim, T., Shen, L.-Y., Zheng, G., Wei, X.-J., Zhang, C.-Q., et al.: Well-dispersed platelet lysate entrapped nanoparticles incorporate with injectable PDLLA-PEG-PDLLA triblock for preferable cartilage engineering application. Biomaterials. **268**, 120605 (2021)
49. Tamai, N., Myoui, A., Hirao, M., Kaito, T., Ochi, T., Tanaka, J., et al.: A new biotechnology for articular cartilage repair: subchondral implantation of a composite of interconnected porous hydroxyapatite, synthetic polymer (PLA-PEG), and bone morphogenetic protein-2 (rhBMP-2). Osteoarthr. Cartil. **13**(5), 405–417 (2005)
50. Narayanan, L.K., Huebner, P., Fisher, M.B., Spang, J.T., Starly, B., Shirwaiker, R.A.: 3D-bioprinting of polylactic acid (PLA) nanofiber–alginate hydrogel bioink containing human adipose-derived stem cells. ACS Biomater Sci. Eng. **2**(10), 1732–1742 (2016)
51. Malikmammadov, E., Tanir, T.E., Kiziltay, A., Hasirci, V.: PCL and PCL-based materials in biomedical applications. J. Biomater. Sci. Polym. Ed. **29**, 863 (2018)
52. Ko, C., Ku, K., Yang, S., Lin, T., Peng, S., Peng, Y.: In vitro and in vivo co-culture of chondrocytes and bone marrow stem cells in photocrosslinked PCL—PEG—PCL hydrogels enhances cartilage formation. J. Tissue Eng. Regen. Med. **10**(10), E485–E496 (2013)
53. Schipani, R., Scheurer, S., Florentin, R., Critchley, S.E., Kelly, D.J.: Reinforcing interpenetrating network hydrogels with 3D printed polymer networks to engineer cartilage mimetic composites reinforcing interpenetrating network hydrogels with 3D printed polymer networks to engineer cartilage mimetic composites. Biofabrication. **12**(3), 035011 (2020)
54. Tang, Y., Wang, H., Sun, Y., Jiang, Y., Fang, S., Kan, Z., et al.: Using platelet-rich plasma hydrogel to deliver mesenchymal stem cells into three-dimensional PLGA scaffold for cartilage tissue engineering. ACS Appl. Bio. Mater. **4**, 8607–8614 (2021)

Chapter 14
Hydrogels Used in Bone Treatment

Ali Deniz Dalgic and Dilek Keskin

Abstract The most widely used approach for treating critical-sized bone loss is the application of "autologous bone grafts", which involves using the patient's own bone tissue. Autologous bone grafts are preferred due to their strong osteoconductive and osteoinductive properties. However, when the patient serves as the donor, several issues can arise, including significant tissue loss at the donor site, limitations in available donor areas, and an increased risk of postoperative infections. As alternative approaches, "allografting"—which involves using bone tissue from another person—and "xenografting"—which uses bone grafts derived from animal tissues—can also be employed. However, these methods may lead to complications such as tissue incompatibility, heightened immune responses, and the risk of pathogen transmission. For these reasons, there is a growing need to develop novel tissue scaffolds using tissue engineering approaches. Synthetic materials and biologically derived hydrogels are among the most commonly used materials in such applications and treatments. In recent years, hydrogels have attracted increasing interest in regenerative therapies due to their biocompatibility, versatility, and hydrophilic network structure, which resembles the extracellular matrix of natural tissues. As a result, hydrogels hold significant potential as biomaterials for various bone treatment applications, including local drug delivery and bone tissue engineering.

Keywords Bone · Bone Tissue Engineering · Calcium phosphate · Bioglass · Bioactive ions · Demineralized bone matrix · Hybrid hydrogel · Interconnected pores · Mechanical strength

---

A. D. Dalgic
Department of Genetics and Bioengineering, Istanbul Bilgi University, Istanbul, Türkiye

D. Keskin (✉)
Department of Engineering Sciences, Middle East Technical University, Ankara, Türkiye

BIOMATEN Center of Excellence in Biomaterials and Tissue Engineering, Middle East Technical University, Ankara, Türkiye
e-mail: dkeskin@metu.edu.tr

N. Hasirci, V. Hasirci (eds.), *Hydrogels and Bioinks in Tissue Engineering*, Biological and Medical Physics, Biomedical Engineering,
https://doi.org/10.1007/978-3-031-90966-5_14

14.1 Hydrogels in Bone Treatment Applications

Hydrogels that are formed by physical and chemical cross-linking of hydrophilic polymers have the ability to swell in biological fluids without being dissolved easily. Owing to this property, hydrogels provide advantages in the requirements of tissue scaffolds like cell loading and differentiation. In addition to this, hydrogels are known for their good performance in loading and local controlled release of drugs and biological agents like growth factors. In hydrogel preparation, polymers from natural sources: (1) proteins (collagen, gelatin, fibroin, etc.); (2) glycosaminoglycans (hyaluronic acid, sGAG, chondroitin sulfate, etc.); (3) carbohydrates (chitosan, alginate, pullulan cellulose acetate, xanthan, etc.); and (4) polymers of synthetic origin (polyvinyl alcohol, polyethylene oxide, polyacrylic acid, etc.) are among the most investigated ones, some of which are being used as biomaterials in clinical applications.

The advantageous properties of hydrogels as biomaterials can be summarized as follows:

1. Hydrogels can be injected in liquid form and then transform into a gel at the target site, allowing for minimally invasive application and localized retention.
2. They exhibit a high rate of biodegradation and can be engineered for controllable degradation through various cross-linking approaches.
3. Hydrogels often contain the same or similar components as natural biological tissues, such as collagen and hyaluronic acid, enhancing their compatibility with the body.
4. They are capable of carrying and releasing therapeutic agents at the site of application in a controlled manner.
5. Due to their porous, three-dimensional, cross-linked polymer networks and high water-holding capacity, hydrogels closely mimic the structure of native tissue.

Bone tissue damage can result not only from trauma or accidents, but can also arise from diseases or treatments that lead to tissue loss at the affected site. In such cases, the application of biomaterials instead of biological grafts—while simultaneously supporting tissue regeneration and disease treatment—can be of great importance (Fig. 14.1).

Hydrogels are improvable for the purpose of bone tissue engineering and thus can form structures that have the ability of cell entrapment. Hydrogels, which can mimic the native extracellular matrix (ECM) microarchitecture by high water holding capacity and 3D polymer network, can be prepared with specialty enabling high porosity, which provides nutrient transfer necessary for cell viability [1]. This interconnected porous structure, while enhancing cell attachment with cell-caffold interaction and cell migration, can also enable the transfer of nutrients, oxygen, and waste in an efficient way. All these properties play an important role in cell proliferation, tissue regeneration, and new blood vessel formation [2].

Hydrogels can be generated in different shapes and dimensions owing to 3D cross-linking of the polymer networks. The cross-linking of polymer chains, besides

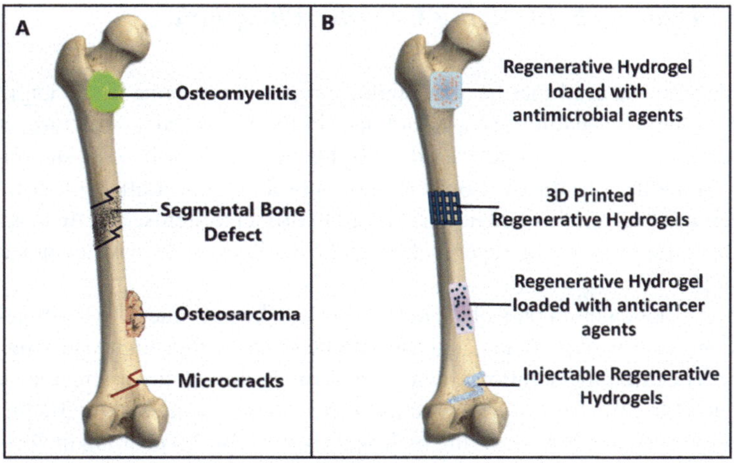

Fig. 14.1 (**a**) Bone diseases and damages that can cause bone tissue loss. (**b**) Hydrogel systems-based treatment approaches for regeneration of bone tissue

providing absolute control of biodegradation, is effective in the optimization of mechanical strength. Biomaterials and hydrogels with biocompatibility, controllable biodegradation, and mechanical strength have a large spectrum of potential in bone tissue engineering applications. Therefore, their cross-linking has critical importance for improving their mechanical strength and enabling them to stay at the tissue site for a longer period. In addition to being cross-linked with physical means like light and temperature, hydrogels can also be cross-linked by chemical means using various cross-linker compounds [3].

14.2 Reinforcement Materials in Hydrogels Developed for Bone Tissue Engineering

Hydrogels, in general, are structures that can support cell viability. However, these polymers are usually brought together with various different materials as composites to develop successful bone tissue engineering scaffolds that can speed up the tissue regeneration. The most widely used approach is the incorporation of calcium phosphate bioceramic materials into hydrogel networks to mimick the bone tissue structure, which is itself a native composite. Another approach for this purpose is the entrapment of bioactive materials or ions into hydrogel structure to trigger the bone cell differentiation process. Composite hydrogels with reinforcement/supportive materials for bone tissue engineering are given with some examples in the following sections.

14.2.1 Hydrogels Involving Calcium Phosphate

Reinforcement material that has the highest proportion in bone tissue engineering research is hydroxyapatite (HA), found also in the native bone structure; it is the most used material in hydrogel-based scaffolds designed for bone tissue regeneration [4]. In addition to this, various calcium phosphate bioceramics other than HA have been used as reinforcement material in hydrogel systems, and their improvement effects on bone tissue regeneration and osteointegration have been shown in literature [5].

HA and other calcium phosphate derivatives can be introduced to hydrogel structures as micro or nanoparticles, and this can be done by thoroughly mixing before gelation. The mixture solution, then, can form hydrogel involving the additive ingredient as the final form or can be used as material in injectable, 3D printable, and gelation systems. HA involving hydrogels formed by 3D printing methods have been observed as widely used approaches for the production of bone tissue scaffolds according to studies in the literature. As an example of such studies, Bendtsen and coauthors have prepared 3D printed alginate and polyvinyl alcohol polymer solution including HA and reported that it provided a suitable environment for bone formation by means of the osteoconductive properties of the HA component of the hydrogels [6]. Similarly, in the case of injectable hydrogel systems that are developed for bone tissue regeneration purposes, mixing and homogenous distribution of HA particles within a polymer solution is usually applied [7].

The inorganic calcium phosphate micro/nanoparticles present in hydrogels behave as the seeds/grains for stimulating mineral deposition and better connection with the bone tissue occurs by these accumulated minerals [8, 9]. Furthermore, bioactive ions like Ca and P can be released from the hydrogel matrix to promote the osteogenic activity of bone cells [10]. The presence of HA also strengthens the hydrogen bonding with the newly formed tissue around the biomaterial, thereby improving cell attachment and integration to the scaffold. Additionally, the enhancement of mechanical strength, protein adsorption, and osteoconduction properties of hydrogels by integrated bioceramic nanoparticles has also been identified. In the research conducted by Sareethammanuwat et al., beta-tricalciumphosphate (β-TCP) was incorporated into hydrogels prepared with chitosan and collagen and has been shown to improve the compressive mechanical strength of hydrogels [9]. Similarly, β-TCP addition to dextran-based hydrogels has been shown to improve compressive strength and cell proliferation while β-TCP nanoparticle containing gelatin/sodium alginate-based hydrogels has been shown to promote osteogenic differentiation [11, 12].

14.2.2 Hydrogels Incorporated with Bioglass

Bioglasses, having silicon dioxide, calcium oxide, and phosphorus pentoxide in their structure, are widely used biomaterial components of scaffolds for bone tissue engineering applications due to their osteoconductive and osteoinductive properties [13]. Hydrogels involving bioglasses are evaluated as potential cell scaffolds for bone regeneration and bone tissue engineering purposes with their bioactivity, biodegradability, and hydrophilicity characteristics. In a study by Han et al., composite hydrogels were prepared using calcium silicate and sodium alginate with the addition of d-gluconic acid δ-lactone; gelation was achieved by in situ (at the site) release of calcium from the calcium silicates. Investigations of these composite hydrogels with bone marrow stem cells (BMSC) and human umbilical cord endothelial cells (HUVEC) showed that hydrogels have improved bone mineral formation and successfully induced both osteogenesis and vascularization [14].

14.2.3 Hydrogels Involving Bioactive Ions

In order to elevate the bioactivity of hydrogels to increase the rate of bone regeneration and bone tissue differentiation via effecting bone cells' activities, different elements can be introduced to hydrogels. It is known that bone inorganic composition does not occur only from HA minerals but also involves many different elements in small quantities. While the main elements, defined as macro minerals, in bone tissue are calcium, phosphorus, and magnesium, other elements in smaller amounts can be ordered as boron, iron, zinc, copper, and selenium, which also play important roles in bone metabolism. They can affect osteoblast and osteoclast cells directly or can modify bone tissue properties by their presence in the bone matrix. Additionally, they can organize the resorptive and reshaping processes of the cells by interacting with the key enzymes as cofactors. Among these enzymes, lysyl hydroxylase and pyrolyl hydroxylase are involved in collagen synthesis, and 25-hydroxylase, a member of cytochrome P450 enzyme family that hydroxylates 25th carbon as the first step of vitamin Vitamin D activation, can be counted and bioactive elements such as iron and copper act as cofactors for these enzymes [15–17].

Biologically active elements can also be used as supportive agents in bone tissue engineering scaffold structure. As an example, magnesium (Mg) is naturally substituted into hydroxyapatite crystals in bone tissue matrix. This has an important effect on bone density and thus bone mechanical properties. Additionally, magnesium ion (Mg^{2+}), when used in biomaterials, has been shown to increase mesenchymal cell proliferation and osteogenic differentiation [18, 19]. In the research conducted by Zang et al., magnesium-enriched nanocomposite hydrogel system was produced by photopolymerization using active Mg^{2+} and gelatin methacryloyl (GELMA), thiolated chitosan (TCS), and modified polyhedral oligomeric silsesquioxan (POSS) nanoparticles. Magnesium has been shown to enhance osteogenic differentiation of mesenchymal stem cells and stimulate angiogenesis both in in vitro and in vivo conditions [20]. As an example of other elements, in the study of Cal et al., silica-based boron-incorporated collagen/keratin hybrid cryogels have been synthesized and investigated for bone formation potential [21]. In another research, strontium-doped bioglass was added to chitosan and alginate hydrogels, which proved to increase the bone regeneration capacity [22]. Due to the stimulating effects of zinc and iron ions on stem cell proliferation and differentiation, they have been reported as efficient additive materials that can be used in hydrogel structures related to bone tissue engineering applications [23]. In addition to these, doping more than one type of ion into hydroxyapatite structure has been suggested to support bone formation within hydrogel structure and can be used in tissue engineering applications [24].

14.2.4 Cell Carrying Hydrogel Applications

The involvement of cells/stem cells and bioactive agents in hydrogel structures provides improvement toward successive bone tissue formation. The hydrogel systems carrying stem cells that have the capacity to differentiate into bone cells can enable new bone tissue formation and fast regeneration at the site by populating the scaffold owing to their better proliferation rate in a porous hydrogel network structure. For example, mesenchymal stem cells (MSCs) can be obtained easily and added to sterile hydrogel solutions before gelation, and cell-carrying hydrogels can be attained by safe gelation methods that will not harm the loaded cells. Thus, with the biocompatible and appropriate cell-carrying properties of hydrogels, the viability of the cells within these systems can be achieved [25].

14.2.5 Hydrogels Containing Demineralized Bone Matrix (DBM)

Demineralized bone matrix (DBM) is obtained by removing the inorganic mineral part of the bone tissue and is a material that basically contains the organic part that forms the bone matrix. DBM that is obtained from allograft bone tissue has high biological activity thanks to the natural bone proteins and growth factors contained in its structure and DBM can regulate osteogenic differentiation of cells and plays an important role in bone regeneration. When blended in hydrogels, DBM will release growth factors to surrounding tissue, which will recruit osteoprogenitor cells and promote osteogenic differentiation [26]. One of these factors is bone morphogenetic proteins (BMPs) in DBM, which are known to promote osteogenic differentiation of mesenchymal stem cells to osteoblast lineage [27, 28]. The growth factors that DBM carries include but not limited to transforming growth factor-beta (TGF-β), fibroblast growth factor (FGF), and insulin-like growth factor (IGF), which also stimulate proliferation and differentiation of bone cells [26, 29].

In the study of Zheng et al., a hydrogel system was formed with covalently cross-linked polyethylene glycol diacrylate (PEGDA) and ionically cross-linked sodium alginate, and demineralized bone matrix was added to this system. It has been shown that groups containing demineralized bone matrix have higher alkaline phosphatase (ALP) activity and increased vascular endothelial growth factor expression [30]. In another study by Wu et al., it was stated that the addition of DBM significantly increased the elastic modulus of the hydrogel in GelMA hydrogels containing vascular endothelial growth factor (VEGF)-loaded PLGA microspheres and DBM [31].

14.2.6 Hybrid Hydrogel Structures

Hydrogels prepared with more than one hydrogel component are defined as hybrid hydrogels. Combining the positive properties of different hydrogels in such structures improves the ability to control desired properties of hydrogels and can enhance bone tissue regeneration. A study by Lu et al. has reported the production of heat-sensitive hybrid hydrogels by cross-linking collagen and chondroitin sulfate with genipin. Chondroitin sulfate, which has a sulfated glycosaminoglycan structure, is a macromolecule that is mainly found in cartilage tissue and is also known to play a role in bone formation and mineralization. Collagen Type I is the major protein of the bone matrix. In a hybrid hydrogel formed with these two molecules, osteogenic differentiation of mesenchymal stem cells and precursor-osteoblast cells have shown to increase; (i) expression of osteogenic genes (OSX, RUNX2, OCN, and ALP) in both Stat3-deficient (Stat 3 deficient *) and normal cells, (2) OSX, RUNX2, and OCN protein levels of cells, and (3) biomineralization level. In addition, in vivo experiments have shown that after monocortical bone damage, this hybrid hydrogel was successful in improving bone regeneration compared to the hydrogel-free groups in experiments with normal and osteogenesis-suppressed transgenic mice [32].

*It has been shown in previous studies that cells lacking Stat 3 have less osteogenic differentiation (impaired osteogenesis) [33, 34].

14.3 Improving the Bone Like Properties of Hydrogels

In the use of hydrogels for bone tissue engineering, some properties may need to be improved to match the mechanical and structural properties of bone. To achieve appropriate hardness and strength, the structure of hydrogel should be improved, formation of structure with a certain average pore size is necessary to enable cell growth as well as material transfer, and bioactivity needs to be improved to accelerate tissue regeneration by activating bone cells. The techniques used to improve hydrogels to be used in bone tissue engineering are summarized as follows.

14.3.1 Improving Mechanical Strength and Biodegradation Time with Cross-Linking

By adding active groups to polymer chains in hydrogel systems, new chemical bonds can be created between polymer chains and this method can be used to strengthen the mechanical properties of hydrogels. Cross-linking of hydrogels can be achieved by two approaches: chemical cross-linking and physical cross-linking [35]. Chemical cross-linking forms new covalent bonds between active groups of polymer chains and gives the ability to tune mechanical and structural properties [36]. On the other hand, chemical cross-linking can have drawbacks to hydrogel structure such as high brittleness and toxicity emerging from unreacted cross-linkers [37]. To prevent these problems, different cross-linking chemistries have been engineered, which creates a less brittle final structure. As an example, in the study of Zerbinati et al., polyethylene glycol (PEG) was used to cross-link hyaluronic acid hydrogels, which resulted in high resistance to degradation with more elastic structure [38]. Physical cross-linking of hydrogels depends on hydrogen bonding, ionic interactions, and van der Waals forces and can be used to achieve less brittle and biocompatible hydrogel networks. The physical interaction of bonds enables the formation of self-healing and stimuli-responsive hydrogels [39, 40]. Additionally, the self-healing properties of physically cross-linked hydrogels have been reported to be effective in 3D printing fabrication of hydrogels [41]. As an example, cross-linked hydrogels can be produced using polymer/peptide chains that can bond with metal ions. By using amino acid sequences that have the ability to bind to metal ions, polymer chains can be cross-linked and hydrogels with high mechanical strength can be obtained. In a previous study, polymer micelles were formed with polymer chains containing peptide motifs that can bind to Zn^{+2} ions and cross-linked hydrogel structures were obtained by the interaction of peptides with Zn^{+2} ions [42]. Furthermore, chemical and physical cross-linking can be combined to achieve hydrogels with appropriate bioactivity, mechanical strength, and self-healing properties [39].

14.3.2 Formation of Interconnected Pores by Using Porogen Materials

Hydrogels are structures that can be produced in different shapes and hardness levels with various controllable methods. The production method will determine important properties like porosity, ultimate strength, and elastic modulus that will affect bone tissue response according to how they mimic the physical and mechanical properties of bone (Fig. 14.2). Previous studies have reported that porosity between 35% to 70% was successful in promoting bone tissue regeneration by enabling the transfer of materials and tissue ingrowth [43]. The optimum pore size for cell attachment and migration has been reported to be approximately 100 to

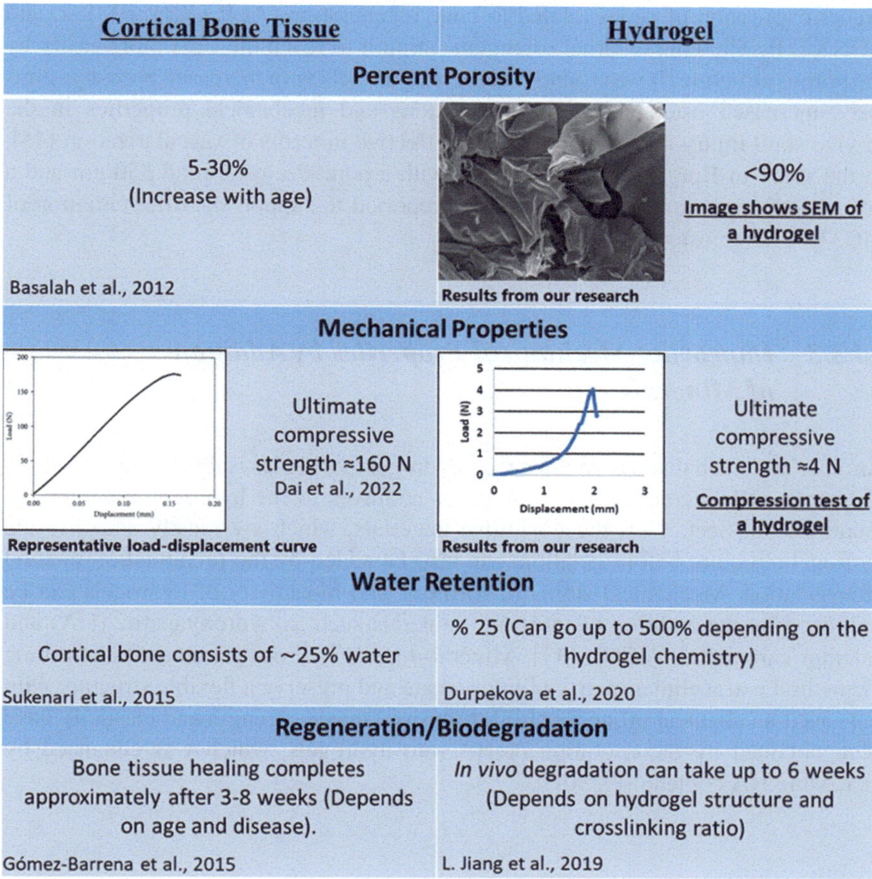

Fig. 14.2 Comparison of physical and mechanical properties of bone (cortical) tissue and hydrogel structures [62–68]

400 μm to allow proper osteogenesis and vascularization [44]. Wu et al. have reported the production of a gelatin methacryloyl (GelMA) hydrogel loaded with VEGF (Vascular Endothelial Growth Factor) carrying PLGA microspheres to provide vascularization. The hydrogel was fabricated by creating an emulsion of GELMA and polyethylene oxide (PEO) and cross-linking GELMA with UV light. Porosity was induced by the porogen leaching method as polyethylene oxide (PEO) was removed by rinsing with water to create porous hydrogel [31]. Results have shown that increased porosity has decreased the elastic modulus while increasing the swelling ratio and degradation rate. Studies have reported that despite carrying the same bioactive substances (DBM and VEGF), hydrogels with increased porosity are more successful in terms of cell viability and cell spreading. Additionally, porous hydrogels have been shown to be more effective than standard hydrogels in terms of osteogenic differentiation of BMSCs. This situation is explained by the

higher expression of genes related to bone regeneration (ALP, OCN, BMP-2, and RUNX2) by allowing external osteogenic stimuli to reach the cells more easily by the porous structure. It was found that these hydrogels with increased porosity, similarly, increased osteogenic gene expressions and mechanical properties in the in vivo skull injury model, but were less effective in terms of vascularization [45]. In the study of Burger et al., a hydrogel with a pore size of around 350 μm and a porosity of approximately 70% has been reported to support the differentiation of MC3T3 cells to osteoblasts [46].

14.3.3 Improving Mechanical Properties by Addition of Minerals

Another approach that has an impact on hydrogel mechanics is the addition of materials such as bioceramics and bioglass as additives to the hydrogel structure. As mentioned in Sect. 14.1, these additive materials, which are mostly added to the hydrogel structure before gelation, can also be added by the precipitation method after gelation. Mechanical strength, stiffness, and bioactivity of hydrogels can be improved by the addition of inorganic minerals such as hydroxyapatite (HA) and calcium carbonate ($CaCO_3$) [47]. Mineral-dopped hydrogel structures can mimic the natural extracellular matrix of bone tissue and preserve a flexible structure with improved mechanical properties [48]. Improved tensile strength and elasticity have been obtained by incorporation of HA into hydrogels, which was enhanced by increasing HA content [49, 50].

14.3.4 Improving Bone Tissue Regeneration with Biologically Active Agents

To improve the bone regeneration potential of hydrogels, biological agents such as cytokines, antibiotics and growth factors can be loaded into the hydrogel. The most used growth factors in clinical applications are bone morphogenetic proteins (BMP) and recombinant human bone morphogenic proteins (BMP-2 and BMP-7 factors) [51]. However, conflicting results were found due to the complications associated with the use of these factors in the clinic, and therefore it was stated that more detailed and comprehensive studies were needed to define the best delivery system model and comparison criteria [52].

One of the bone conditions that require bioactive agents for treatment is bone infection (osteomyelitis) which can occur after a trauma, fracture, or tumor-related operations. These infections may require antibiotic treatment for up to 4–6 weeks [53, 54]. In such infections, especially bacterial film-plaque formation reduces tissue healing and the success of systemic antibiotic treatment. The most common and difficult-to-treat bacterial infection is infection caused by "methicillin-resistant

Staphylococcus aureus" (MRSA) [53]. During treatment with high doses of systemic antibiotics, side effects such as nephrotoxicity and gastrointestinal disturbances may occur [54]. Therefore, the use of hydrogels and/or biomaterials that provide locally controlled antibiotic release is the most used treatment approach for osteomyelitis. There is a lot of research in the literature on the treatment of osteomyelitis with local release of antibiotics from antibiotic-loaded hydrogel or hydrogel-containing composite structures. One of the reasons that the research has focused on the use of biodegradable hydrogels is the necessity of removing nonbiodegradable bone cements like gentamicin-loaded polymethylmethacrylate (PMMA) beads prepared for osteomyelitis treatment [55]. The results of an example study have shown that by loading vancomycin into silk fibroin hydrogels containing halloysite (a type of kaolin with alumina silicate structure) nanotubes, a long-term bacterial inhibition zone was obtained by creating a higher water retention and controlled release profile [56]. However, the decrease in mechanical properties and increased water retention due to halloysite nanotubes have been stated as features of these hydrogels that need to be improved.

14.4 Encountered Challenges and Expected Impact

Hydrogels are successful biomaterials used for bone regeneration with their inherent advantages such as porosity, water retention, and biodegradability (Fig. 14.2). Interconnected pores in the three-dimensional structures of hydrogels increase the coordinated interaction of bioactive molecules and cells that can be loaded into the hydrogel. In addition to these advantages, there are still some problems that need to be overcome for the use of hydrogels in the field of bone tissue engineering [57].

The main problems encountered are the inability of hydrogels to maintain their structural integrity during application and their weak mechanical strength. Hydrogels can be inherently flexible, and their mechanical strength is much lower compared to the mechanical strength of bone (Fig. 14.2). Additionally, when natural polymers are used, the mechanical strength of the produced 3D porous structures will be weaker. Biomaterials to be used in bone tissue engineering applications need to have a strength close to the mechanical properties of bone and should resist the mechanical forces encountered by the bone in its natural state. A hydrogel that does not have sufficient strength and/or degrades rapidly may not support the area of tissue damage and may cause a delay in bone tissue healing or failure to heal completely.

In tissue engineering applications, attempts are often made to increase the regenerative effect of the material by entrapping bioactive agents and/or stem cells in the structure of hydrogels [58]. Since the ultimate purpose of hydrogels for bone tissue engineering is to provide new bone formation, osteointegration can be increased by loading growth factors. As an example, the rate of new tissue formation has been increased by using angiogenic growth factors and providing vascularization, which is important in the hydrogel scaffolds used for large-volume tissue healing [59, 60].

Besides that, if hydrogel does not possess enough mechanical strength, premature and sudden release of bioactive molecules can occur, and the cells in the structure may be damaged and lose their viability, due to the deterioration of hydrogel under mechanical forces [57]. Additionally, during the extrusion-based bioprinting method, cells loaded into the hydrogel may be exposed to shear forces during printing and lose their integrity. In particular, in 3D printing techniques, being able to print with high resolution while preserving cell viability is a challenge that must be overcome [2].

One of the most direct ways to increase mechanical strength in polymer-structured biomaterials is the use of synthetic polymers. It is expected that the mechanical strength of hydrogels produced by adding metal and ceramic particles or synthetic polymers will be higher. However, hydrogels produced in this way may have less biocompatibility, tissue bonding, and osteointegration. Increasing the polymer concentration in the hydrogel structure can be considered as a solution to provide mechanical strength. However, although this method increases the mechanical strength of the hydrogel, it may reduce the biodegradation rate and biocompatibility [2]. When selecting materials to be used in hydrogel production, different combinations and ratios of natural and synthetic polymers that will best promote bone regeneration should be taken into account. The best approach to increase the mechanical strength of the hydrogel to the desired extent is to select the correct cross-linkers and optimize the cross-linking method [61].

Another difficulty encountered in hydrogel systems is controlling the release of bioactive molecules loaded into the hydrogel. Growth factors that are released in the dynamic body environment may quickly lose their activity or move away from the applied area. In addition, local and uncontrolled release of growth factors in high doses may cause ectopic ossification (abnormally located bone formation) and carcinogenic effects [69, 70]. Therefore, the release of bioactive molecules in therapeutic doses should be ensured. Hydrogels, which are flexible and can hold large amounts of water, can stretch during application or under the mechanical forces they are exposed to in the tissue and cause the sudden and high dose release of the molecules. To prevent this problem, smart hydrogel systems with optional release features have been developed. In these systems, bioactive molecules are contained in a capsule or chemically bonded to the polymer chains that form the hydrogel system, and when there is a stimulus, the release of active molecules occurs by changing the shape of the hydrogel or breaking the chemical bonds. Smart hydrogel systems can be designed to respond to stimuli such as temperature, light, and pH and can be enhanced to control bioactive agent release [57].

14.5 Conclusion

Currently, only a few hydrogels are approved by the US Food and Drug Administration (FDA) for use in clinical applications. This shows that innovative developments are needed in the field of hydrogels. To overcome the mechanical limitations of hydrogels, they can be combined with other mechanically strong scaffolds to create hybrid tissue scaffolds. One innovative production technique in this context is four-dimensional (4D) printing, which enables the fabrication of hydrogels that can change their shape, function, or properties over time in response to specific predetermined external stimuli. In the future, innovative production methods can successfully combine hydrogel systems with biological agents, and smart hydrogels can overcome many limitations.

References

1. Naahidi, S., Jafari, M., Logan, M., Wang, Y., Yuan, Y., Bae, H., Dixon, B., Chen, P.: Biocompatibility of hydrogel-based scaffolds for tissue engineering applications. Biotechnol. Adv. **35**(5), 530–544 (2017). https://doi.org/10.1016/J.BIOTECHADV.2017.05.006
2. Lin, H., Yin, C., Mo, A., Hong, G.: Applications of hydrogel with special physical properties in bone and cartilage regeneration. Materials. **14**(1), 235 (2021). https://doi.org/10.3390/MA14010235
3. Xue, X., Hu, Y., Wang, S., Chen, X., Jiang, Y., Su, J.: Fabrication of physical and chemical crosslinked hydrogels for bone tissue engineering. Bioact. Mater. **12**, 327–339 (2022). https://doi.org/10.1016/J.BIOACTMAT.2021.10.029
4. Suvarnapathaki, S., Wu, X., Lantigua, D., Nguyen, M.A., Camci-Unal, G., Suvarnapathaki, S., Wu, X., Lantigua, D., Nguyen, M.A., Camci-Unal, G.: Hydroxyapatite-incorporated composite gels improve mechanical properties and bioactivity of bone scaffolds. Macromol. Biosci. **20**(10), 2000176 (2020). https://doi.org/10.1002/MABI.202000176
5. Pei, B., Wang, W., Dunne, N., Li, X.: Applications of carbon nanotubes in bone tissue regeneration and engineering: superiority, concerns, current advancements, and prospects. Nanomaterials. **9**(10), 1501 (2019). https://doi.org/10.3390/NANO9101501
6. Bendtsen, S.T., Quinnell, S.P., Wei, M.: Development of a novel alginate-polyvinyl alcohol-hydroxyapatite hydrogel for 3D bioprinting bone tissue engineered scaffolds. J. Biomed. Mater. Res. A. **105**(5), 1457–1468 (2017). https://doi.org/10.1002/JBM.A.36036
7. Bendtsen, S.T., Wei, M.: Synthesis and characterization of a novel injectable alginate–collagen–hydroxyapatite hydrogel for bone tissue regeneration. J. Mater. Chem. B. **3**(15), 3081–3090 (2015). https://doi.org/10.1039/C5TB00072F
8. Ma, Y., Kwon, J., Ji, R., Espinosa-Marzal, R.M.: Polymer–mineral interaction influences the mineralization of hydroxyapatite in hydrogels. Cryst. Growth Des. **24**(21), 8970–8988 (2024). https://doi.org/10.1021/acs.cgd.4c01002
9. Sareethammanuwat, M., Boonyuen, S., Arpornmaeklong, P.: Effects of beta-tricalcium phosphate nanoparticles on the properties of a thermosensitive chitosan/collagen hydrogel and controlled release of quercetin. J. Biomed. Mater. Res. A. **109**(7), 1147–1159 (2021). https://doi.org/10.1002/JBM.A.37107
10. Douglas, T.E.L., Schietse, J., Zima, A., Gorodzha, S., Parakhonskiy, B.V., KhaleNkow, D., Shkarin, R., Ivanova, A., Baumbach, T., Weinhardt, V., Stevens, C.V., Vanhoorne, V., Vervaet, C., Balcaen, L., Vanhaecke, F., Slośarczyk, A., Surmeneva, M.A., Surmenev, R.A., Skirtach, A.G.: Novel self-gelling injectable hydrogel/alpha-tricalcium phosphate composites for bone regeneration: physiochemical and microcomputer tomographical characterization. J. Biomed. Mater. Res. A. **106**(3), 822–828 (2018). https://doi.org/10.1002/jbm.a.36277
11. Ghaffari, R., Salimi-Kenari, H., Fahimipour, F., Rabiee, S.M., Adeli, H., Dashtimoghadam, E.: Fabrication and characterization of dextran/nanocrystalline β-tricalcium phosphate nanocomposite hydrogel scaffolds. Int. J. Biol. Macromol. **148**, 434–448 (2020). https://doi.org/10.1016/J.IJBIOMAC.2020.01.112

12. Zhuang, P., Wu, X., Dai, H., Yao, Y., Qiu, T., Han, Y., Li, S.: Nano β-tricalcium phosphate/hydrogel encapsulated scaffolds promote osteogenic differentiation of bone marrow stromal cells through ATP metabolism. Mater. Des. **208**, 109881 (2021). https://doi.org/10.1016/J.MATDES.2021.109881
13. Al-Harbi, N., Mohammed, H., Al-Hadeethi, Y., Bakry, A.S., Umar, A., Hussein, M.A., Abbassy, M.A., Vaidya, K.G., Al Berakdar, G., Mkawi, E.M., Nune, M.: Silica-based bioactive glasses and their applications in hard tissue regeneration: a review. Pharmaceuticals. **14**(2), 75 (2021). https://doi.org/10.3390/PH14020075
14. Han, Y., Zeng, Q., Li, H., Chang, J.: The calcium silicate/alginate composite: preparation and evaluation of its behavior as bioactive injectable hydrogels. Acta Biomater. **9**(11), 9107–9117 (2013). https://doi.org/10.1016/J.ACTBIO.2013.06.022
15. Cimmino, L., Neel, B.G., Aifantis, I.: Vitamin C in stem cell reprogramming and cancer. Trends Cell Biol. **28**(9), 698–708 (2018). https://doi.org/10.1016/j.tcb.2018.04.001
16. Gaffney-Stomberg, E.: The impact of trace minerals on bone metabolism. Biol. Trace Elem. Res. **188**(1), 26–34 (2019). https://doi.org/10.1007/S12011-018-1583-8/TABLES/1
17. Song, J., Li, L., Chen, B., Shan, L., Yuan, S., Yu, H.: Dietary copper requirements of postlarval coho salmon (*Oncorhynchus kisutch*). Aquacult. Nutr. **27**(6), 2084–2092 (2021). https://doi.org/10.1111/anu.13342
18. Grigolato, R., Pizzi, N., Brotto, M.C., Corrocher, G., Desando, G., Grigolo, B.: Magnesium-enriched hydroxyapatite as bone filler in an ameloblastoma mandibular defect. Int. J. Clin. Exp. Med. **8**(1), 288 (2015)
19. Qi, T., Weng, J., Yu, F., Zhang, W., Li, G., Qin, H., Tan, Z., Zeng, H.: Insights into the role of magnesium ions in affecting osteogenic differentiation of mesenchymal stem cells. Biol. Trace Elem. Res. **199**(2), 559–567 (2021). https://doi.org/10.1007/S12011-020-02183-Y/TABLES/1
20. Zhang, X., Huang, P., Jiang, G., Zhang, M., Yu, F., Dong, X., Wang, L., Chen, Y., Zhang, W., Qi, Y., Li, W., Zeng, H.: A novel magnesium ion-incorporating dual-crosslinked hydrogel to improve bone scaffold-mediated osteogenesis and angiogenesis. Mater. Sci. Eng. C. **121**, 111868 (2021). https://doi.org/10.1016/J.MSEC.2021.111868
21. Cal, F., Sezgin Arslan, T., Derkus, B., Kiran, F., Cengiz, U., Arslan, Y.E.: Synthesis of silica-based boron-incorporated collagen/human hair keratin hybrid Cryogels with the potential bone formation capability. ACS Appl. Bio Mater. **4**(6), 7266–7279 (2021). https://doi.org/10.1021/ACSABM.1C00805/ASSET/IMAGES/LARGE/MT1C00805_0008.JPEG
22. Manoochehri, H., Ghorbani, M., Moosazadeh Moghaddam, M., Nourani, M.R., Makvandi, P., Sharifi, E.: Strontium doped bioglass incorporated hydrogel-based scaffold for amplified bone tissue regeneration. Sci. Rep. **12**(1), 1–18 (2022). https://doi.org/10.1038/s41598-022-14329-0
23. Chen, Z., Zhang, W., Wang, M., Backman, L.J., Chen, J.: Effects of zinc, magnesium, and iron ions on bone tissue engineering. ACS Biomater Sci. Eng. **8**(6), 2321–2335 (2022). https://doi.org/10.1021/ACSBIOMATERIALS.2C00368/ASSET/IMAGES/LARGE/AB2C00368_0004.JPEG
24. Liu, T., Jin, M., Zhang, Y., Weng, W., Wang, T., Yang, H., Zhou, L.: K+/Sr2+/Na+ triple-doped hydroxyapatites/GelMA composite hydrogel scaffold for the repair of bone defects. Ceram. Int. **47**(21), 30929–30937 (2021). https://doi.org/10.1016/J.CERAMINT.2021.07.277
25. Zhang, Y., Yu, T., Peng, L., Sun, Q., Wei, Y., Han, B.: Advancements in hydrogel-based drug sustained release systems for bone tissue engineering. Front. Pharmacol. **11**, 622 (2020). https://doi.org/10.3389/FPHAR.2020.00622/BIBTEX
26. Shirai, M., Yamamoto, R., Chiba, T., Komatsu, K., Shimoda, S., Yamakoshi, Y., Oida, S., Ohkubo, C.: Bone augmentation around a dental implant using demineralized bone sheet containing biologically active substances. Dent. Mater. J. **35**(3), 470–478 (2016). https://doi.org/10.4012/dmj.2016-026
27. Jang, C.H., Cho, G.W., Song, A.: Effect of bone powder/mesenchymal stem cell/BMP2/fibrin glue on osteogenesis in a mastoid obliteration model. In Vivo. **34**(3), 1103–1110 (2020). https://doi.org/10.21873/invivo.11881

28. Solomon, R., Hommen, J.P., Travascio, F.: Effects of platelet-rich osteoconductive–osteoinductive allograft compound on tunnel widening of ACL reconstruction: a randomized blind analysis study. Pathophysiology. **29**(3), 394–404 (2022). https://doi.org/10.3390/pathophysiology29030031
29. Saito, H., Chiba-Ohkuma, R., Yamakoshi, Y., Karakida, T., Yamamoto, R., Shirai, M., Ohkubo, C.: Characterization of bioactive substances involved in the induction of bone augmentation using demineralized bone sheets. Int. J. Implant Dentistry. **8**(1), 49 (2022). https://doi.org/10.1186/s40729-022-00449-9
30. Zheng, Y., Huang, K., You, X., Huang, B., Wu, J., Gu, Z.: Hybrid hydrogels with high strength and biocompatibility for bone regeneration. Int. J. Biol. Macromol. **104**, 1143–1149 (2017). https://doi.org/10.1016/J.IJBIOMAC.2017.07.017
31. Wu, X., Huo, Y., Ci, Z., Wang, Y., Xu, W., Bai, B., Hao, J., Hu, G., Yu, M., Ren, W., Zhang, Y., Hua, Y., Zhou, G.: Biomimetic porous hydrogel scaffolds enabled vascular ingrowth and osteogenic differentiation for vascularized tissue-engineered bone regeneration. Appl. Mater. Today. **27**, 101478 (2022). https://doi.org/10.1016/J.APMT.2022.101478
32. Lu, J., Xu, R., Chen, Y., Chan, L., Feng, X., Lin, L., Yao, Y., Hu, X., Zhang, X.: Injectable col-I/CS hydrogel enhances bone regeneration in mice tibial mono-cortical defect with impaired osteogenesis. Mater. Today Commun. **32**, 104070 (2022). https://doi.org/10.1016/J.MTCOMM.2022.104070
33. Huang, Z., Feng, J., Feng, X., Chan, L., Lu, J., Lei, L., Huang, Z., Zhang, X.: Loss of signal transducer and activator of transcription 3 impaired the osteogenesis of mesenchymal progenitor cells in vivo and in vitro. Cell Biosci. **11**(1), 172 (2021). https://doi.org/10.1186/S13578-021-00685-3/FIGURES/7
34. Lei, L., Huang, Z., Feng, J., Huang, Z., Tao, Y., Hu, X., Zhang, X.: Loss of receptor tyrosine kinase-like orphan receptor 2 impairs the osteogenesis of mBMSCs by inhibiting signal transducer and activator of transcription 3. Stem Cell Res. Ther. **11**(1), 137 (2020). https://doi.org/10.1186/S13287-020-01646-2
35. de Siqueira, E.C., de França, J.A.A., de Souza, R.F.M., da Leoterio, D.M., Cordeiro, J.N., Doboszewski, B.: Recent advances in the development of the physically crosslinked hydrogels and their biomedical applications. Res. Soc. Dev. **12**(8), e18212843073 (2023). https://doi.org/10.33448/rsd-v12i8.43073
36. Gao, Y., Peng, K., Mitragotri, S.: Covalently crosslinked hydrogels via step-growth reactions: crosslinking chemistries, polymers, and clinical impact. Adv. Mater. **33**(25) (2021). https://doi.org/10.1002/adma.202006362
37. Hu, X., Zhang, L., Yan, L., Tang, L.: Recent advances in polysaccharide-based physical hydrogels and their potential applications for biomedical and wastewater treatment. Macromol. Biosci. **22**(9), e2200153 (2022). https://doi.org/10.1002/mabi.202200153
38. Zerbinati, N., Esposito, C., Cipolla, G., Calligaro, A., Monticelli, D., Martina, V., Golubovic, M., Binic, I., Sigova, J., Gallo, A.L., D'Este, E., Jafferany, M., Pratosoni, M., Tirant, M., Van Thuong, N., Sangalli, F., Rauso, R., Lotti, T.: Chemical and mechanical characterization of hyaluronic acid hydrogel cross-linked with polyethylen glycol and its use in dermatology. Dermatol. Ther. **33**(4), e13747 (2020). https://doi.org/10.1111/dth.13747
39. Jiang, Y., Jeng, J., Wu, P., Chien, H.: A rapidly and highly self-healing poly(Sulfobetaine methacrylate) hydrogel with stretching properties, adhesive properties, and biocompatibility. Macromol. Biosci. **23**(3), e2200368 (2023). https://doi.org/10.1002/mabi.202200368
40. Shao, J., Zhang, Z., Zhao, S., Wang, S., Guo, Z., Xie, H., Hu, Y.: Self-healing hydrogel of poly(vinyl alcohol)/agarose with robust mechanical property. Starch. **71**, 1800281 (2019). https://doi.org/10.1002/star.201800281
41. Alruwaili, M., Lopez, J.A., McCarthy, K., Reynaud, E.G., Rodriguez, B.J.: Liquid-phase 3D bioprinting of gelatin alginate hydrogels: influence of printing parameters on hydrogel line width and layer height. Bio-Des. Manuf. **2**(3), 172–180 (2019). https://doi.org/10.1007/s42242-019-00043-w

42. Ghoorchian, A., Simon, J.R., Bharti, B., Han, W., Zhao, X., Chilkoti, A., Lõpez, G.P.: Bioinspired reversibly cross-linked hydrogels comprising polypeptide micelles exhibit enhanced mechanical properties. Adv. Funct. Mater. **25**(21), 3122–3130 (2015). https://doi.org/10.1002/ADFM.201500699
43. Bastos, A.R., da Silva, L.P., Maia, F.R., Pina, S., Rodrigues, T., Sousa, F., Oliveira, J.M., Cornish, J., Correlo, V.M., Reis, R.L.: Lactoferrin-hydroxyapatite containing spongy-like hydrogels for bone tissue engineering. Materials. **12**(13), 2074 (2019). https://doi.org/10.3390/ma12132074
44. Yuan, L., Li, X., Ge, L., Jia, X., Lei, J., Mu, C., Li, D.: Emulsion template method for the fabrication of gelatin-based scaffold with a controllable pore structure. ACS Appl. Mater. Interfaces. **11**(1), 269–277 (2019). https://doi.org/10.1021/acsami.8b17555
45. Qiao, Y., Liu, X., Zhou, X., Zhang, H., Zhang, W., Xiao, W., Pan, G., Cui, W., Santos, H.A., Shi, Q.: Gelatin templated polypeptide co-cross-linked hydrogel for bone regeneration. Adv. Healthc. Mater. **9**(1), e1901239 (2020). https://doi.org/10.1002/adhm.201901239
46. Burger, D., Beaumont, M., Rosenau, T., Tamada, Y.: Porous silk fibroin/cellulose hydrogels for bone tissue engineering via a novel combined process based on sequential regeneration and Porogen leaching. Molecules. **25**(21), 5097 (2020). https://doi.org/10.3390/molecules25215097
47. Strasser, V., Matijaković, N., Mihelj Josipović, T., Kontrec, J., Lyons, D.M., Kralj, D., Dutour Sikirić, M.: Factors affecting calcium phosphate mineralization within bulk alginate hydrogels. J. Polym. Res. **26**(12), 1–11 (2019). https://doi.org/10.1007/S10965-019-1942-Y/FIGURES/7
48. Wu, X., Zhang, T., Hoff, B., Suvarnapathaki, S., Lantigua, D., McCarthy, C., Wu, B., Camci-Unal, G.: Mineralized hydrogels induce bone regeneration in critical size cranial defects. Adv. Healthc. Mater. **10**(4), e2001101 (2021). https://doi.org/10.1002/adhm.202001101
49. Fan, Z., Chen, Z., Zhang, H., Nie, Y., Xu, S.: Gradient mineralized and porous double-network hydrogel effectively induce the differentiation of BMSCs into osteochondral tissue in vitro for potential application in cartilage repair. Macromol. Biosci. **21**(3), e2000323 (2021). https://doi.org/10.1002/mabi.202000323
50. Yang, C., Li, G.: A biomimetic mineralized collagen hydrogel containing uniformly distributed and highly abundant dopamine-modified hydroxyapatite particles for bone tissue engineering. J. Appl. Polym. Sci. **141**(26) (2024). https://doi.org/10.1002/app.55567
51. Nauth, A., Ristevski, B., Li, R., Schemitsch, E.H.: Growth factors and bone regeneration: how much bone can we expect? Injury. **42**(6), 574–579 (2011). https://doi.org/10.1016/j.injury.2011.03.034
52. Krishnakumar, G.S., Roffi, A., Reale, D., Kon, E., Filardo, G.: Clinical application of bone morphogenetic proteins for bone healing: a systematic review. Int. Orthop. **41**(6), 1073–1083 (2017). https://doi.org/10.1007/S00264-017-3471-9/TABLES/3
53. Ahadi, F., Khorshidi, S., Karkhaneh, A.: A hydrogel/fiber scaffold based on silk fibroin/oxidized pectin with sustainable release of vancomycin hydrochloride. Eur. Polym. J. **118**, 265–274 (2019). https://doi.org/10.1016/J.EURPOLYMJ.2019.06.001
54. Popat, K.C., Eltgroth, M., LaTempa, T.J., Grimes, C.A., Desai, T.A.: Decreased staphylococcus epidermis adhesion and increased osteoblast functionality on antibiotic-loaded titania nanotubes. Biomaterials. **28**(32), 4880–4888 (2007). https://doi.org/10.1016/J.BIOMATERIALS.2007.07.037
55. Van De Belt, H., Neut, D., Uges, D.R.A., Schenk, W., Van Horn, J.R., Van Der Mei, H.C., Busscher, H.J.: Surface roughness, porosity and wettability of gentamicin-loaded bone cements and their antibiotic release. Biomaterials. **21**(19), 1981–1987 (2000). https://doi.org/10.1016/S0142-9612(00)00082-X
56. Avani, F., Damoogh, S., Mottaghitalab, F., Karkhaneh, A., Farokhi, M.: Vancomycin loaded halloysite nanotubes embedded in silk fibroin hydrogel applicable for bone tissue engineering. Int. J. Polym. Mater. Polym. Biomater. **69**(1), 32–43 (2019). https://doi.org/10.1080/00914037.2019.1616201

57. Bai, X., Gao, M., Syed, S., Zhuang, J., Xu, X., Zhang, X.Q.: Bioactive hydrogels for bone regeneration. Bioact. Mater. **3**(4), 401–417 (2018). https://doi.org/10.1016/J.BIOACTMAT.2018.05.006
58. Chuah, Y.J., Peck, Y., Lau, J.E.J., Hee, H.T., Wang, D.A.: Hydrogel based cartilaginous tissue regeneration: recent insights and technologies. Biomater. Sci. **5**(4), 613–631 (2017). https://doi.org/10.1039/C6BM00863A
59. Champeau, M., Alves Heinze, D., Nunes Viana, T., Rodrigues de Souza, E., Cristine Chinellato, A., Titotto, S., Champeau, M., Heinze, D.A., Viana, T.N., de Souza, E.R., Chinellato, A.C., Titotto, S.: 4D printing of hydrogels: a review. Adv. Funct. Mater. **30**(31), 1910606 (2020). https://doi.org/10.1002/ADFM.201910606
60. He, B., Ou, Y., Zhou, A., Chen, S., Zhao, W., Zhao, J., Li, H., Zhu, Y., Zhao, Z., Jiang, D.: Functionalized d-form self-assembling peptide hydrogels for bone regeneration. Drug Des. Dev. Ther. **10**, 1388 (2016). https://doi.org/10.2147/DDDT.S97530
61. Short, A.R., Koralla, D., Deshmukh, A., Wissel, B., Stocker, B., Calhoun, M., Dean, D., Winter, J.O.: Hydrogels that allow and facilitate bone repair, remodeling, and regeneration. J. Mater. Chem. B Mater. Biol. Med. **3**(40), 7830 (2015). https://doi.org/10.1039/C5TB01043H
62. Basalah, A., Shanjani, Y., Esmaeili, S., Toyserkani, E.: Characterizations of additive manufactured porous titanium implants. J. Biomed. Mater. Res. B Appl. Biomater. **100B**(7), 1970–1979 (2012). https://doi.org/10.1002/jbm.b.32764
63. Czerner, M., Fellay, L.S., Suárez, M.P., Frontini, P.M., Fasce, L.A.: Determination of elastic modulus of gelatin gels by indentation experiments. Procedia Mater. Sci. **8**, 287–296 (2015). https://doi.org/10.1016/J.MSPRO.2015.04.075
64. Dai, X., Fan, R., Wu, H., Jia, Z.: Investigation on the differences in the failure processes of the cortical bone under different loading conditions. Appl. Bionics Biomech. **2022**, 1–9 (2022). https://doi.org/10.1155/2022/3406984
65. Durpekova, S., Filatova, K., Cisar, J., Ronzova, A., Kutalkova, E., Sedlarik, V.: A novel hydrogel based on renewable materials for agricultural application. Int. J. Polym. Sci. **2020**, 1–13 (2020). https://doi.org/10.1155/2020/8363418
66. Gómez-Barrena, E., Rosset, P., Lozano, D., Stanovici, J., Ermthaller, C., Gerbhard, F.: Bone fracture healing: cell therapy in delayed unions and nonunions. Bone. **70**, 93–101 (2015). https://doi.org/10.1016/j.bone.2014.07.033
67. Jiang, L., Su, D., Ding, S., Zhang, Q., Li, Z., Chen, F., Ding, W., Zhang, S., Dong, J.: Salt-assisted toughening of protein hydrogel with controlled degradation for bone regeneration. Adv. Funct. Mater. **29**(26), 1901314 (2019). https://doi.org/10.1002/adfm.201901314
68. Sukenari, T., Horii, M., Ikoma, K., Kido, M., Hayashi, S., Hara, Y., Yamasaki, T., Matsuda, K., Kawata, M., Kubo, T.: Cortical bone water changes in ovariectomized rats during the early postoperative period: objective evaluation using sweep imaging with Fourier transform. J. Magn. Reson. Imaging. **42**(1), 128–135 (2015). https://doi.org/10.1002/jmri.24765
69. Carragee, E.J., Hurwitz, E.L., Weiner, B.K.: A critical review of recombinant human bone morphogenetic protein-2 trials in spinal surgery: emerging safety concerns and lessons learned. Spine J. **11**(6), 471–491 (2011). https://doi.org/10.1016/j.spinee.2011.04.023
70. de Melo Pereira, D., Habibovic, P.: Biomineralization-inspired material design for bone regeneration. Adv. Healthc. Mater. **7**(22), 1800700 (2018). https://doi.org/10.1002/ADHM.201800700

Chapter 15
Hydrogels Used in Wound Healing

Elif Conger-Onder, Merve Gültekinoğlu, and Kezban Ulubayram

Abstract The skin, which is the body's largest organ, plays a crucial role in protecting the body from external influences. In addition, the skin contributes to the prevention of fluid loss, thermoregulation, immune surveillance, and sensory perception. Impairment of the integrity and functionality of skin tissue, whether due to mechanical forces, surgical procedures, burns (chemical, electrical, or radiation), frostbite, genetic disorders (such as epidermolysis bullosa), or ulcers from chronic conditions, has a profound impact on the patient's quality of life and may even lead to death. Chronic wounds, particularly those associated with diabetes mellitus, represent a significant health burden, with a considerable economic and social impact on the patient. Therefore, understanding the fundamental mechanisms of wound healing is crucial for selecting an appropriate wound dressing and developing an effective treatment plan. This section will present information regarding the development of hydrogels for use in wound healing. Prior to addressing this topic, it would be beneficial to discuss the structure and functions of the skin, as well as the mechanisms of wound healing, to gain a deeper understanding of the subject.

E. Conger-Onder
Department of Basic Pharmaceutical Sciences, Faculty of Pharmacy, Hacettepe University, Ankara, Türkiye

Department of Bioengineering, Institute of Science, Hacettepe University, Ankara, Türkiye

M. Gültekinoğlu
Department of Nanotechnology and Nanomedicine, Institute of Science, Hacettepe University, Ankara, Türkiye

K. Ulubayram (✉)
Department of Basic Pharmaceutical Sciences, Faculty of Pharmacy, Hacettepe University, Ankara, Türkiye

Department of Bioengineering, Institute of Science, Hacettepe University, Ankara, Türkiye

Department of Nanotechnology and Nanomedicine, Institute of Science, Hacettepe University, Ankara, Türkiye
e-mail: ukezban@hacettepe.edu.tr

N. Hasirci, V. Hasirci (eds.), *Hydrogels and Bioinks in Tissue Engineering*, Biological and Medical Physics, Biomedical Engineering,
https://doi.org/10.1007/978-3-031-90966-5_15

Keywords Skin · Wound healing · Hydrogel · Bioactive · Antibacterial · Hemostatic · Anti-oxidative · Anti-inflammatory · Smart · Four-dimensional

15.1 Structure and Functions of the Skin

The skin is a protective covering system that envelops the human body. It serves to safeguard muscles and organs from external factors, including mechanical, thermal, and physical injuries. Additionally, it acts as a barrier to prevent the entry of microorganisms and harmful chemicals into the body. Furthermore, the skin plays a pivotal role in regulating fluid levels and body temperature. The skin is the largest vital organ in the human body, comprising approximately 16% of body weight in adults and an average surface area of 1.6 m^2 [1, 2].

The skin consists of layers: epidermis, dermis, and subcutaneous layer. Hair, nails, sebaceous, sweat, and apocrine glands are derivatives of the skin. The outermost layer of the skin, the epidermis, is in a state of constant renewal and is composed of multiple layers. Approximately 95% of the epidermal cells are keratinocytes. The remaining 5% of the epidermis is composed of melanocytes, Langerhans cells, dendritic α cells, and Merkel cells, which are responsible for melanin production, antigen creation, and sensation, respectively. The epidermis layer has a mean thickness of approximately 0.2 mm. The dermis is situated beneath the epidermis. The papillary layer is composed of fine fibers, a multitude of capillaries, nerves, and cells with a substantial amount of cytoplasm. Furthermore, projections extending into the epidermis are observed. The subpapillary layer is composed of the same components as the papillary layer. Consequently, some sources exclude this layer and define the dermis as comprising only two layers. The reticular layer is the largest part of the dermis. It consists of fibers and dense connective tissue. The dermis contains fibroblasts, which are responsible for the production of collagen, elastin, and structural proteoglycans, as well as mast cells and macrophages. Approximately, 70% of the dermis is made up of collagen fibers, which provide the tissue with structural integrity. Elastin endows the tissue with elasticity, whereas proteoglycans enhance viscosity and hydration. The dermis also contains a network of blood and lymphatic vessels, nerves, sweat glands, hair follicles, and a small amount of striated muscle. The subcutaneous layer, situated between the dermis and fascia, is predominantly constituted of adipose tissue and fulfills a number of functions, including the storage of fat, an increase in resistance to physical impact or pressure, the retention of moisture, and the provision of heat. The thickness of this tissue is subject to variation depending on the specific body region, age, and other contributing factors.

One of the most pivotal aspects in the formation and sustenance of the skin barrier function is the lipid layer situated between the cells. The lipids penetrate the intercellular spaces from the granular layer, thereby forming a bilayer barrier. The water content retained in the upper layer of the skin is typically around 30%. Free amino acids, lactic acid, urea, and certain salts serve as natural moisturizing factors, thereby maintaining the skin's hydration levels. Melanocytes located in the basal layer contribute to the skin's barrier function by synthesizing melanin, which serves to mitigate the detrimental effects of UV radiation.

15.2 Wound Healing

A wound is typically defined as a disruption of the normal integrity of the skin. This may result from a number of causes, including burns (due to thermal, chemical, electrical, or radiation causes), frostbite, mechanical trauma, surgical interventions, genetic disorders (such as epidermolysis bullosa), or physiological conditions [3].

Injuries compromise the integrity of the skin, thereby exposing it to environmental factors and potentially resulting in the temporary or permanent loss of existing functions. Wounds may be classified as either acute or chronic. Acute wounds result from surgical procedures, burns, trauma, and other causes and are expected to undergo a healing process over time, thereby restoring the integrity of the skin. In contrast, a chronic wound is defined as one that does not respond to appropriate treatment within a timeframe of 6–8 weeks. This may result in either a prolonged healing process or a failure to heal at all [4]. Chronic wounds frequently have underlying systemic diseases such as diabetes, venous hypertension, arterial insufficiency, and peripheral vascular diseases. A subsequent section will address the subject of chronic wounds, whereas this section will concentrate on the acute wound healing process.

The term "wound healing" is defined as the process of repairing damaged tissue through a series of cellular and biochemical interactions. The repair process is dynamic and complex, comprising a series of sequential and partially overlapping stages [5]. These stages are referred to as hemostasis, inflammation, proliferation, maturation, and remodeling (Fig. 15.1).

15.2.1 Hemostasis

The process of hemostasis is initiated in response to an injury or other disruption to the vascular integrity of the body. In the aftermath of an injury, damage to the blood vessels gives rise to a disruption of vascular integrity. Platelets interact with collagen within the damaged vessel wall, forming a provisional clot that serves to reduce bleeding. Subsequently, a permanent thrombus is formed through the development of fibrin. The clot is composed of collagen, platelets, thrombin, and fibronectin. While this tissue serves to protect the wound from contamination, it also provides a matrix for cell migration and adhesion. Subsequently, serotonin released from platelets and the activation of other systems result in the dilation of constricted blood vessels, thereby increasing the passage of blood cells to the damaged tissue. The aggregation of the platelets and the formation of a fibrin clot create a hemostatic effect, which rapidly transitions the wound healing process into the inflammatory phase [6].

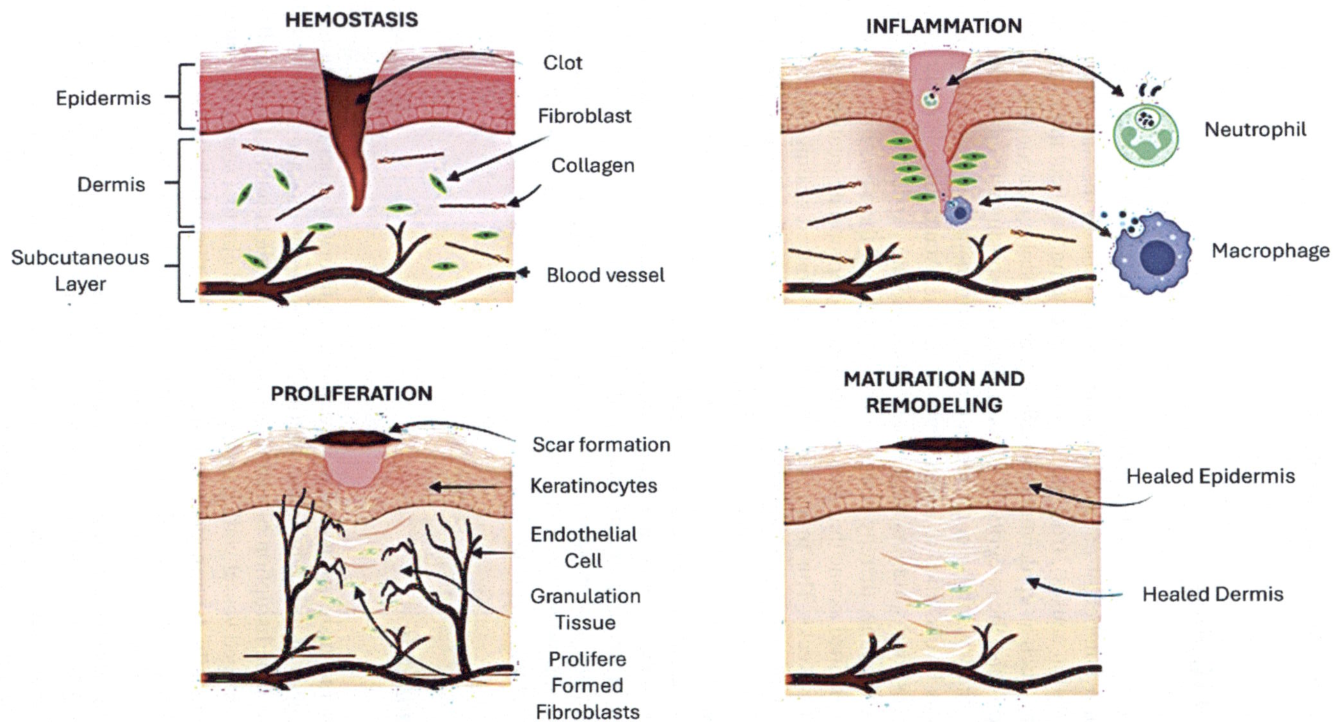

Fig. 15.1 The stages of wound healing: (**a**) Hemostasis: In an injured tissue, blood vessels are opened, which results in platelet aggregation and clot formation. (**b**) Inflammation: This is the process by which inflammatory cells migrate to the wound site, accompanied by the release of proteolytic enzymes and cytokines. (**c**) Proliferation: This stage involves angiogenesis and the development of granulation tissue, which ultimately leads to the completion of epithelialization. (**d**) Maturation and Remodeling: This phase encompasses collagen synthesis (types I and III), cross-link formation, and the regeneration of hair follicles, sweat glands, sebaceous glands, and some skin appendages

15.2.2 *Inflammation*

The increase in vascular permeability and dilation allows for the migration of immune cells, primarily neutrophils, to the wound site. Neutrophils play a pivotal role in the prevention of infection by eliminating foreign substances and bacteria from damaged tissue. Subsequently, monocytes that have arrived in the affected area differentiate into macrophages. Macrophages secrete a range of growth factors that promote the migration and proliferation of fibroblasts, epithelial cells, and endothelial cells to the wound site, thereby instigating the proliferative phase of the healing process [7].

15.2.3 *Proliferation*

This phase begins 2–3 days following the injury and persists until complete epithelialization, which occurs approximately 2 weeks later. During the proliferation phase, the number of inflammatory cells in the wound area declines, while fibroblasts and keratinocytes migrate to the wound area and commence proliferation. The secretion of growth factors instigates the synthesis of the extracellular matrix and the formation of new blood vessels. Concurrently, granulation tissue develops in the wound area, presenting as pale pink tissue, while basal cells continue to multiply around the damaged tissue, migrating to the wound site for the formation of new epithelium. Once the process of epithelialization is complete and granulation tissue has fully occupied the region of the wound, the formation of scar tissue will ensue. At this stage, the production of collagen (in particular, type III collagen) exceeds its degradation, resulting in an accumulation of collagen in the wound area.

15.2.4 *Maturation and Remodeling*

In this phase, the production and degradation of collagen gradually reach equilibrium, with type III collagen being converted into type I collagen in a progressive manner. Collagen bundles form interconnections with one another and with the underlying dermis layer. Because of these interconnections, the tissue undergoes contraction. By the third week, the strength of the newly formed tissue increases as a result of remodeling and contraction. The newly formed capillaries undergo maturation, ultimately developing into small arteries and veins. Regeneration takes place in the hair follicles, the sweat glands, the sebaceous glands, and certain appendages of the skin. Subsequently, the newly formed tissue is replaced by functional skin tissue.

The process of healing in acute wounds is facilitated by cell proliferation, angiogenesis, and the synthesis of the extracellular matrix. In contrast, chronic wounds exhibit stalling in the inflammatory phase, accompanied by the presence of exudate, severe infection, pain, and necrosis [8]. The definition of chronic wounds varies between sources. Some sources define it as a wound in which the healing process does not progress normally after 3–4 weeks, while other sources define this timeframe as 6–8 weeks [4, 8]. The defining characteristics of chronic wounds include the presence of excessive inflammation, persistent infection, drug-resistant biofilms, and insufficient epidermal cells. The chronicity of a wound is influenced not only by the width and depth of the lesion but also by several additional factors, including age, nutritional status, stress levels, ischemia, obesity, immunosuppression, smoking habits, and nutritional intake of the patient. It is not uncommon for chronic leg and foot ulcers to be observed in adults with vascular disorders or diabetes. Such ulcers have the potential to result in functional or organ loss in patients. It is estimated that ~1% of the adult population has chronic wounds, with a prevalence of 3–5% in people over 65. In Western societies, the prevalence of chronic wounds, including burns and large skin defects, is increasing due to an aging population, and an increase in the incidence of obesity and diabetes. The treatments necessitate the administration of multiple medical interventions and a protracted period of healing.

Synthetic or biological wound dressings have been employed in wound treatment from the past to the present as a means of accelerating healing, protecting the wound from external factors, and maintaining optimal moisture levels following the initial stages of intervention. These dressings come in a variety of forms, including film, foam, sponge, hydrocolloid, and hydrogel. A plethora of wound dressings are currently available on the market [9, 10]. The selection of an appropriate wound dressing is contingent upon the specific type and condition of the wound in question. The following section will address the topic of hydrogel wound dressings.

15.3 Hydrogels Used in Wound Healing Treatments

Hydrogels play an important role in the wound healing process. They provide a hydrophilic structure, adjustable physical and mechanical properties, a three-dimensional (3D) structure that resembles the extracellular matrix, and the ability to avoid causing secondary wound damage, which is a common issue with autografting procedures [11, 12]. Hydrogels are synthesized through the physical or chemical cross-linking of hydrophilic polymers. These structures facilitate the healing of wounds by absorbing exudate, which accumulates in the wound area due to the high water retention capacity, thereby allowing for the diffusion of oxygen. The accumulation of exudate can facilitate the proliferation of bacteria and the production of malodorous compounds within the wound environment. Hydrogels provide a soothing and cooling effect by lowering the temperature of the wound site and act as a barrier against external infections in the wound area [12]. A variety of natural and synthetic polymers and polypeptides are used as hydrogel wound dressings, including chitosan, gelatin, alginate, silk fibroin, hyaluronic acid, poly(ethylene glycol), poly(vinyl alcohol), polyacrylamide, and poly(acrylic acid), among others [13–19]

(Fig. 15.2). While hydrogel wound dressings were initially developed with the primary objective of protecting the wound from environmental factors and maintaining moisture, they have subsequently been employed in a broader range of functional roles in response to increasing clinical demands. Hydrogel wound dressings are classified according to their functions, which include antibacterial, hemostatic, anti-inflammatory, antioxidant, bioactive, and smart properties.

15.3.1 Antibacterial Hydrogels

While acute wounds have a relatively short healing period, chronic wounds require a longer time frame due to the colonization of the wound site by bacteria or fungi [4]. Complications such as sepsis, amputation, and death may ensue from chronic wound infections. The loss of the epidermis' protective effect allows for the proliferation of microorganisms in the wound area. Furthermore, inadequate blood flow and a hypoxic environment impede the process of wound healing. The wound area is populated by a variety of microorganisms, with the presence of 3–10 dominant species, including *Staphylococcus aureus, Streptococcus, Pseudomonas*, anaerobic bacteria, and others. In addition, a multitude of microbial organisms are embedded within and attached to the extracellular matrix of the biofilm layer. The effective process of wound healing necessitates the initial elimination of infection at the site of the wound. Despite the availability of numerous treatments for infected wounds, antibacterial hydrogels are the preferred option due to their capacity to mimic the properties of skin, absorb exudate, and be applied easily. To this end, hydrogels can be loaded with antibacterial agents. The ability to adjust the amount and release kinetics of these agents enhances the potential for the utilization of such hydrogels in wound healing. In general, the objective of achieving an antibacterial effect is accomplished by incorporating antibiotics or inorganic nanoparticles into hydrogels or by utilizing antibacterial polymers.

Antibiotics are chemical compounds that impede the growth and proliferation of infectious microorganisms at the site of an injury or infection. This is achieved through two different mechanisms: biocidal and biocytidal activity. The most used antibiotics for this purpose include gentamicin, ciprofloxacin, ampicillin, amoxicillin, vancomycin, and tetracycline, which are administered orally or topically. Antibiotics are currently considered the gold standard for the treatment of infection. However, the uncontrolled use of antibiotics may result in the emergence of bacterial resistance. Moreover, the systemic use of antibiotics may result in adverse effects on the liver, kidney, and intestines. Additionally, the inadequate tissue penetration of antibiotics limits their use. Consequently, the topical application of antibiotics to the wound site via hydrogel delivery overcomes these limitations by providing an efficacious dose with an extended/controlled-release profile.

In light of the increasing prevalence of antibiotic-resistant bacterial strains, there has been a notable surge in interest in metallic nanoparticles. A variety of metallic nanoparticles, including silver, gold, cobalt, copper, and numerous metal oxide

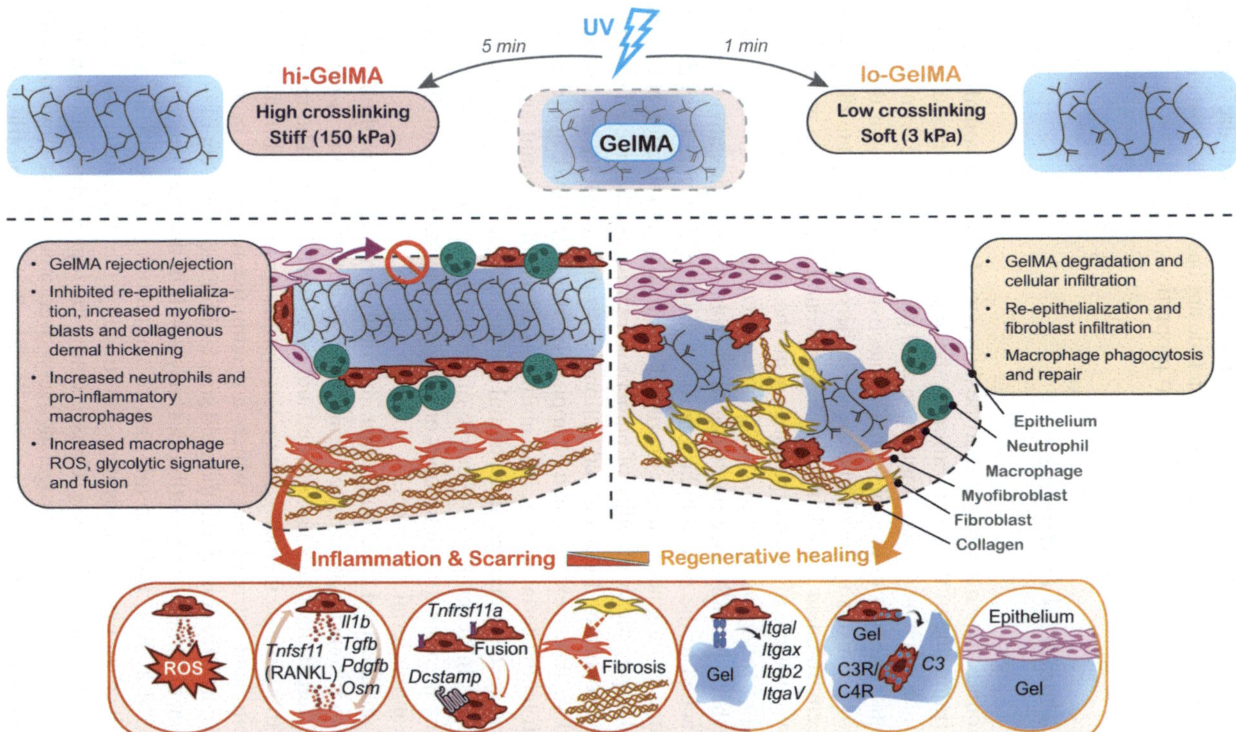

Fig. 15.2 The schematic representation depicts the impact of cross-linking rates on the wound healing effects of methacrylated gelatin (GelMA) hydrogels. UV irradiation time has a direct impact on the cross-linking and degradation profile. The cross-linking ratio is also effective in determining the pathway to inflammation and scarring, or regenerative healing. NonCommercial-NoDerivs 4.0 International license. (https://creativecommons.org/licenses/by-nc-nd/4.0/deed.en) [19]

nanoparticles (such as zinc oxide (ZnO), titanium dioxide (TiO_2), and nickel oxide (NiO)), have been loaded into hydrogels in order to provide an antibacterial effect [20–23]. The precise mechanism by which metallic nanoparticles interact with bacteria remains unclear. It is hypothesized that they disrupt membrane integrity and provide an antibacterial effect by stopping the metabolic activities of organelles [24]. Among metallic nanoparticles, silver nanoparticles (AgNPs) are widely used due to their low cost, broad-spectrum antimicrobial activity, and nontoxicity in tissues.

Another prominent strategy is the utilization of polymers with intrinsic antibacterial activity to prevent the emergence of antibiotic resistance. Cationic polymers, including chitosan, polyethyleneimine, and antibacterial peptides, have been employed in the development of hydrogel wound dressings [13, 18, 25]. Antimicrobial peptides (AMPs) are natural antibiotics derived from a variety of sources, including animals, plants, and protozoa. The antibacterial mechanism of action of AMPs is the disruption of bacterial cell membranes by electrostatic interactions, which ultimately results in the lysis of the bacteria. Among natural polymers, chitosan has been demonstrated to exhibit antimicrobial and antioxidant properties due to its polycationic structure. At low pH values, the free amine groups present in chitosan undergo a transformation into quaternary ammonium salts, thereby acquiring a polycationic structure. Consequently, an antibacterial hydrogel is frequently utilized in the fabrication of wound dressings [13]. In addition to its antibacterial properties, it is employed in isolation or in conjunction with a variety of active ingredients and hydrogels due to its derivation from sustainable natural sources, including shellfish, insect exoskeletons, and fungal cell walls. It has been demonstrated to possess antioxidant and hemostatic properties, facilitate wound healing, and minimize scar formation.

15.3.2 Hemostatic Hydrogels

The term "hemostasis" is used to describe the process by which blood flow is stopped because of the formation of a clot. This occurs due to the interaction between platelets and coagulation factors [26]. In the human body, hemostasis is a complex and dynamic process, comprising several sequential stages. In the natural coagulation mechanism, the acute response to tissue damage, vessel rupture, and subsequent bleeding results in regional vasoconstriction, which, in conjunction with the activation of coagulation factors and platelets, serves to minimize blood loss. In individuals with no underlying health issues, hemorrhage typically stops on its own without the need for intervention. However, certain diseases (such as thrombocytopenia, Von Willebrand disease, and hemophilia) have the potential to disrupt the coagulation mechanism. Moreover, individuals who are taking anticoagulant and antiplatelet medications are at an elevated risk of prolonged bleeding and hemorrhage during injuries or surgical procedures. In such cases, the use of hemostatic agents and/or hemostatic wound dressings is often employed. The current range of commercially available hemostatic products includes chitosan, alginate, gelatin, collagen, oxidized cellulose, polyethylene glycol (PEG), fibrin, fibrinogen, cyanoacrylate, thrombin, zeolites, and clays. The efficacy of each of these varies considerably.

It is essential that hemostatic biomaterials effectively mimic the natural coagulation mechanism. Consequently, an efficacious hemostatic wound dressing must be capable of rapidly stopping bleeding and effectively stimulating the healing process. It must demonstrate robust mechanical properties to maintain adhesion in wet and mobile areas, thereby ensuring its retention at the bleeding site. Moreover, the material should exhibit rapid gelling ability to prevent being washed away by blood flow. Furthermore, these properties must be optimally balanced. The swelling process of hydrogels in contact with blood is dependent on the density and structure of the cross-links present. This results in an increase in the concentrations of blood components in the wound. Additionally, they facilitate passive hemostasis through their rapid gelation properties. Furthermore, their injectability and fluidity allow them to assume the shape of the wound area, exerting a hemostatic effect. Most hemostatic hydrogels are derived from polysaccharides, proteins, or synthetic sources.

Polysaccharides, which are formed by the combination of multiple and different glucoses obtained from sustainable sources in nature, are widely used in hemostatic hydrogels due to their high biocompatibility, antibacterial, and hemostatic properties. The principal polysaccharide polymers among these include chitosan, hyaluronic acid, alginate, and carrageenan. Chitosan plays a role in hemostasis by promoting platelet aggregation [27]. In addition, robust red blood cell aggregation at the wound site occurs as the positively charged amino groups of chitosan interact with the negatively charged red blood cell membranes and proteins. It is recommended that chitosan hydrogels be applied topically, as intravenous administration carries an increased risk of thrombosis. Another polysaccharide that is employed in hemostatic hydrogels is hyaluronic acid (HA). Hydrogels based on HA are frequently the preferred option for wound healing due to their physicochemical and biological properties. Such properties include good gelation ability, high biocompatibility, biodegradability, the induction of cell adhesion and migration, stimulation of angiogenesis, and support for tissue regeneration [28]. Nevertheless, the weak mechanical properties and rapid degradation of these materials limit their potential applications.

Alginate is a linear polysaccharide derived from algae. Due to its robust water retention, biocompatibility, low toxicity, adjustable gelation properties, and cost-effectiveness, it is a frequently preferred polymer in the context of wound healing. Several alginate-based hemostatic products have been developed. These include hemostatic needles, transcatheter interventional embolic preparations, microneedles, and injectable hydrogels. The carboxyl groups present in alginate have the capacity to activate coagulation reactions, promote platelet aggregation, accelerate the process of hemostasis, and prevent bacterial infection while simultaneously absorbing fluid from the wound site. Moreover, it initiates the rapid formation of epithelial tissue and granulation tissue, thereby accelerating the healing of wounds

[29]. However, calcium alginate hydrogel does not undergo natural degradation within the body, and its adhesion to the wound site is relatively weak. Oxidized cellulose-based hemostatic dressings represent the most employed topical dressing for the purpose of controlling bleeding. These were primarily employed for the purpose of tamponade and did not contain any coagulation agents. Despite the advantages of cellulose-based hemostatic agents, some complications have been reported, including localized drying, which can lead to compression-related ischemia or sepsis [26].

Protein-based hydrogels are frequently employed in the context of wound healing and hemostasis due to their capacity to mimic the structural characteristics of the skin, their biocompatibility, biodegradability, and flexibility, as well as their compatibility for combined use with a range of biomaterials. The most used products in this field are collagen and gelatin hydrogels. Collagen plays a role in forming a clot with platelets to facilitate hemostasis; however, its effectiveness is reduced in patients with severe thrombocytopenia due to weak tissue adhesion [30]. Gelatin is a safe, biocompatible, and biodegradable polymer that is obtained through either an acid (Type A gelatin) or alkaline (Type B gelatin) process. Gelatin can be used as a hemostatic gel, with the capacity to absorb over 40 times its weight in blood. It can be used in conjunction with or in the absence of thrombin on a surface exhibiting bleeding. Gelatin exhibits hemostatic effects by facilitating platelet activation and aggregation, a property shared by other hydrogels. Moreover, it expands upon contact with blood, creating a tamponing effect at the wound site [31]. However, the high cost of thrombin, which facilitates rapid hemostasis, and the challenges associated with its storage represent factors that limit its use. Silk is a natural fiber protein produced by arthropods. Silk fibroin, commonly obtained from silkworms, has been demonstrated to activate the coagulation stages, reduce the time required for fibrin formation, and play a pivotal role in hemostasis by enhancing clot strength and forming a natural barrier against blood loss [32]. Additionally, silk fibroin has the capacity to facilitate the proliferation and migration of diverse cell types through the various phases of wound healing. However, silk fibroin is lacking in mechanical strength and demonstrates inadequate adhesion in wet and dynamic environments. Elastin is a significant protein present within the extracellular matrix of various tissues, including the vascular system, skin, and lungs. Biologic derivatives such as fibrin, fibrinogen, and thrombin are implicated in the process of blood coagulation. Fibrin is a fibrous protein that plays a role in the process of blood clotting. Fibrinogen is the precursor of fibrin, which is formed during the process of blood coagulation. Fibrinogen is converted to fibrin by thrombin and ionized calcium, thereby forming the clot. However, they have a limited shelf life and can carry the risk of transmitting disease.

The utilization of synthetic polymers for hemostatic purposes is more circumscribed than that of natural polymers. For example, although polyethylene glycol (PEG) exhibits a high water retention capacity, biocompatibility, and structural flexibility, it lacks intrinsic hemostatic properties. Consequently, it is employed in conjunction with natural polysaccharides, including chitosan and hyaluronic acid. For example, Zhu et al. developed hybrid hydrogel designs by aminoethyl methacrylate hyaluronic acid (HA-AEMA) and methacrylated methoxy polyethylene glycol (mPEG-MA) with chlorhexidine diacetate (CHX)-loaded nanogels. These hybrid hydrogels exhibited rapid hemostasis and antibacterial properties [33].

15.3.3 Anti-Oxidative Hydrogels

The high concentration of free radicals, known as reactive oxygen species (ROS), within the wound area exerts an additional influence on the wound healing process. In the aftermath of tissue injury, the immune system instigates the generation of superoxide radical ($O2^{\bullet-}$), hydrogen peroxide (H_2O_2), and hydroxyl radical ($OH^{\bullet}$) in response to bacterial pathogens. These act as secondary messengers in the wound healing process. Normally, ROS are essential for the immune system's defense against pathogenic microorganisms and for maintaining normal vascular functions in the wound area. However, the excessive production of these radicals and their transformation into other oxidant structures results in oxidative stress, which hinders the healing of tissue damage in the wound area. Hydrogen peroxide (H_2O_2) is regarded as a nonradical entity that nevertheless serves as a source of ROS. It has been demonstrated to impede fibroblast cell damage and the migration of keratinocytes by interfering with epidermal growth factor (EGF) signaling communication [34]. A role in the development of chronic wounds is played by oxidative stress resulting from increased levels of reactive oxygen species. To maintain intracellular redox balance and prevent oxidative stress and metabolic disorders during the process of wound healing, antioxidant hydrogels have been developed.

The use of antioxidants is pervasive, with a plethora of forms currently in use. These include natural antioxidants, such as vitamin C, phenolic compounds, carotenoids, and enzymes; antioxidant substances found in mammals, such as glutathione and coenzyme Q10; and those derived from plants, including phenolic acids, flavonoids, tannins, curcumin, gallic acid, and jujube. The utilization of curcumin-loaded gelatin microsphere hydrogels has been demonstrated to facilitate the healing process by reducing the levels of ROS present in diabetic wounds [35]. However, the use of curcumin necessitates meticulous monitoring due to the potential for adverse effects associated with high doses, including mitochondrial dysfunction. Furthermore, there are limitations related to rapid metabolic activation and stability. Thi et al. developed an injectable gallic acid-gelatin-based antioxidant hydrogel, which benefits from the advantages of gallic acid in free radical scavenging and collagenase inhibition. It has been demonstrated that gallic acid possesses antioxidant properties and promotes angiogenesis, as well as the formation of organized collagen fiber structures and hair follicles [36]. Tannic acid, a type of polyphenol, has been employed in silk fibroin hydrogels, wherein it was observed that elevated tannic acid concentrations resulted in enhanced cross-linking, diminished pore sizes, and superior adhesive properties of the hydrogel. Moreover, these hydrogels have been shown to facilitate wound healing by displaying antioxidant and antibacterial characteristics in comparison to silk fibroin hydrogels [37].

Another approach to wound healing that utilizes an antioxidant effect is the use of ROS-driven hydrogels. Given the high production of ROS in the wound area, these can be employed as a trigger for the release of an antioxidant agent. Exosomes are naturally occurring biological particles that carry proteins, microRNAs, and small interfering RNAs. They have recently gained popularity in the scientific community. In a new study, exosomes were used to obtain an antioxidant effect. They designed a chitosan and gelatin-based hydrogel and loaded it with exosomes carrying a specific microRNA (miR-4500) that shows antioxidant effects. The release of the exosomes was triggered by the ROS concentration through reversible boronic ester bonds in chitosan. The positive outcomes in terms of wound healing demonstrated in this study indicate that anti-oxidative hydrogels displaying a ROS concentration-dependent activity may be an effective approach [38].

15.3.4 Anti-Inflammatory Hydrogels

Disruptions to the wound healing process may impede the normal healing trajectory and prolong the inflammatory response. Macrophages typically play a pivotal role in the clearance of apoptotic neutrophils and the regulation of inflammation through phenotype transition. However, in chronic wounds, abnormal macrophage differentiation and the anomalous metabolic effects of neutrophils result in macrophages persisting in the M1 phenotype, thereby leading to heightened inflammation due to the excessive release of pro-inflammatory factors [39]. As previously stated, the initial stage of the typical wound healing process is the elimination of necrotic and infected tissue debris from the wound site. In chronic wounds, which are characterized by persistent inflammation, it is of the utmost importance to prevent bacterial contamination and biofilm formation during this phase. In order to achieve this, a variety of antiseptic hydrogels are employed with the objective of preventing microbial colonization of the area in question. Antiseptics, including povidone-iodine, iodine, cefazolin, chlorhexidine, octenidine, zinc oxide, hydrogen peroxide, and antimicrobial peptides, are frequently employed for this purpose [40]. In more advanced systems, hydrogels containing antiseptic-loaded liposomes or nanocapsules have been demonstrated to be more effective in the wound area due to their prolonged antiseptic release. In instances of severe infection where antiseptic therapy proves inadequate, hydrogels containing antibiotics, including vancomycin, tetracycline, ciprofloxacin, and gentamicin, are utilized. Another nanomaterial approach for anti-inflammatory effects in wound healing is the nanoenzymes (NZs) that show enzyme-like activity. NZs preserve their activities in extreme temperature and pH conditions, unlike natural enzymes. This field gained attention with a study investigating ferromagnetic magnetite (Fe_3O_4) nanoparticles, which exhibit peroxidase (POD)-like activity. A comprehensive investigation has been conducted on a range of nanomaterials, including metallic, metal oxide-based, and carbon-based nanomaterials, with the objective of elucidating their intrinsic enzymatic activities, particularly those associated with peroxidase (POD), oxidase (OXD), catalase

(CAT), and superoxide dismutase (SOD) enzymes. Hydrogels are effective NZs carriers for enhanced and targeted release. There are several studies introducing NZ@ hydrogels that have anti-inflammatory effects in wound healing. In particular, Cu^{2+}-based NZs loaded into hydrogels have been studied for their anti-inflammatory effects through the clearance of ROS (Fig. 15.3) [41, 42].

Hyaluronic acid, chitosan, alginate, gelatin, and silk are commonly used as natural components in hydrogels, functioning as drug delivery matrices with anti-inflammatory properties [43]. To illustrate, ibuprofen, a hydrophobic and nonsteroidal anti-inflammatory pharmaceutical agent, has been loaded onto alginate via an adsorption precipitation method, resulting in controlled ibuprofen release over a 12–24 h period with minimal inflammatory response [44]. Liu et al. developed a self-healing hydrogel with anti-inflammatory properties using chitosan and hyaluronic acid. In this study, chitosan was modified with tannic acid with the objective of imparting adhesive and anti-inflammatory characteristics to the hydrogel, while also extending its degradation time. The developed hydrogel was observed to demonstrate anti-inflammatory effects, thereby accelerating the process of wound healing [45].

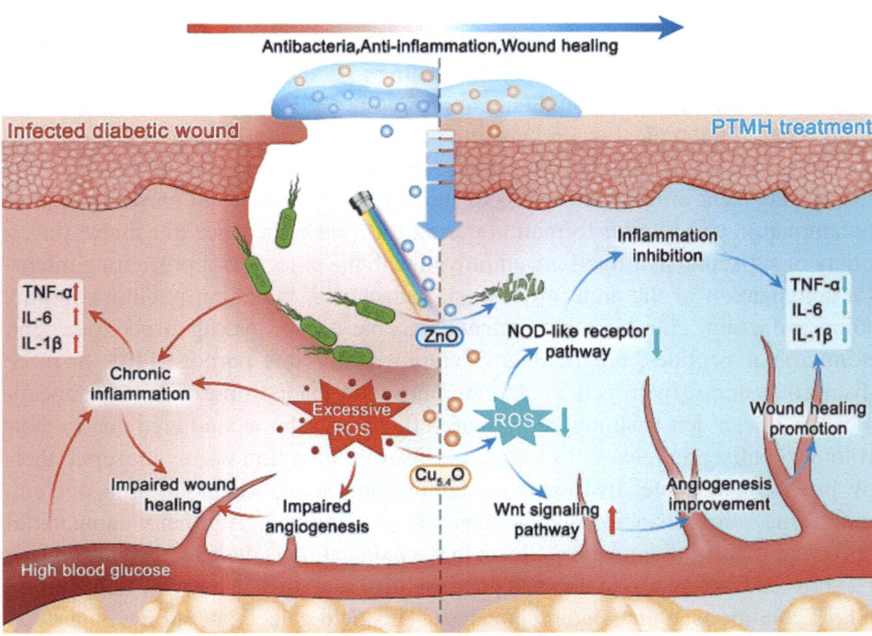

Fig. 15.3 The schematic illustrates the dual action of PTMH on impaired wound healing. PTMH is a hydrogel with antibacterial and anti-inflammatory effects that triggers tissue repair and promotes impaired wound healing. While ZnO shows an antibacterial effect, $Cu_{5.4}O$ NZs show an anti-inflammatory effect. The underlying mechanisms are explained (*IL*: interleukin, *PTMH*: programmed time-release multifunctional hydrogel, *TNF-α*: tumor necrosis factor-α) [42]

15.3.5 Bioactive Hydrogels

Chronic wounds remain predominantly in the inflammatory phase, frequently accompanied by exudate, severe infections, pain, and tissue necrosis. These factors have the potential to prolong the healing process for several months or even years. The management of chronic wounds frequently necessitates a combination of treatments and prolonged hospitalization, as well as regular dressing changes. For example, it has been documented that venous leg ulcers can produce exudate volumes of up to 12 L/m^2 per day [14, 46]. It is therefore imperative that active intervention be employed in the management of chronic or complex wounds. Moreover, strategies to promote angiogenesis and cell proliferation are necessary. The development of bioactive hydrogels incorporating signaling molecules, such as bioactive agents and growth factors, or stem cells, has become imperative due to the inadequacy of traditional hydrogel dressings in the treatment of chronic wounds. In-depth research into the complex stages of wound healing has identified cell proteins that are involved in the healing process. These include epidermal growth factor (EGF), fibroblast growth factor (FGF), keratinocyte growth factor (KGF), vascular endothelial growth factor (VEGF), transforming growth factor β (TGF-β), and platelet-derived growth factor (PDGF) [14]. It has been demonstrated that the loading of growth factors into hydrogels can result in enhanced wound healing. To illustrate, the influence of EGF on wound healing was demonstrated by Ulubayram et al. using EGF-loaded gelatin hydrogels. In this study, EGF was stabilized in gelatin microspheres and loaded onto gelatin sponges to stimulate cell proliferation in the wound area. Subsequently, the impact of EGF on wound healing was assessed in a full-thickness wound model. The study indicated that the effects of EGF were dose dependent. The administration of 15 μg EGF/cm^2 gelatin sponges to the full-thickness wound area resulted in complete wound healing within 10 days, whereas the control group exhibited a 12% reduction in wound area. It has been demonstrated that EGF-gelatin sponges promote angiogenesis, epithelialization, and granulation tissue formation in the wound area, thereby accelerating the healing of both acute and chronic wounds [17]. These findings have been corroborated by clinical studies, resulting in the commercialization of EGF-gelatin sponges under the brand name Neoderm® (Ars Arthro Biotechnology) in 2008 for the treatment of burns and chronic wounds. The high activity demonstrated by growth factors and hydrogels has also attracted the interest of commercial firms, resulting in the development of Regranex® (Smith & Nephew), which contains 0.01% PDGF-BB for chronic wounds [47]. Nevertheless, the commercialization of these products is impeded by several significant obstacles, including the high costs of growth factors, their relatively short shelf lives, the potential loss of activity, and the associated cancer risk from excessive dosing. It is therefore imperative to confirm the dose effect of growth factors in wound healing and to avoid overdosing.

An alternative approach to restoring tissue and function lost due to chronic or complex wounds or burns is the development of skin substitutes through a tissue engineering method referred to as advanced therapeutic products [48]. In this method, autologous, allogeneic, or stem cells are seeded onto hydrogel tissue scaffolds and cultured under appropriate conditions to produce three-dimensional (3D) skin substitutes [49]. Skin substitutes are composed of dermal–epidermal structures, including the epidermal, dermal, or both layers. Currently, a plethora of products are available for the treatment of chronic wounds and burns. Examples of such products include Apligraft® (USA), OrCell (USA), Theraskin (USA), Strata Graft (USA), DenovaSkin (Switzerland), Epidex (Switzerland), MySkin (UK), BioSeed-S (Germany), and Laserskin (Italy). To illustrate, Apligraft® (Organogenesis, USA) was the inaugural skin tissue engineering product to obtain FDA approval in 1998. Apligraft® is developed using bovine-derived Type I collagen, fibroblast, and keratinocyte cells derived from human neonatal foreskin, resulting in a dermal–epidermal structure. Another tissue engineering product, Dermagraf®, is a dermal product developed from fibroblast cells cultured in polyglycolic acid or polygactin-910 meshes derived from the neonatal foreskin. These products are utilized in the treatment of partial and full-thickness wounds, burn wounds, acute and chronic wounds, and epidermolysis bullosa. Furthermore, additional dermal–epidermal skin substitutes currently utilized in clinical practice include OrCel®, PolyActive™, and the TissueTech Autograft System®. While the PolyActive and TissueTech products are based on autologous fibroblast and keratinocyte cells, the Apligraft and OrCel products are developed using allogeneic fibroblast cells. Stem cells (embryonic, induced pluripotent, and adult) play a significant role in cellular therapies and represent a promising cell source. Nevertheless, their utilization in advanced therapy products remains a topic of investigation, given the intricate regulatory frameworks governing their utilization and production [50]. It is of the utmost importance to accurately observe the differentiation processes of stem cells to achieve the desired effects and prevent uncontrolled differentiation, which could potentially lead to cancer. Consequently, the accessibility of dermal substitutes is constrained by elevated production costs, extended manufacturing processes, and rigorous regulatory standards. Furthermore, the precise shelf life of these products is not defined.

15.3.6 Smart Hydrogels

Smart hydrogels are three-dimensional, porous, networked polymers that exhibit a high water retention capacity, sensitivity to stimuli, and the ability to form a response. In response to changes in their surrounding environment, smart hydrogels undergo a volume change, either swelling or shrinking, in response to a variety of stimuli, including pH, ionic strength, light, and heat. The stimuli for smart hydrogels can be classified into three main categories: chemical (e.g., pH, glucose, oxidative), physical (e.g., temperature, light, magnetic field, and electrical sensitivity), or biological (e.g., enzymes and reactive oxygen species) [51]. Upon the removal of the stimulus from the environment, smart hydrogels revert to their original state. The biomedical field has identified smart hydrogels as a significant area of interest, with several potential applications including biosensors, tissue engineering, cellular therapies, and drug delivery systems. This is due to the highly versatile specialization potential of these materials. The selection of an appropriate stimulus and smart polymer is crucial for the success of a treatment, considering the physiology of the application site and the protocol to be applied. Furthermore, smart hydrogels are of considerable significance in the field of wound healing, exhibiting a broad range of potential applications. Changes in pH and temperature within the wound area serve as stimuli for smart polymers. This enables the design of diagnostic and therapeutic approaches that encompass a range of applications, from biosensors to drug delivery systems. For example, when a smart hydrogel comprising a pH indicator is applied to the wound area, in addition to its primary functions, such as forming a barrier in the wound area and maintaining moisture balance, it also enables in situ observation of whether infection develops, as indicated by a yellow-red color change observed when the pH reaches 7 [52].

Temperature-sensitive hydrogels can be composed of natural and/or synthetic polymers. These hydrogels display amphiphilic characteristics resulting from the combination of hydrophilic and hydrophobic monomers, comprising two distinct segments [53]. The temperature-dependent swelling or shrinking of hydrogels is contingent upon whether the temperature change exceeds or falls below a specified limit value. The temperature at which this limit value is reached is referred to as the "low critical solution temperature" (LCST). At temperatures below the "upper critical solution temperature" (UCST), hydrogel structures are formed; at temperatures above this value, the reduction in folding regions results in an increase in solubility. The LCST and UCST can be controlled by differentiating the ratios of the segments in the structure [51]. Temperature-sensitive hydrogels are employed in a variety of applications, including controlled drug delivery systems, where they facilitate the release of active pharmaceutical ingredients in a controlled manner, thus preventing the attainment of toxic doses; in the transportation of biomolecules for tissue regeneration; and in the removal of cell layers on polymer surfaces in cell layer tissue engineering. In the context of wound healing, hydrogel systems are designed to facilitate the controlled delivery of cell proliferation-inducing biomolecules or anti-inflammatory and anti-infectious agents to the wound site [54]. As an example, the study conducted by Zhang et al. resulted in the synthesis of GelMA-PDA-ASP nanofiber hydrogel through the reaction of amine groups from methacrylated gelatin (GelMA) with Michael addition and catechol groups from polydopamine (PDA) encapsulated with fatty acids/aspirin (ASP) under alkaline conditions. It was demonstrated that a twofold more efficient release profile was achieved by heating the hydrogels to 40 °C in comparison to room conditions. Furthermore, it was demonstrated that nanofiber hydrogels loaded with fatty acids/aspirin (ASP) are biocompatible and facilitate cell regeneration and wound healing [54].

The term "pH-sensitive hydrogels" is used to describe a type of polymeric material that contains weak acidic and basic groups or ions, depending on the specific composition. They can be prepared from natural polymers, including alginate (cationic), chitosan (anionic), albumin, and gelatin, or from synthetic polymers such as poly (acrylic acid) (PAA), poly (methacrylic acid) (PMAA), and polyacrylamide (PAM) [51]. The shrinkage or swelling response of pH-sensitive hydrogels is dependent on the pKa value at which the functional groups in the polymer source ionize. The response enables the regulated release of pharmaceutical agents through alterations in pH. For example, Haidari et al. prepared pH-sensitive hydrogels loaded with silver nanoparticles for the treatment of infected wounds. In this study, the effective antibacterial activity and release profile were determined to be higher in the pH 7.4–10 range, which mimics acute-chronic infections, than in the pH 4–5.5 range, which is the pH value of healthy skin, when infection treatment was required [55]. Polymers with carboxyl groups lack the capacity to undergo swelling at low pH values. The hydrophobic interactions between the polymer chains result in a reduction in volume as the acidic group undergoes protonation. As the pH value increases, the hydrogel undergoes a swelling process due to the ionization of the carboxyl groups, resulting in an increase in volume. As an example, carboxymethyl chitosan (CMC) swells in high-pH solutions due to the ionization of acidic groups [56]. Conversely, if the polymer has a basic structure, ionization increases in acidic conditions at low pH values, resulting in hydrogel swelling. Cationic polymers, including chitosan and polyethyleneimine, exhibit swelling behavior at low pH values. In the preparation of pH-sensitive hydrogels for use in wound healing, the pH value of the wound site is also a key factor in studies targeting drug release by the swelling shrinkage effect. In the study conducted by Omidi et al., pH-sensitive chitosan, amine-containing nanowhisker, and amine-containing graphene hydrogels were prepared for the dual release of curcumin (CUR) and doxorubicin (DOX). The Schiff base reaction, which employs synthetic aldehyde groups (2P), resulted in the successful synthesis of pH-sensitive hydrogels capable of transferring active substances in a pH-sensitive manner within a PBS medium. The release profiles for DOX and CUR at pH 5.4 and pH 7.4 were evaluated individually and in pairs. It was found that the percentage cumulative release values were found to be higher for both DOX and CUR in acidic media conditions (pH 5.4) [57].

The examination of light-sensitive hydrogels is conducted in accordance with two headings: ultraviolet (UV) and visible light sensitivity. The application of UV radiation results in the transformation of the polymer structures into their cationic form, thereby inducing a swelling effect due to the action of electrostatic repulsion caused by ionization. Conversely, hydrogel formation is facilitated by the introduction of chromophore groups in the visible region by the light source. Several factors affect the properties of photosensitive hydrogels, including the characteristics of the light source, the intensity of the light, and the concentration of chromophore moieties. The majority of literature on light-sensitive hydrogels pertains to ophthalmic applications [51]. Additionally, they are utilized in the context of dermal wound healing. For example, Kim et al. utilized light-sensitive hydrogels induced by a light source in the visible region to develop transdermal patches [58].

Hydrogels that are sensitive to electric and magnetic fields are composed of polyelectrolyte polymers with a high ionization capacity, which are similar in composition to those that are sensitive to pH. In these hydrogels, which employ a combination of natural and synthetic polymers, the degree of swelling or shrinkage can be directly observed in response to the applied electric or magnetic field. The application of an electrical potential difference applied to the hydrogel with an anode and cathode electrode results in a change in volume, thereby facilitating the release of the active drug substance. Furthermore, the rate of drug release can be regulated by modifying the intensity of the electrical field. Di Luca et al. prepared electric/magnetic field-sensitive hydrogels consisting of curcumin-loaded graphene oxide, gelatin, or trypsin and investigated the release profiles of curcumin from hydrogels with an electric field applied in the range of 0–48 V. The results demonstrated that the release of curcumin was slow at 0 V, moderate at 12 and 48 V, and fast at 24 V [59]. Consequently, the controlled application of the electric field enabled the achievement of precise release profiles, which proved effective in both the antibacterial effect and wound healing.

15.3.7 Four-Dimensional (4D) Hydrogels

Concurrent with the acceleration of technological development, multidisciplinary studies are also advancing rapidly. In recent years, three-dimensional (3D) bioprinting technologies have begun to give way to four-dimensional (4D) printing technologies. In the context of hydrogels, the most notable distinction between 3D and 4D materials is their capacity to alter their physical configuration or functional characteristics in response to external stimuli, including heat, light, moisture, magnetic fields, and pH changes [51]. The adaptation of stimulus-responsive hydrogels to 4D bioprinting technologies offers several notable advantages, including biocompatibility, ease of production, rapid manufacturing, low production costs, and high elasticity. The polymers most commonly used in 4D bioprinting processes are alginate, gelatin, poly(N-isopropylacrylamide) (PNIPAm), poly(N, N-dimethylacrylamide) (PDMAAm), and hyaluronic acid [60]. The most common printing techniques employed with these polymers include extrusion-based fused deposition modeling (FDM), direct ink writing (DIW), stereolithography (SLA), digital light synthesis (DLS), and digital light processing (DLP). As a next-generation technology, 4D hydrogels are expected to facilitate significant advancements in the field of tactile sensors and other related disciplines [61].

15.4 Conclusions

By 2016, approximately 20 million patients worldwide were afflicted with chronic wounds [58, 59]. It is projected that the global market for wound dressings will reach 11.2 billion USD by 2023, driven largely by the increasing number of accidents and the rising incidence of chronic, surgical, traumatic, and burn injuries. It is anticipated that this growth will be further accelerated by the aging population. Notwithstanding the considerable technological advancements that have been made in the field of wound treatment, a few significant challenges remain, particularly in the context of chronic and complex wound care. When evaluating hydrogel wound dressings, although the benefits of hemostatic/antimicrobial/antioxidant drug-carrying hydrogels for promoting wound healing have been demonstrated, their highly hydrophilic nature results in low loading efficiencies for hydrophobic drugs. It is thus imperative to devise novel methodologies to guarantee the delivery of an efficacious dose of hydrophobic drugs at the site of the wound. Similarly, while antibiotic-loaded hydrogels offer a local alternative to systemic antibiotic administration for infected wounds, ensuring the safe and effective delivery of extended/controlled-release antibiotics to the wound site represents another critical issue that must be addressed for hydrogels. Furthermore, given that rapid fluctuations in pH are commonly observed in chronic wounds and can impede the healing process, the development of new systems to regulate and maintain the wound's pH at physiological levels represents a promising avenue for chronic wound treatment. The creation of novel bioactive hydrogels for chronic wounds, especially those comprising bioactive molecules that stimulate cell proliferation and angiogenesis, requires comprehensive investigation and assessment in terms of dose-dependent effects, cost, and safety. This is with the aim of enhancing treatment efficacy. Furthermore, given that the wound microenvironment in chronic wounds frequently exhibits disparate symptoms due to metabolic disorders, it is essential to consider personalized treatments as a crucial element to enhance efficacy.

In conclusion, hydrogel wound dressings have the potential to play a more functional role in addressing the clinical needs of chronic/complex wounds and extensive burns. It is therefore crucial to develop smart hydrogel systems capable of delivering combined therapies (such as antibacterial, anti-inflammatory, and angiogenic agents) and being selectively activated at different stages of the wound healing process.

Acknowledgments The authors would like to thank Soghrat Salamati for drawing the figure included in this book chapter.

References

1. Sontheimer, R.D.: Skin is not the largest organ. J. Invest. Dermatol. **134**(2), 581 (2014)
2. Shimizu, H.: Shimizu's Dermatology. Wiley, Hoboken, NJ (2017)
3. Diegelmann, R.F., Evans, M.C.: Wound healing: an overview of acute, fibrotic and delayed healing. Front. Biosci. **9**, 283–289 (2004)
4. Lazarus, G.S., et al.: Definitions and guidelines for assessment of wounds and evaluation of healing. Wound Repair Regen. **2**(3), 165–170 (1994)
5. Clark, R.A.F., Henson, P.M.: The Molecular and Cellular Biology of Wound Repair, p. 597. Plenum Press, New York (1988)
6. Robson, M.C., Steed, D.L., Franz, M.G.: Wound healing: biologic features and approaches to maximize healing trajectories. Curr. Probl. Surg. **38**(2), 72–140 (2001)
7. Eming, S.A., Krieg, T., Davidson, J.M.: Inflammation in wound repair: molecular and cellular mechanisms. J. Invest. Dermatol. **127**(3), 514–525 (2007)
8. Enoch, S., Leaper, D.J.: Basic science of wound healing. Surgery (Oxford). **23**(2), 37–42 (2005)

9. Shi, C., et al.: Selection of appropriate wound dressing for various wounds. Front. Bioeng. Biotechnol. **8**, 182 (2020)
10. Rezvani Ghomi, E., et al.: Wound dressings: current advances and future directions. J. Appl. Polym. Sci. **136**(27), 47738 (2019)
11. Tavakoli, S., Klar, A.S.: Advanced hydrogels as wound dressings. Biomolecules. **10**(8), 1169 (2020)
12. Liang, Y., He, J., Guo, B.: Functional hydrogels as wound dressing to enhance wound healing. ACS Nano. **15**(8), 12687–12722 (2021)
13. Ribeiro, M.P., et al.: Development of a new chitosan hydrogel for wound dressing. Wound Repair Regen. **17**(6), 817–824 (2009)
14. Koehler, J., Brandl, F.P., Goepferich, A.M.: Hydrogel wound dressings for bioactive treatment of acute and chronic wounds. Eur. Polym. J. **100**, 1–11 (2018)
15. Gupta, A., et al.: The production and application of hydrogels for wound management: a review. Eur. Polym. J. **111**, 134–151 (2019)
16. Bostanci, N.S., Büyüksungur, S., Hasirci, N., Tezcaner, A.: Bioprinted scaffolds assembled as synthetic skin grafts by natural hydrogels containing fibroblasts and bioactive agents. Int. J. Polym. Mater. Polym. Biomater. **73**(11), 927–945 (2024)
17. Ulubayram, K., Çakar, N., Korkusuz, P., Ertan, C., Hasirci, N.: EGF containing gelatin-based wound dressings. Biomaterials. **22**(11), 1345–1356 (2001)
18. Calamak, S., et al.: Silk fibroin based antibacterial bionanotextiles as wound dressing materials. Mater. Sci. Eng. C Mater. Biol. Appl. **43**, 11–20 (2014)
19. Butenko, S., et al.: Hydrogel crosslinking modulates macrophages, fibroblasts, and their communication, during wound healing. Nat. Commun. **15**(1), 6820 (2024)
20. Mahmoud, N.N., et al.: Gold nanoparticles loaded into polymeric hydrogel for wound healing in rats: effect of nanoparticles' shape and surface modification. Int. J. Pharm. **565**, 174–186 (2019)
21. Tao, B., et al.: Copper-nanoparticle-embedded hydrogel for killing bacteria and promoting wound healing with photothermal therapy. J. Mater. Chem. B. **7**(15), 2534–2548 (2019)
22. Sudheesh Kumar, P.T., et al.: Flexible and microporous chitosan hydrogel/Nano ZnO composite bandages for wound dressing: in vitro and in vivo evaluation. ACS Appl. Mater. Interfaces. **4**(5), 2618–2629 (2012)
23. Calamak, S., et al.: Ag/silk fibroin nanofibers: effect of fibroin morphology on ag+ release and antibacterial activity. Eur. Polym. J. **67**, 99–112 (2015)
24. Li, S., et al.: Antibacterial hydrogels. Adv. Sci. **5**(5), 1700527 (2018)
25. Li, G., Lai, Z., Shan, A.: Advances of antimicrobial peptide-based biomaterials for the treatment of bacterial infections. Adv. Sci. **10**(11), 2206602 (2023)
26. Colak, S., Altan, A., Akbulut, N.: Lokal Hemostatik Ajanlar. J. Int. Dent. Sci. **2018**, 147–152 (2018)
27. Chou, T.C., et al.: Chitosan enhances platelet adhesion and aggregation. Biochem. Biophys. Res. Commun. **302**(3), 480–483 (2003)
28. Hussain, Z., et al.: Hyaluronic acid-based biomaterials: a versatile and smart approach to tissue regeneration and treating traumatic, surgical, and chronic wounds. Polym. Rev. **57**(4), 594–630 (2017)
29. Aderibigbe, B.A., Buyana, B.: Alginate in wound dressings. Pharmaceutics. **10**(2), 42 (2018)
30. Abbott, W.M., Austen, W.G.: The effectiveness and mechanism of collagen-induced topical hemostasis. Surgery. **78**(6), 723–729 (1975)
31. Hickman, D.A., et al.: Biomaterials and advanced technologies for hemostatic management of bleeding. Adv. Mater. **30**(36), 1804635 (2018)
32. Lei, C., et al.: Preparation and hemostatic property of low molecular weight silk fibroin. J. Biomater. Sci. Polym. Ed. **27**(5), 403–418 (2016)
33. Zhu, J., et al.: Hyaluronic acid and polyethylene glycol hybrid hydrogel encapsulating Nanogel with hemostasis and sustainable antibacterial property for wound healing. ACS Appl. Mater. Interfaces. **10**(16), 13304–13316 (2018)

34. Hensley, K., et al.: Reactive oxygen species, cell signaling, and cell injury. Free Radic. Biol. Med. **28**(10), 1456–1462 (2000)
35. Liu, J., et al.: Encapsulation of curcumin nanoparticles with MMP9-responsive and thermos-sensitive hydrogel improves diabetic wound healing. ACS Appl. Mater. Interfaces. **10**(19), 16315–16326 (2018)
36. Thi, P.L., et al.: In situ forming and reactive oxygen species-scavenging gelatin hydrogels for enhancing wound healing efficacy. Acta Biomater. **103**, 142–152 (2020)
37. Jing, J., et al.: Fabrication of hybrid hydrogels from silk fibroin and tannic acid with enhanced gelation and antibacterial activities. ACS Biomater Sci. Eng. **5**(9), 4601–4611 (2019)
38. Yang, H., et al.: ROS-responsive injectable hydrogels loaded with exosomes carrying miR-4500 reverse liver fibrosis. Biomaterials. **314**, 122887 (2025)
39. Xu, Z., et al.: Anti-inflammation biomaterial platforms for chronic wound healing. Biomater. Sci. **9**(12), 4388–4409 (2021)
40. da Silva, L.P., et al.: Hydrogel-based strategies to advance therapies for chronic skin wounds. Annu. Rev. Biomed. Eng. **21**, 145–169 (2019)
41. Kurian, A.G., et al.: Nanozyme-engineered hydrogels for anti-inflammation and skin regeneration. Nano Micro Lett. **16**(1), 110 (2024)
42. Peng, Y., et al.: Construction of programmed time-released multifunctional hydrogel with antibacterial and anti-inflammatory properties for impaired wound healing. J. Nanobiotechnol. **22**(1), 126 (2024)
43. Lebaudy, E., et al.: Recent advances in Antiinflammatory material design. Adv. Healthc. Mater. **10**(1), e2001373 (2021)
44. Johnson, K.A., et al.: Drug-impregnated, pressurized gas expanded liquid-processed alginate hydrogel scaffolds for accelerated burn wound healing. Acta Biomater. **112**, 101–111 (2020)
45. Liu, S., et al.: Injectable and self-healing hydrogel based on chitosan-tannic acid and oxidized hyaluronic acid for wound healing. ACS Biomater Sci. Eng. **8**(9), 3754–3764 (2022)
46. Cutting, K.F.: Wound exudate: composition and functions. Br. J. Community Nurs. **8**(9 Suppl), suppl 4–suppl 9 (2003)
47. Sun, B.K., Siprashvili, Z., Khavari, P.A.: Advances in skin grafting and treatment of cutaneous wounds. Science. **346**(6212), 941–945 (2014)
48. Chouhan, D., et al.: Emerging and innovative approaches for wound healing and skin regeneration: current status and advances. Biomaterials. **216**, 119267 (2019)
49. Borena, B.M., et al.: Regenerative skin wound healing in mammals: state-of-the-art on growth factor and stem cell based treatments. Cell. Physiol. Biochem. **36**(1), 1–23 (2015)
50. Isakson, M., et al.: Mesenchymal stem cells and cutaneous wound healing: current evidence and future potential. Stem Cells Int. **2015**, 831095 (2015)
51. Aliakbar Ahovan, Z., et al.: Antibacterial smart hydrogels: new hope for infectious wound management. Mater. Today Bio. **17**, 100499 (2022)
52. Stan, D., et al.: Wound healing applications of creams and "smart" hydrogels. Exp. Dermatol. **30**(9), 1218–1232 (2021)
53. Yu, Y., et al.: Recent advances in thermo-sensitive hydrogels for drug delivery. J. Mater. Chem. B. **9**(13), 2979–2992 (2021)
54. Zhang, K., et al.: Nanofibrous hydrogels embedded with phase-change materials: temperature-responsive dressings for accelerating skin wound healing. Compos. Commun. **25**, 100752 (2021)
55. Haidari, H., et al.: Eradication of mature bacterial biofilms with concurrent improvement in chronic wound healing using silver nanoparticle hydrogel treatment. Biomedicine. **9**(9), 1182 (2021)
56. Rizwan, M., et al.: pH sensitive hydrogels in drug delivery: brief history, properties, swelling, and release mechanism, material selection and applications. Polymers (Basel). **9**(4), 137 (2017)
57. Omidi, S., Pirhayati, M., Kakanejadifard, A.: Co-delivery of doxorubicin and curcumin by a pH-sensitive, injectable, and in situ hydrogel composed of chitosan, graphene, and cellulose nanowhisker. Carbohydr. Polym. **231**, 115745 (2020)

58. Kim, H., et al.: Visible light-triggered on-demand drug release from hybrid hydrogels and its application in transdermal patches. Adv. Healthc. Mater. **4**(14), 2071–2077 (2015)
59. di Luca, M., et al.: Electro-responsive graphene oxide hydrogels for skin bandages: the outcome of gelatin and trypsin immobilization. Int. J. Pharm. **546**(1–2), 50–60 (2018)
60. Shie, M.-Y., et al.: Review of polymeric materials in 4D printing biomedical applications. Polymers. **11**(11), 1864 (2019)
61. Tang, Y., et al.: Recent advances of 4D printing technologies toward soft tactile sensors. Front. Mater. **8**, 658046 (2021)

Chapter 16
Hydrogels Used for the Treatment of Cornea

Gozde Ervin Kole, Deniz Basoz, and Deniz Yucel

Abstract The cornea is a transparent, dome-shaped tissue located at the front of the outer (fibrous) protective layer of the eye, and it has the ability to transmit light. In severe corneal damage where more than one layer of the cornea is affected, it is necessary to replace the damaged area and/or the entire cornea. In this case, although the transplantation of healthy corneal tissue from a donor to the diseased cornea, known as keratoplasty, is commonly preferred, issues such as donor limitations and allograft rejection may arise. To overcome these challenges, artificial corneal implants developed using biomaterials and tissue engineering techniques offer an alternative approach for corneal treatment. Due to their biocompatibility, high water retention capacity, tailorable shape, and flexible structure, hydrogel-based corneal

G. E. Kole
Graduate School of Health Sciences, Medical Biotechnology, Acıbadem Mehmet Ali Aydinlar University, Istanbul, Türkiye

Biomaterials Application and Research Center, Acıbadem Mehmet Ali Aydinlar University, Istanbul, Türkiye

D. Basoz
Biomaterials Application and Research Center, Acıbadem Mehmet Ali Aydinlar University, Istanbul, Türkiye

Graduate School of Natural and Applied Sciences, Biomaterials, Acıbadem Mehmet Ali Aydinlar University, Istanbul, Türkiye

D. Yucel (✉)
Graduate School of Health Sciences, Medical Biotechnology, Acıbadem Mehmet Ali Aydinlar University, Istanbul, Türkiye

Biomaterials Application and Research Center, Acıbadem Mehmet Ali Aydinlar University, Istanbul, Türkiye

Graduate School of Natural and Applied Sciences, Biomaterials, Acıbadem Mehmet Ali Aydinlar University, Istanbul, Türkiye

Department of Histology and Embryology, School of Medicine, Acıbadem Mehmet Ali Aydinlar University, Istanbul, Türkiye
e-mail: deniz.yucel@acibadem.edu.tr

N. Hasirci, V. Hasirci (eds.), *Hydrogels and Bioinks in Tissue Engineering*, Biological and Medical Physics, Biomedical Engineering,
https://doi.org/10.1007/978-3-031-90966-5_16

tissue substitutes provide significant advantages. In cases of severe corneal damage, three-dimensional (3D) printing and bioprinting technologies are commonly used to develop a layered, organized, three-dimensional tissue substitute that mimics the light transmitting stroma layer. Hydrogels are the most frequently used bioinks, and cells can be easily loaded into hydrogels. Hydrogel-based constructs are commonly produced by 3D bioprinting or formed as thick slabs to develop tissue substitutes for corneal treatment.

Keywords Cornea · Hydrogel · Collagen · Gelatin · Chitosan · Decellularized extracellular matrix · Alginate · Polyethylene glycol · Polyvinyl alcohol

16.1 Structure of Cornea

The cornea, located at the front of the eye, has the ability to transmit light and is the first structure where light entering the eye is refracted. In the thickest layer of the cornea, the stroma, the tissue-specific organization of collagen fibers ensures the cornea's. The horizontal diameter of the human cornea is 11–12 mm, and the vertical diameter is 9–11 mm, with a central thickness of 550 μm and peripheral thickness ranging from 610 to 650 μm [1, 2]. Although the cornea is an avascular tissue without blood vessels, it is rich in peripheral nerves that perceive pain, irritation, heat, and chemical changes. Nutrition and oxygen requirements are provided by the aqueous humor, limbal blood vessels, and tear secretion [1]. The cornea, with a refractive index of 1.376, is the structure where light entering the eye is refracted first and most intensely [3].

This special tissue, which protects the eye from external factors and focuses light, consists of five layers, three cellular layers, and two membranes separating them. These layers, from the external environment to the anterior chamber of the eye, are the epithelium, Bowman's membrane, stroma, Descemet's membrane, and endothelium, respectively (Fig. 16.1).

The outermost, frontal surface of the cornea is lined with a non-keratinized, stratified squamous epithelium that is approximately 50 μm thick and consists of 5–7 cell layers [3]. The basal cells of the corneal epithelium, which are short and columnar, adhere to the basal membrane through hemidesmosomes. Basal epithelial cells, differentiated from limbal stem cells, proliferate and enable the regeneration of the corneal epithelium in cases of abrasion and injury. The superficial epithelial cells have a microvilli structure and help maintain the tear film on the surface, preventing the cornea from drying out. The corneal epithelium, which contains free nerve endings, is sensitive to touch.

Bowman's membrane, located beneath the corneal epithelium, is a homogeneous, acellular layer with a thickness of 8–10 μm, composed mainly of irregular

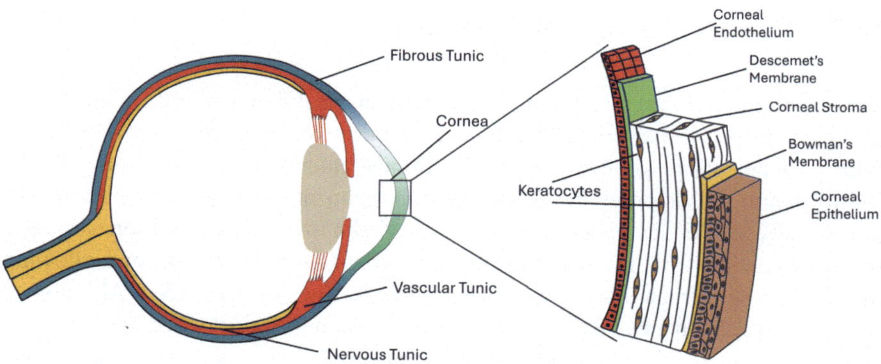

Fig. 16.1 The general structure of the eye and the layers of the corneal tissue

collagen fibers and proteoglycans [3]. Bowman's membrane is important for maintaining the shape and mechanical strength of the cornea [4]. Additionally, this membrane serves as a barrier against the spread of potential infection. This membrane, lacking self-renewal property, forms opaque scar tissue when damaged, which can impair vision.

The stroma, the thickest layer of the cornea, is located between Bowman's and Descemet's membranes. The stroma layer, which makes up about 90% of the corneal tissue, is approximately 450 μm in thickness. The stroma contains 78% water, 15% collagen, 5% protein, 1% glycosaminoglycans, and 1% salt. The corneal stroma is composed of approximately 200–250 thin layers called lamellae. Each lamella contains regularly organized collagen fibers that are parallel to one another, and the lamellae are arranged orthogonally to each other [3, 5]. The stroma, which is predominantly composed of type I collagen, also contains proteoglycans such as lumican and decorin, which regulate the formation of collagen fibers and their organization. Additionally, there are thin, flat keratocyte cells between the collagen fibers. This organization in the stroma layer is crucial for the transparency, and optical clarity of the cornea. The stroma provides tensile strength, protecting the cornea against intraocular pressure and trauma.

Descemet's membrane, which separates the stroma from the endothelium, is 10 μm thick, an acellular structure consisting of fibers and pores, forming an interwoven mesh that contains a high amount of type VIII collagen and serves as a basal membrane for endothelial cells [3].

The innermost layer is the corneal endothelium, which consists of a single layer of flat endothelial cells. Endothelial cells are connected to each other by well-developed interconnections, partially permeable tight junctions, and desmosomes. The endothelial layer, which allows the passage of water, oxygen, and metabolites, regulates the hydration of the stroma, and is important for the transparency of the cornea.

16.2 Corneal Diseases and Damages

Various corneal diseases or traumas can cause corneal damage at different levels, and depending on the severity of the damage, there are different treatment approaches such as medication, gene therapy, and surgical interventions. According to the World Report on Vision by the World Health Organization, at least 2.2 billion people globally experience vision impairment or vision loss, and 4.2 million cases due to corneal opacity [6]. The cornea is the entry point for light into the eye, allowing the light to be refracted and a portion of the UV light to be filtered. Although it serves as a protective barrier, factors such as abrasion, physical, chemical, thermal, and photokeratitis-related factors can cause varying levels of damage to the cornea. In addition, corneal diseases such as corneal dystrophies, dry eye, keratitis, and allergies can also cause varying degrees of damage to the corneal tissue.

Allergies in the cornea can arise due to pollen, various medications, contact lenses, and cosmetics, and may present symptoms such as redness, itching, burning, stinging, tearing, and watering. An inability to continuously produce tears that nourish the cornea, connective tissue diseases, various medications, and long-term use of contact lenses can lead to eye irritation and the development of dry eye. Corneal dystrophies, which are mostly hereditary, can affect different layers of the cornea and may even result in an opaque cornea, leading to a decrease in vision. Keratitis can occur due to corneal infections caused by bacteria, viruses, or fungi and may present symptoms of pain, watering, discharge, and blurred vision. The transparent corneal tissue can become cloudy, thin, and even perforated due to keratitis. Another disease is corneal ulceration caused by a bacterial infection that invades the cornea due to injury or trauma and can lead to vision loss in the later stages. Keratoconus is the disruption of the cornea's structural integrity, characterized by thinning of the corneal tissue, loss of its normal curvature, and protrusion in the shape of a cone. Although the exact cause is unknown, this disease may be hereditary and can lead to moderate to severe astigmatism and blurred vision.

The tissue response following corneal damage involves a highly complex series of events, including cell adhesion, migration, differentiation, proliferation, death, and remodeling of the extracellular matrix (ECM) mediated by growth factors and immunomodulators. Preserving corneal transparency is crucial for visual function. Therefore, any damage to the cornea due to corneal diseases or trauma can lead to blurred vision and even vision loss. Depending on the level of damage to the corneal tissue, different treatment approaches such as medication, gene therapy, and surgical interventions are available. In complicated cases where more than one layer of the cornea is affected, the damaged area and/or the entire cornea need to be replaced. For permanent or severe corneal injuries where clear vision cannot be restored through alternative methods, keratoplasty, which is the transplantation of healthy corneal tissue from a donor to the patient's damaged cornea, is preferred. Keratoplasty, which is widely used, is the best clinical practice with an 80% success rate. However, in keratoplasty, issues such as the unavailability of donors due to increasing demand and graft rejection may arise. To overcome these potential problems, an artificial cornea (keratoprosthesis) implant has been developed as an alternative treatment method for damaged corneas.

It is important that the corneal implant material allows sufficient oxygen and nutrient permeability to preserve the cellular components of the formed tissue and enable interaction with surrounding tissues. It has been found that using polymeric implants would be more suitable for corneal treatment. Biocompatible synthetic and natural polymeric biomaterials are used to better mimic the corneal extracellular matrix (ECM). Besides cell-free biomaterials, tissue engineering, which is based on the principle of developing tissue equivalents by combining biomaterials with cells, is also a frequently studied alternative treatment approach. Cell-free and cell-containing tissue substitutes are produced for the replacement of part or all (full-thickness) of the damaged cornea, enabling the patients to regain their vision.

Considering the functionality of the cornea, a corneal tissue substitute must meet essential requirements, such as protecting the eye from the external environment, allowing light transmission, providing an ideal optical interface, being biocompatible, and integrating well with the natural tissue. In addition, it is important for the corneal tissue substitute to have biodegradation properties that support appropriate tissue regeneration, tensile strength (3.8 MPa) and mechanical properties similar to natural tissue, and a suitable microstructure and permeability for the passage of nutrients [7]. Moreover, considering that the water content of the cornea is close to 80%, which is important for light transmission, the water retention properties of the artificial cornea should be similar. Hydrogels are 3D hierarchical structures consisting of cross-linked polymeric networks that can hold a large amount of water [8]. Hydrogel-based substitutes stand out in the fields of biomaterials, tissue engineering, and regenerative medicine, particularly in corneal tissue substitutes developed for corneal treatment, due to their high water retention capacity, shape-tailoring ability, and flexible structure.

16.3 Hydrogel-Based Tissue Substitutes for Corneal Treatment

The lamellar organization of collagen fibers in the corneal stroma is crucial for light transmittance. An ideal corneal tissue equivalent is also expected to have an organized, 3D architecture that can closely resemble the native cornea. Hydrogel-based corneal tissue substitutes are frequently constructed using 3D (bio)printing or produced as thick unprinted hydrogels (slabs), and their applications in regenerative medicine are steadily increasing. The advancement of 3D printing and bioprinting technology, also known as additive manufacturing, enables the fabrication and stacking of materials layer by layer with the desired microarchitecture to obtain ideal tissue substitutes with complex internal geometries, such as the cornea. 3D (bio)printing provides a platform that can overcome the limitations faced in traditional biomaterial and tissue engineering fabrication methods by enabling the use of heterogeneous components and the production of heterotypic layered architectures that closely resemble native tissue. The physical and chemical properties can be tailored throughout the layers, and constructs specific to the size of the damage can be produced by 3D (bio)printing. In addition, the cells, even different cell types, can be integrated into layers of the tissue substitute via 3D bioprinting [9].

One of the commonly used 3D printing methods is Fused Deposition Modeling (FDM), which is based on the principle of melting a thermoplastic polymer and extruding it under pressure [9]. Another 3D printing method is stereolithography (SLA), which is based on the curing of a liquid photopolymer using UV light. Selective Laser Sintering (SLS) is a 3D printing method in which layers are formed from powdered material by melting and fusing it with a laser. In 3D bioprinting, bioinks composed of hydrogel solutions, cells, and bioactive agents like growth factors, are used to deposit cell-laden layers on top of each other. Hydrogels are preferred in 3D printing and bioprinting for their printability, cross-linking capability, and biocompatibility. Hydrogels are the most commonly used bioinks, and cells can be easily loaded into them. Bioprinting methods are compatible with one or more types of bioinks, which are cross-linked using various approaches. 3D bioprinting methods include extrusion-based, droplet-based, and stereolithography-based approaches (Fig. 16.2). In extrusion-based bioprinting, bioink is continuously printed as cylindrical filaments (struts), forming cell-laden layers and ultimately constructing a 3D tissue equivalent. In droplet-based bioprinting, bioink is printed in droplet form and can be precisely deposited at predefined locations to form 3D cell layers. In the stereolithography-based method, bioprinting is achieved by the gelation of the bioink solution through photo-cross-linking and depositing each layer on top of the previous one.

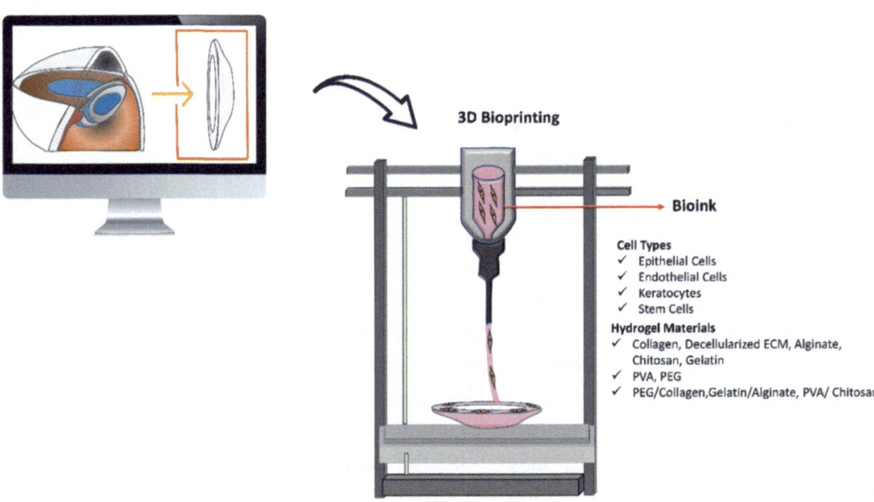

Fig. 16.2 3D bioprinting of corneal tissue using the extrusion method and the types of bioinks used (*ECM* extracellular matrix, *PVA* polyvinyl alcohol, *PEG* polyethylene glycol)

In the development of tissue equivalents for corneal treatment, hydrogels produced by 3D (bio)printing or obtained as hydrogel slabs are commonly used (Table 16.1). Hydrogel-based corneal substitutes are fabricated from various natural or synthetic polymers, such as collagen, gelatin, decellularized ECM, chitosan, hyaluronic acid, alginate, polyvinyl alcohol (PVA), and polyethylene glycol (PEG). Moreover, depending on the level of damage and the treatment strategy, cell sources such as epithelial cells, keratocytes, and/or endothelial cells can be used to enhance corneal healing.

16.3.1 Collagen-Based Hydrogels

Collagen, a protein abundantly found in natural corneal tissue and highly biocompatible, is the most commonly used polymer for scaffold fabrication in corneal tissue substitutes [4, 7]. The mechanical properties, pore size, and biodegradation rate of collagen-based hydrogels can be modified using various cross-linking methods (chemical, physical, enzymatic). Isaacson et al. designed a 3D corneal stromal equivalent in a digital environment and fabricated it with extrusion-based bioprinting using a bioink composed of methacrylated collagen I, alginate, and keratocytes [10]. The fabricated artificial corneal stroma was 500–700 μm thick and had sufficient transparency to transmit light. Additionally, it was reported that keratocytes showed over 80% viability during a 7-day in vitro culture following the bioprinting process. In another study, a bioink including recombinant human laminin and collagen type I was developed to imitate both the stroma and corneal epithelium, and the layered hydrogel was fabricated with a laser-assisted bioprinter [11]. The cell components of the bioink were human embryonic stem cell-derived limbal epithelial stem cells (hESC-LESCs) for the corneal epithelial layer and human adipose-derived mesenchymal stem cells (hASCs) for the corneal stroma. Both cell types conserved their viability and exhibited the characteristic cell morphology of the cornea after bioprinting. It was observed that the LESCs formed a multilayered, epithelium-like layer and expressed corneal epithelium-specific markers such as p63α and CK15. It was also seen that the hASCs were organized in the construct similarly to keratocytes in the corneal stroma and the cells migrated into the surrounding native tissue after implantation into the porcine cornea. In a study by Wang et al., a corneal tissue substitute with high transparency and mechanical strength was produced from a hydrogel containing collagen and decellularized ECM using a vitrification method [12]. A dual-layered corneal equivalent was designed to replace the stromal layer and Bowman's membrane. It was reported the implant retained its transparency after vitrification, with significantly enhanced mechanical properties and suturability. In vivo studies using a rabbit anterior lamellar keratoplasty model showed that this acellular hydrogel achieved full epithelialization within 14 days, integrated with the host stromal tissue within 30 days, and promoted corneal stroma regeneration.

Table 16.1 Hydrogel studies used in various applications for corneal treatment

Application	Form of corneal substitute and production method	Hydrogel material	Cell type	Target tissue	Ref.
In situ	Hydrogel slab/repeated freeze-thaw	PVA-based, composite structure containing iron-doped hydroxyapatite and graphite	Acellular	Cornea	2014 [35]
In situ	3D printed hydrogel/ aerosol jet-based bioprinting/chemical cross-linking with EDC-NHS	Type III collagen	Acellular	Stroma	2021 [36]
In vitro	Hydrogel slab/physical cross-linking with heat	Decellularized porcine ECM	Human corneal epithelial cells and mesenchymal stem cells	Stroma	2021 [18]
In vitro	Hydrogel slab/physical cross-linking with UV	GelMA	Human keratocyte cells	Stroma	2018 [37]
In vitro	Hydrogel slab/chemical cross-linking with genipin	PVA, silk fibroin, nano-hydroxyapatite composite	Human corneal fibroblast cells	Stroma	2019 [38]
In vitro	Hydrogel slab + film/ physical cross-linking with heat	Type I collagen hydrogel sheet and silk-based film	Human corneal stromal cells, epithelial cells, and chicken dorsal root ganglion cells	Corneal epithelium and stroma	2017 [39]
In vitro	Hydrogel slab/chemical and physical cross-linking	GelMA and pHEMA	Human corneal keratocyte cells	Stroma	2019 [40]
In vitro	3D bioprinted hydrogel/ extrusion with chemical cross-linking using calcium chloride	Type I collagen, gelatin, alginate	Human corneal epithelial cells	Corneal epithelium	2016 [28]

(continued)

Table 16.1 (continued)

Application	Form of corneal substitute and production method	Hydrogel material	Cell type	Target tissue	Ref.
In vitro	Injectable, shape-retaining hydrogel/cross-linking using thiol-maleimide click chemistry	Thiol collagen	Human corneal epithelial cells	Corneal epithelium and stroma	2023 [41]
In vitro	3D printed hydrogel/ extrusion	PVA and chitosan	Human adipose-derived mesenchymal stem cells	Stroma	2020 [23]
In vitro	3D bioprinted hydrogel/ extrusion with chemical cross-linking using calcium chloride	Sodium alginate and methacrylated type I collagen	Keratocyte cells	Stroma	2018 [10]
In vitro	3D bioprinted hydrogel/ extrusion with chemical cross-linking using calcium chloride	Sodium alginate and gelatin	Keratocyte cells	Stroma	2020 [42]
In vitro	3D bioprinted hydrogel/ extrusion with physical cross-linking with UV	GelMA	Keratocyte cells	Stroma	2020 [13]
In vitro	3D bioprinted hydrogel/ drop-on-demand inkjet printing (drop-on-demand) with physical cross-linking using heat	Agarose and type I collagen	Keratocyte cells	Stroma	2019 [43]
In vitro	3D bioprinted hydrogel/ digital light processing (DLP) with physical cross-linking with UV	GelMA or HAGM	Limbal stem cells	Cornea	2021 [26]
In vitro	3D bioprinted hydrogel/stereolithography (SLA) with physical cross-linking with UV	GelMA	Corneal stromal cells	Stroma	2020 [15]
In vitro	3D bioprinted hydrogel/ laser-assisted bioprinter (LaBP) with physical cross-linking using laser	Laminin and type I collagen	Human embryonic stem cell-derived limbal epithelial stem cells and adipose-derived mesenchymal stem cells	Corneal epithelium and stroma	2018 [11]

(continued)

Table 16.1 (continued)

Application	Form of corneal substitute and production method	Hydrogel material	Cell type	Target tissue	Ref.
In vitro	3D bioprinted hydrogel/ extrusion and digital light processing (DLP) with chemical cross-linking using calcium chloride and physical cross-linking with UV	Sodium alginate, gelatin, or GelMA	Human corneal epithelial cells	Cornea	2019 [44]
In vitro, ex vivo, in vivo (rabbit model)	Hydrogel slab/chemical and UV cross-linking using CMC-NHS	Decellularized porcine ECM and methacrylated HA	Corneal epithelial cells and corneal stromal cells	Corneal epithelium and stroma	2023 [19]
In vitro, ex vivo, in vivo (pig model)	Hydrogel slab/physical cross-linking with heat	CLP, PEG, and fibrinogen	Corneal epithelial cells	Cornea	2020 [45]
In vitro, ex vivo, in vivo (rabbit model)	Hydrogel slab/vitrification	Decellularized porcine ECM and β-CD-mediated type I collagen	Corneal fibroblast cells	Stroma and Bowman membrane	2020 [46]
In situ, in vivo (rabbit model)	3D printed hydrogel/ extrusion with spontaneous cross-linking over time	hPL, fibrinogen, thrombin	Acellular	Corneal epithelium	2022 [47]
In vitro, in vivo (rabbit model)	3D bioprinted hydrogel/ digital light processing (DLP) with physical cross-linking with UV	GelMA and PEGDA	Rabbit epithelial cells and mesenchymal stem cells from rabbit adipose tissue	Corneal epithelium and stroma	2022 [30]
In vitro, in vivo (rabbit model)	3D bioprinted hydrogel/ physical cross-linking with heat	Decellularized ECM	Keratocyte cells differentiated from human mesenchymal stem cells	Stroma	2019 [21]

(continued)

Table 16.1 (continued)

Application	Form of corneal substitute and production method	Hydrogel material	Cell type	Target tissue	Ref.
In vitro, in vivo (pig model)	Hydrogel slab/chemical cross-linking (carbodiimide, PEG-DBA, and EDC-NHS)	Type I collagen and chitosan	In vitro: Human corneal epithelial cells and chicken embryo dorsal root ganglion cells In vivo: Acellular	Corneal epithelium and stroma	2008 [25]
In vivo (rabbit model)	3D bioprinted hydrogel/ physical cross-linking with heat	Decellularized corneal ECM	Human mesenchymal stem cells differentiated into corneal stromal cells	Stroma	2019 [48]
In vitro, in vivo (rabbit model)	3D bioprinted hydrogel/ physical cross-linking with heat	Decellularized corneal ECM	Keratocytes differentiated from human mesenchymal stem cells	Cornea	2019 [20]
In vitro, in vivo (rat model)	Hydrogel slab/mold-based and physical cross-linking with heat	Chitosan and gelatin	Rat limbal epithelial stem cells and rat mesenchymal stem cells	Corneal epithelium	2017 [24]
In vitro, ex vivo, in vivo (pig model)	Hydrogel slab/physical cross-linking with heat	Type I collagen-based hydrogel and decellularized porcine cornea layers	Human corneal stromal cells	Cornea	2020 [49]
In vitro, in vivo (rabbit and pig model)	Hydrogel slab/chemical cross-linking with EDC and NHS	Collagen type I	In vitro: Human corneal epithelial cells In vivo: Acellular	Corneal epithelium	2006 [50]
In vitro, in vivo (rat model)	Hydrogel slab/cross-linking with UV light	Integration of GelMA hydrogel with PCL-PEG membrane	Keratocytes differentiated from limbal epithelial stem cells	Stroma	2020 [16]

(continued)

Table 16.1 (continued)

Application	Form of corneal substitute and production method	Hydrogel material	Cell type	Target tissue	Ref.
In vitro, in vivo (rabbit model)	Hydrogel slab	Bacterial cellulose and PVA	In vitro: Human corneal stromal cells In vivo: Acellular	Stroma	2020 [34]
In vitro, in vivo (rabbit model)	Hydrogel slab/chemical cross-linking with EDC	HA	Rabbit corneal epithelial cells	Corneal endothelium	2016 [27]
In vitro, in vivo (rabbit model)	Hydrogel slab/using PDMS mold and cross-linking with UV light	GelMA	In vitro: Human keratocytes In vivo: Acellular	Stroma	2019 [14]
In vitro, in vivo (rabbit model)	Hydrogel slab/chemical cross-linking with EDC and NHS	Collagen type I	In vitro: Human mesenchymal stem cells In vivo: Acellular	Stroma	2015 [51]
In vitro, in vivo (mice and rabbit model)	Hydrogel slab/ self-cross-linking	Chitosan and oxidized alginate	Rabbit corneal endothelial cells	Corneal endothelium	2011 [29]
In vitro, in vivo (rabbit model)	Injectable hydrogel/visible light cross-linking using eosin Y and triethanolamine	Thiolated gelatin and methacrylated hyaluronic acid (GelMA-HAMA)	Acellular	Corneal epithelium and stroma	2024 [17]
In vitro, in vivo (rabbit model)	3D bioprinted hydrogel/ digital light processing (DLP) with physical cross-linking using UV	Corneal decellularized extracellular matrix/gelatin methacryloyl (CECM-GelMA)	Human corneal fibroblasts	Corneal epithelium and stroma	2023 [22]
In vitro, in vivo (mice model)	Hydrogel slab/chemical cross-linking with genipin	Chitosan-based thermosensitive hydrogel with long-term release of murine nerve growth factor	Acellular	Corneal epithelium	2023 [52]

(continued)

Table 16.1 (continued)

β-CD beta-cyclodextrin, *CMC-NHS* N-cyclohexyl-N′-(2-morpholinoethyl) carbodiimide metho-p-toluenesulfonate–N-hydroxysuccinimide, *ECM* extracellular matrix, *EDC* 1-ethyl-3-(3-dimethylaminopropyl) carbodiimide hydrochloride, *GelMA* gelatin methacrylate, *HA* hyaluronic acid, *HAGM* hyaluronic acid glycidyl methacrylate, p*HEMA* poly(2-hydroxyethyl methacrylate), *hPL* human platelet lysate, *CLP* collagen-like peptide, *NHS* N-hydroxysuccinimide, *PVA* polyvinyl alcohol, *PEG* polyethylene glycol, *PEGDA* polyethylene glycol diacrylate, *PEG-DBA* poly(ethylene glycol)-dibutyraldehyde, *PCL* polycaprolactone, *PDMS* polydimethylsiloxane

16.3.2 Gelatin-Based Hydrogels

Gelatin, derived from the hydrolysis of collagen, is a biocompatible material that does not exhibit cytotoxic effects [4]. It is also biodegradable, possesses low immunogenicity, and promotes cell adhesion. These properties make gelatin-based hydrogels a promising candidate for corneal treatment approaches. In biomaterials and tissue engineering studies, gelatin is frequently chemically modified with an acrylate reactive group to form methacrylated gelatin (GelMA), which plays a key role in the production of hydrogels.

In a study conducted by Kilic Bektas et al., a corneal equivalent was obtained with an extrusion-based 3D bioprinting method using GelMA-based bioink containing keratocytes [13]. The 3D corneal substitute demonstrated 98% cell viability, exhibited mechanical properties comparable to those of the native cornea, and achieved transparency exceeding 80%, offering a promising approach for stromal damage repair. In another study by the same group, the in vivo performance of cell-free GelMA-based hydrogels in sheet form was evaluated using a rabbit model [14]. After 8 weeks of implantation, it was reported that the corneal stromal equivalent demonstrated high transparency, appropriate mechanical properties, and good integration with the surrounding native tissue, making it a promising candidate for corneal healing. A 3D bioprinted tissue equivalent was developed in another study using the UV light-based stereolithography (SLA) method, with a bioink containing GelMA and human corneal stromal cells (hCSC), to form dome-shaped structures anatomically similar to the corneal stroma [15]. It was found that the corneal tissue equivalent exhibited good mechanical properties and optical transparency comparable to that of the native corneal stroma. Moreover, the stromal cells in the substitute maintained over 80% viability after 7 days of in vitro culture and expressed cornea-specific markers, such as lumican and collagen type I, after 28 days of culture. Kong et al. fabricated membranes composed of orthogonally aligned nanofibers (polycaprolactone (PCL)–PEG) using the electrospinning technique and integrated these membranes into GelMA-based hydrogels to enhance the mechanical properties of the hydrogels, thereby transferring the mechanical load from the hydrogel to the nanofiber structure [16]. In in vivo studies using rat intrastromal keratoplasty models, the implanted construct promoted stromal regeneration and maintained high transparency for up to 3 months.

Agrawal et al. developed an injectable, photo-cross-linkable hydrogel corneal substitute, known as Kuragel, to promote corneal regeneration [17]. Kuragel was composed of thiolated gelatin (Gel-SH) and methacrylated hyaluronic acid (HA-MA), cross-linked using Eosin Y and triethanolamine under visible light (white light). It was reported that the mechanical and optical properties, including a compressive modulus of 346.49 kPa and over 91% transparency, were similar to those of the cornea. In vivo studies using a rabbit corneal injury model demonstrated rapid re-epithelialization within 1 month, as well as stromal and nerve regeneration by 3 months. Its biocompatibility and integration with host tissue supported its potential for use in the treatment of corneal injuries and defects.

16.3.3 Decellularized Extracellular Matrix (ECM)-Based Hydrogels

Using decellularized corneal ECM as a hydrogel material for corneal treatment has gained importance due to its tissue-specific composition. An excessive decellularization process can lead to the loss of mechanical and biological structural integrity of the ECM. Conversely, inadequate decellularization of ECM may provoke an immune response due to the remaining cells. However, when an optimized decellularization procedure is used, the resulting tissue substitute, containing the complex components of natural tissue ECM, significant potential for successful application. In a study by Yazdanpanah et al., a thermoresponsive, transparent hydrogel was obtained by a heat-induced physical cross-linking of decellularized porcine cornea ECM (COMatrix) [18]. The effectiveness of the decellularization protocol and the composition of the resulting ECM were determined by comparing it with the human amniotic membrane, which is commonly used in ophthalmology applications. Proteomic studies showed that the decellularized ECM-based hydrogel was rich in proteins, such as lumican, laminin, and keratocan, with their regenerative effects on the cornea. In vitro results demonstrated that corneal epithelial and mesenchymal cells proliferated in the hydrogels, maintaining their viability at over 90%.

In a study by Shen et al., a hydrogel with an interpenetrating network (IPN) structure was developed using decellularized porcine corneal stroma matrix and methacrylated hyaluronic acid (HAMA) [19]. A heat-induced dual-cross-linking was carried out using N-cyclohexyl-N′-(2-morpholinoethyl) carbodiimide metho-p-toluenesulfonate (CMC-NHS) and UV light. The decellularized ECM component provided biological cues, while HAMA contributed to the hydrogel's shape, integrity, and structural stability. In situ characterization studies showed that the developed hydrogel exhibited remarkable physical properties essential for ophthalmic applications, such as high transparency, low swelling ratio, slow degradation rate, and increased mechanical strength. Additionally, the formation of the interpenetrating polymer network and stable bonds with the surrounding tissue suggest that this hydrogel could serve as a potential suture-free treatment method, demonstrating excellent adhesion. In vitro results confirmed that the hydrogel effectively supported cell adhesion, viability, and proliferation, and the stromal cells retained the keratocyte phenotype. In vivo studies showed that the hydrogel adhered to the native stroma, maintained transparency, and promoted both corneal epithelialization and stromal healing. In another study, Kim et al. developed a bioink composed of decellularized bovine corneal ECM and keratocytes differentiated from mesenchymal stem cells [20]. The corneal tissue equivalent, produced by 3D bioprinting with minimal impact on cell viability, demonstrated optical transparency comparable to that of native human cornea. In vivo studies showed that keratocan expression in cells within the ECM-based hydrogel was 21 times higher than in cells within collagen hydrogels. In another study, a functional corneal equivalent was produced by 3D bioprinting using a bioink composed of decellularized bovine corneal ECM and differentiated keratocytes [21]. In this study, a shear stress was induced during the bioprinting to align collagen type I fibers of the ECM to better imitate the stroma architecture. A rabbit model was used for in vivo studies, and a lattice structure similar to that of the native cornea was observed by week 4 post-implantation. Zhang et al. developed a novel bioink composed of corneal decellularized extracellular matrix (CECM) and GelMA and produced corneal stromal substitutes with digital light processing (DLP) 3D bioprinting [22]. The 3D-printed hydrogels exhibited high transparency and tunable mechanical properties, closely mimicking the natural cornea. In vitro studies with human corneal fibroblasts demonstrated that hydrogels supported cell viability and expression of corneal-specific markers. In vivo studies showed successful epithelial regeneration, reduced scarring, and restored corneal clarity within 8 weeks in a rabbit model.

16.3.4 Chitosan-Based Hydrogels

Chitosan is a cationic polysaccharide derived from the deacetylation of chitin, the primary component of shellfish exoskeletons. Its biocompatibility, antibacterial properties, biodegradability, and low cost make chitosan a promising polymer in biomaterials and tissue engineering fields, particularly for corneal treatment. Chitosan has been combined with synthetic and other natural polymers, such as PVA, gelatin, and collagen, to develop hydrogels for corneal tissue substitutes [23–25]. Ulag et al. developed a corneal stroma substitute using an extrusion-based 3D printing with a polymer blend of chitosan and PVA [23]. The cell-free chitosan/PVA construct had a thickness of 400 μm and was capable of maintaining the required shape and size. Additionally, the hydrogels exhibited suitable mechanical properties and good optical transparency. In in vitro studies, human adipose tissue-derived mesenchymal stem cells (hASCs) were integrated into the printed scaffolds, and it was shown that the cells maintained over 95% viability within the 3D hydrogels. In another study, a hydrogel composed of chitosan and collagen was produced using 1-ethyl-3-(3-dimethylaminopropyl) carbodiimide hydrochloride (EDC)-NHS and poly(ethylene glycol) dibutyraldehyde (PEG-DBA) cross-linkers [25]. The hydrogels demonstrated suitable optical and mechanical properties, suturability, and good permeability to glucose and albumin. In in vitro studies, human corneal epithelial cells and dorsal root ganglion cells from chick embryos adhered and proliferated within the hydrogel. For in vivo studies, the hydrogels were implanted into pig corneas, and 12 months post-implantation results showed successful regeneration of the corneal epithelium, stroma, and nerves, with seamless host-graft integration.

16.3.5 Hyaluronic Acid-Based Hydrogels

Hyaluronic acid (HA), also known as hyaluronan, is an anionic natural polymer composed of repeating D-glucuronic acid and N-acetyl-D-glucosamine units. Hyaluronic acid is a high molecular weight polysaccharide and one of the main components of the extracellular matrix (ECM). The features of hyaluronic acid, including biodegradability, biocompatibility, non-thrombogenicity, non-immunogenicity, and high flexibility, make it a promising polymer for biomaterials and tissue engineering applications, particularly for corneal tissue substitutes. Zhong et al. produced microscale hydrogels for corneal epithelium mimicry using Digital Light Processing (DLP) bioprinting with two different bioinks containing either GelMA or HA glycidyl methacrylate (HAGM) combined with limbal stem cells [26]. The cells showed viability ratios of 86% and 92% in GelMA- and HAGM-based hydrogels, respectively. The immunocytochemistry and transcriptome analysis results demonstrated that limbal stem cells were active within GelMA-based hydrogels and passive within HAGM-based hydrogels, indicating that the ECM could regulate and reprogram the fate of stem cells. This study provided valuable insights into regenerative medicine and stem cell therapy approaches for corneal reconstruction, demonstrating that both active and passive states of limbal stem cells can be supported by two different hydrogels. In another study, a hyaluronic acid-based hydrogel was produced to support corneal endothelial cells [27]. Assessing various HA concentrations (0.05%–1.25%) revealed that the hydrogel with 0.25% HA concentration provided the best results in terms of achieving a porous microstructure, biocompatibility, and high nutrient permeability. The corneal endothelial cell-loaded hydrogels were implanted in a rabbit model with corneal endothelial dysfunction. At week 4 post-implantation, the cell-loaded hydrogel implant group compared to the untreated damaged model group demonstrated better outcomes in terms of tissue healing and corneal transparency.

16.3.6 Alginate-Based Hydrogels

Alginate, a linear anionic polysaccharide, is composed of β-D-mannuronic acid and α-L-guluronic acid building blocks and is derived from the cell walls of brown algae. Alginate-based hydrogels, known for their transparency, rapid gelation, high biocompatibility, and stability properties, also have a water content comparable to that of natural corneal tissue, making them a preferable material for corneal treatment. Moreover, the biophysical properties of alginates can be modulated in a controlled manner, enabling the development of specific therapeutic strategies for the treatment of corneal diseases. Furthermore, the biophysical properties of alginates can be precisely adjusted, allowing for the production of specific therapeutic approaches for corneal diseases. Wu et al. developed a bioink using a collagen/gela tin/alginate (Col/Gel/Alg) polymer blend that was suitable for extrusion-based 3D bioprinting [28]. The layer-by-layer printed Col/Gel/Alg hydrogel provided a favorable, stable, macroporous 3D environment for cell infiltration and proliferation. In addition to its outstanding mechanical and optical properties, the hydrogel supported 95% cell viability for human corneal epithelial cells. Furthermore, the expression of cytokeratin 3, a specific marker for corneal epithelium, increased from day 1 to day 5 within the hydrogel during in vitro culture. Liang et al. developed a hydrogel using a polymer blend of alginate and chitosan, which is capable of self-cross-linking when applied to tissue [29]. In in vivo studies, the endothelial layer of the rabbit cornea was removed, and then the rabbit corneal endothelial cell containing bioink was applied to the rabbit eyes. The polymer blend was self-cross-linked in 5 min upon application and formed a hydrogel structure. It was observed that two-thirds of the rabbits were found to retain corneal transparency 10 days after application.

16.3.7 Polyethylene Glycol-Based Hydrogels

Alone or in combination with other synthetic polymers, and also with natural polymers like collagen, gelatin, and silk to improve their mechanical properties, Polyethylene glycol (PEG) is a hydrophilic and biocompatible synthetic polymer widely used in the biomaterials field, with U.S. Food and Drug Administration (FDA) approval. PEG, which can have varying molecular weights, is frequently utilized in hydrogel form for corneal tissue regeneration studies. He et al. developed a cell-laden bilayer hydrogel to replace both the epithelium and stroma using a bioink composed of GelMA and long-chain poly(ethylene glycol) diacrylate (PEGDA), fabricated via Digital Light Processing (DLP) 3D bioprinting [30]. In this study, rabbit corneal epithelial cells were utilized to print an epithelium-like layer with a thickness of 50 μm, while rabbit adipose-derived mesenchymal stem cells (rASCs) were used in the stroma-like structure, which had a thickness of 150 μm. The stroma part was printed with fibrous layers aligned orthogonally, mimicking the architecture of native tissue. The PEGDA-GelMA hydrogel was shown to promote cell adhesion and proliferation and also provide excellent light transmittance, a suitable degradation rate, and porosity that facilitates nutrient diffusion. In in vivo studies, the dome-shaped, cell-laden hydrogel was implanted into a rabbit model, and it successfully promoted corneal wound closure, supporting re-epithelialization and stromal regeneration.

16.3.8 Polyvinyl Alcohol-Based Hydrogels

Polyvinyl alcohol (PVA), a water-soluble synthetic polymer, has been used in ophthalmic applications for a long time. Numerous studies, in particular, have been conducted on PVA hydrogel-based soft contact lenses. PVA hydrogels can be crosslinked using physical methods such as cryogelation, chemical methods like glutaraldehyde, and irradiation techniques such as UV [31–33]. PVA hydrogels have a low friction coefficient and exhibit mechanical properties similar to collagen. PVA-based hydrogel crystallizes at low temperatures and is a potential polymer for corneal treatment due to its transparency, nutrient permeability, and stability at body temperature for long-term implantation. In hydrogels for corneal treatment, PVA is commonly used in combination with polymers such as chitosan and cellulose [23, 34]. For example, a corneal stromal substitute was developed as a hydrogel using a mixture of bacterial cellulose (BC) and PVA [34]. The hydrogel prepared with BC/PVA demonstrated high biocompatibility and exhibited better water retention and light transmission properties compared to the hydrogel prepared with BC alone. In vivo studies using a rabbit model showed that the hydrogels maintained their transparency and did not induce inflammation by week 4 after implantation. Jiang et al. fabricated a PVA-based hydrogel integrated with graphite and iron (Fe)-doped nano-hydroxyapatite (n-Fe-HA), suggesting that the transparent optical center of this hydrogel could exhibit properties similar to those of the natural cornea, while the porous peripheral regions could enhance integration with the surrounding tissue [35].

16.4 Conclusion

In vitro and in vivo preclinical studies demonstrate that hydrogels can be a potential treatment strategy for corneal injuries due to their biocompatibility, native tissue-like viscoelastic properties, light transmittance, and ability to support tissue healing and regeneration. To more accurately mimic the complex architecture of the stroma layer that provides light transmittance in cornea, corneal tissue substitutes, produced in layers with appropriate geometry via 3D printing/bioprinting technologies, present a promising approach in corneal treatments due to their high transparency properties. One of the greatest challenges in corneal treatments is producing a full-thickness artificial cornea by precisely reconstructing the complex, orthogonal, multilayered structure of the stroma. In this respect, the development of hydrogel-based bioinks for use in 3D bioprinting to produce a corneal tissue equivalent that reflects the complex architecture of the stroma and other corneal layers, including cells, will contribute to significant advances in corneal treatment. Studies are ongoing to develop an ideal corneal tissue equivalent that has improved mechanical properties, the desired shape, structure, and thickness, high transparency, sterility, biomimetic properties, and the potential to be applied in the clinic for corneal treatments.

References

1. Sridhar, M.S.: Anatomy of cornea and ocular surface. Indian J. Ophthalmol. **66**, 190–194 (2018)
2. Feizi, S., Jafarinasab, M.R., Karimian, F., Hasanpour, H., Masudi, A.: Central and peripheral corneal thickness measurement in normal and keratoconic eyes using three corneal pachymeters. J. Ophthalmic Vis. Res. **9**(3), 296–304 (2014)
3. Ross, P., Histology, A.: Text and Atlas with Correlated Cell and Molecular Biology. Lippincott Williams & Wilkins, Philadelphia, PA (2011)
4. Khosravimelal, S., Mobaraki, M., Eftekhari, S., Ahearne, M., Seifalian, A.M., Gholipour-Malekabadi, M.: Hydrogels as emerging materials for cornea wound healing. Small. **17**(30), 2006335 (2021)
5. Gartner, L.P., Hiatt, J.L.: Color Textbook of Histology Ebook. Elsevier Health Sciences, Amsterdam (2006)
6. Organization WH: World Report on Vision, p. 2020. World Health Organization, Geneva (2019)
7. Kishore, V., Alapan, Y., Iyer, R., McLay, R., Gurkan, U.A.: Application of Hydrogels in Ocular Tissue Engineering. Gels Handbook, pp. 137–164. World Scientific, Singapore (2016)
8. Patel, A., Mequanint, K.: Hydrogel biomaterials. In: Biomedical Engineering—Frontiers and Challenges. IntechOpen, London (2011)
9. Tamay, D.G., Dursun Usal, T., Alagoz, A.S., Yucel, D., Hasirci, N., Hasirci, V.: 3D and 4D printing of polymers for tissue engineering applications. Front. Bioeng. Biotechnol. **7**, 164 (2019)
10. Isaacson, A., Swioklo, S., Connon, C.J.: 3D bioprinting of a corneal stroma equivalent. Exp. Eye Res. **173**, 188–193 (2018)
11. Sorkio, A., Koch, L., Koivusalo, L., Deiwick, A., Miettinen, S., Chichkov, B., et al.: Human stem cell based corneal tissue mimicking structures using laser-assisted 3D bioprinting and functional bioinks. Biomaterials. **171**, 57–71 (2018)
12. Wang, F., Shi, W., Li, H., Wang, H., Sun, D., Zhao, L., et al.: Decellularized porcine cornea-derived hydrogels for the regeneration of epithelium and stroma in focal corneal defects. Ocul. Surf. **18**(4), 748–760 (2020)
13. Bektas, C.K., Hasirci, V.: Cell loaded 3D bioprinted GelMA hydrogels for corneal stroma engineering. Biomater. Sci. **8**(1), 438–449 (2020)
14. Kilic Bektas, C., Burcu, A., Gedikoglu, G., Telek, H.H., Ornek, F., Hasirci, V.: Methacrylated gelatin hydrogels as corneal stroma substitutes: in vivo study. J. Biomater. Sci. Polym. Ed. **30**(18), 1803–1821 (2019)
15. Mahdavi, S.S., Abdekhodaie, M.J., Kumar, H., Mashayekhan, S., Baradaran-Rafii, A., Kim, K.: Stereolithography 3D bioprinting method for fabrication of human corneal stroma equivalent. Ann. Biomed. Eng. **48**(7), 1955–1970 (2020)
16. Kong, B., Chen, Y., Liu, R., Liu, X., Liu, C., Shao, Z., et al.: Fiber reinforced GelMA hydrogel to induce the regeneration of corneal stroma. Nat. Commun. **11**(1), 1435 (2020)
17. Agrawal, P., Tiwari, A., Chowdhury, S.K., Vohra, M., Gour, A., Waghmare, N., et al.: Kuragel: a biomimetic hydrogel scaffold designed to promote corneal regeneration. Iscience. **27**, 27(5) (2024)
18. Yazdanpanah, G., Jiang, Y., Rabiee, B., Omidi, M., Rosenblatt, M.I., Shokuhfar, T., et al.: Fabrication, rheological, and compositional characterization of thermoresponsive hydrogel from cornea. Tissue Eng. Part C Methods. **27**(5), 307–321 (2021)
19. Shen, X., Li, S., Zhao, X., Han, J., Chen, J., Rao, Z., et al.: Dual-crosslinked regenerative hydrogel for sutureless long-term repair of corneal defect. Bioact. Mater. **20**, 434–448 (2023)
20. Kim, H., Park, M.N., Kim, J., Jang, J., Kim, H.K., Cho, D.W.: Characterization of cornea-specific bioink: high transparency, improved in vivo safety. J. Tissue Eng. **10**, 2041731418823382 (2019)
21. Kim, H., Jang, J., Park, J., Lee, K.-P., Lee, S., Lee, D.-M., et al.: Shear-induced alignment of collagen fibrils using 3D cell printing for corneal stroma tissue engineering. Biofabrication. **11**(3), 035017 (2019)

22. Zhang, M., Yang, F., Han, D., Zhang, S.-y., Dong, Y., Li, X., et al.: 3D bioprinting of corneal decellularized extracellular matrix: GelMA composite hydrogel for corneal stroma engineering. Int. J. Bioprint. **9**(5), 774 (2023)
23. Ulag, S., Ilhan, E., Sahin, A., Yilmaz, B.K., Ekren, N., Kilic, O., et al.: 3D printed artificial cornea for corneal stromal transplantation. Eur. Polym. J. **133**, 109744 (2020)
24. Tang, Q., Luo, C., Lu, B., Fu, Q., Yin, H., Qin, Z., et al.: Thermosensitive chitosan-based hydrogels releasing stromal cell derived factor-1 alpha recruit MSC for corneal epithelium regeneration. Acta Biomater. **61**, 101–113 (2017)
25. Rafat, M., Li, F., Fagerholm, P., Lagali, N.S., Watsky, M.A., Munger, R., et al.: PEG-stabilized carbodiimide crosslinked collagen–chitosan hydrogels for corneal tissue engineering. Biomaterials. **29**(29), 3960–3972 (2008)
26. Zhong, Z., Balayan, A., Tian, J., Xiang, Y., Hwang, H.H., Wu, X., et al.: Bioprinting of dual ECM scaffolds encapsulating limbal stem/progenitor cells in active and quiescent statuses. Biofabrication. **13**(4), 044101 (2021)
27. Lai, J.-Y.: Hyaluronic acid concentration-mediated changes in structure and function of porous carriers for corneal endothelial cell sheet delivery. Mater. Sci. Eng. C. **59**, 411–419 (2016)
28. Wu, Z., Su, X., Xu, Y., Kong, B., Sun, W., Mi, S.: Bioprinting three-dimensional cell-laden tissue constructs with controllable degradation. Sci. Rep. **6**(1), 24474 (2016)
29. Liang, Y., Liu, W., Han, B., Yang, C., Ma, Q., Song, F., et al.: An in situ formed biodegradable hydrogel for reconstruction of the corneal endothelium. Colloids Surf. B Biointerfaces. **82**(1), 1–7 (2011)
30. He, B., Wang, J., Xie, M., Xu, M., Zhang, Y., Hao, H., et al.: 3D printed biomimetic epithelium/stroma bilayer hydrogel implant for corneal regeneration. Bioact. Mater. **17**, 234–247 (2022)
31. Yokoyama, F., Masada, I., Shimamura, K., Ikawa, T., Monobe, K.: Morphology and structure of highly elastic poly (vinyl alcohol) hydrogel prepared by repeated freezing-and-melting. Colloid Polym. Sci. **264**(7), 595–601 (1986)
32. Zu, Y., Zhang, Y., Zhao, X., Shan, C., Zu, S., Wang, K., et al.: Preparation and characterization of chitosan–polyvinyl alcohol blend hydrogels for the controlled release of nano-insulin. Int. J. Biol. Macromol. **50**(1), 82–87 (2012)
33. Hennink, W.E., van Nostrum, C.F.: Novel crosslinking methods to design hydrogels. Adv. Drug Deliv. Rev. **64**, 223–236 (2012)
34. Han, Y., Li, C., Cai, Q., Bao, X., Tang, L., Ao, H., et al.: Studies on bacterial cellulose/poly(vinyl alcohol) hydrogel composites as tissue-engineered corneal stroma. Biomed. Mater. **15**(3), 035022 (2020)
35. Jiang, H., Zuo, Y., Zhang, L., Li, J., Zhang, A., Li, Y., et al.: Property-based design: optimization and characterization of polyvinyl alcohol (PVA) hydrogel and PVA-matrix composite for artificial cornea. J. Mater. Sci. Mater. Med. **25**, 941–952 (2014)
36. Gibney, R., Patterson, J., Ferraris, E.: High-resolution bioprinting of recombinant human collagen type III. Polymers. **13**(17), 2973 (2021)
37. Kilic Bektas, C., Hasirci, V.: Mimicking corneal stroma using keratocyte-loaded photopolymerizable methacrylated gelatin hydrogels. J. Tissue Eng. Regen. Med. **12**(4), e1899–ee910 (2018)
38. Zhou, H., Wang, Z., Cao, H., Hu, H., Luo, Z., Yang, X., et al.: Genipin-crosslinked polyvinyl alcohol/silk fibroin/nano-hydroxyapatite hydrogel for fabrication of artificial cornea scaffolds—a novel approach to corneal tissue engineering. J. Biomater. Sci. Polym. Ed. **30**(17), 1604–1619 (2019)
39. Wang, S., Ghezzi, C.E., Gomes, R., Pollard, R.E., Funderburgh, J.L., Kaplan, D.L.: In vitro 3D corneal tissue model with epithelium, stroma, and innervation. Biomaterials. **112**, 1–9 (2017)
40. Kilic Bektas, C., Hasirci, V.: Cell loaded GelMA: HEMA IPN hydrogels for corneal stroma engineering. J. Mater. Sci. Mater. Med. **31**, 1–15 (2020)
41. Rosenquist, J., Folkesson, M., Höglund, L., Pupkaite, J., Jn, H., Samanta, A.: An injectable, shape-retaining collagen hydrogel cross-linked using thiol-maleimide click chemistry for sealing corneal perforations. ACS Appl. Mater. Interfaces. **15**(29), 34407–34418 (2023)

42. Kutlehria, S., Dinh, T.C., Bagde, A., Patel, N., Gebeyehu, A., Singh, M.: High-throughput 3D bioprinting of corneal stromal equivalents. J. Biomed. Mater. Res. B Appl. Biomater. **108**(7), 2981–2994 (2020)
43. Duarte Campos, D.F., Rohde, M., Ross, M., Anvari, P., Blaeser, A., Vogt, M., et al.: Corneal bioprinting utilizing collagen-based bioinks and primary human keratocytes. J. Biomed. Mater. Res. A. **107**(9), 1945–1953 (2019)
44. Zhang, B., Xue, Q., Hu, H.-y., Yu, M.-f., Gao, L., Luo, Y.-c., et al.: Integrated 3D bioprinting-based geometry-control strategy for fabricating corneal substitutes. J. Zhejiang Univ. Sci. B. **20**(12), 945 (2019)
45. McTiernan, C.D., Simpson, F.C., Haagdorens, M., Samarawickrama, C., Hunter, D., Buznyk, O., et al.: LiQD cornea: pro-regeneration collagen mimetics as patches and alternatives to corneal transplantation. Sci. Adv. **6**(25), eaba2187 (2020)
46. Wang, X., Majumdar, S., Soiberman, U., Webb, J.N., Chung, L., Scarcelli, G., et al.: Multifunctional synthetic Bowman's membrane-stromal biomimetic for corneal reconstruction. Biomaterials. **241**, 119880 (2020)
47. You, J., Frazer, H., Sayyar, S., Chen, Z., Liu, X., Taylor, A., et al.: Development of an in-situ printing system with human platelet lysate-based bio-adhesive to treat corneal perforation. In: bioRxiv, p. 2021.09.13.460167 (2021)
48. Park, J., Lee, K.-P., Kim, H., Park, S., Wijesinghe, R.E., Lee, J., et al.: Biocompatibility evaluation of bioprinted decellularized collagen sheet implanted in vivo cornea using swept-source optical coherence tomography. J. Biophotonics. **12**(11), e201900098 (2019)
49. Fernández-Pérez, J., Madden, P.W., Ahearne, M.: Engineering a corneal stromal equivalent using a novel multilayered fabrication assembly technique. Tissue Eng. Part A. **26**(19–20), 1030–1041 (2020)
50. Liu, Y., Gan, L., Carlsson, D.J., Fagerholm, P., Lagali, N., Watsky, M.A., et al.: A simple, cross-linked collagen tissue substitute for corneal implantation. Invest. Ophthalmol. Vis. Sci. **47**(5), 1869–1875 (2006)
51. Koulikovska, M., Rafat, M., Petrovski, G., Veréb, Z., Akhtar, S., Fagerholm, P., et al.: Enhanced regeneration of corneal tissue via a bioengineered collagen construct implanted by a nondisruptive surgical technique. Tissue Eng. Part A. **21**(5–6), 1116–1130 (2015)
52. Wu, J., Huang, Y., Yu, H., Li, K., Zhang, S., Qiao, G., et al.: Chitosan-based thermosensitive hydrogel with longterm release of murine nerve growth factor for neurotrophic keratopathy. Neural Regen. Res. **19**, 680 (2023)

Chapter 17
Hydrogels Used in Contact Lenses and Intraocular Lenses

Cemile Kilic Bektas

Abstract Maintaining vision is crucial for healthy aging. Numerous studies focus on vision impairments caused by disease, aging, and ocular trauma, leading to the development of a wide range of lenses, implants, and biomedical devices aimed at correcting vision. Contact lenses are the most commonly used biomedical devices for individuals who prefer not to wear glasses, correcting mild ametropia (a general term for refractive errors such as myopia, hyperopia, and astigmatism, where the eye fails to focus images sharply on the retina). However, in cases such as severe anisometropia (differences in refractive power between the two eyes), regular astigmatism, and irregular astigmatism, contact lenses are the only option for vision correction. Similarly, intraocular lenses are essential for treating cataracts, a condition that can lead to blindness if left untreated. Hydrogels form the fundamental biomaterial in ophthalmology, with applications ranging from contact lenses to flexible intraocular lenses and ophthalmic drug delivery systems. This chapter examines the hydrogels used in the manufacture of contact lenses and intraocular lenses.

Keywords Contact lens · Rigid contact lenses · Soft contact lenses · Silicone-based · HEMA · Dihydroxy methacrylates · Methacrylic acid · Acrylamides · N-Vinyl-2-Pyrrolidone · Intraocular lens

17.1 Contact Lenses

The human eye is a complex organ essential for a person's healthy aging. To address functional deficiencies caused by diseases, aging, and ocular trauma, implants and biomedical devices are widely used. Contact lenses are transparent, hemispherical devices placed on the cornea, the outermost layer of the eye. By altering the direction of light to focus it onto a single point on the retina, contact lenses are primarily

C. Kilic Bektas (✉)
Department of Chemistry and Chemical Biology, Rutgers University, Piscataway, NJ, USA

N. Hasirci, V. Hasirci (eds.), *Hydrogels and Bioinks in Tissue Engineering*, Biological and Medical Physics, Biomedical Engineering,
https://doi.org/10.1007/978-3-031-90966-5_17

used to correct vision impairments, often as the only viable option in certain cases. Some scenarios where contact lenses are preferred or the only option to achieve satisfactory vision include: (1) severe ametropia, where using glasses is mechanically or optically unacceptable for correcting extreme nearsightedness or farsightedness, (2) severe anisometropia, where the patient's eyes have significantly different refractive powers, (3) corneal transplant cases followed by regular postoperative astigmatism, and (4) irregularities in the corneal surface caused by keratoconus, corneal scarring, or irregular astigmatism [1].

The contact lens market is rapidly growing, with over 140 million users worldwide, 45 million of whom are in the United States [2]. In 2021, the global market size was estimated at $9.94 billion, with projections to reach $14.80 billion by 2029 [3]. For such a market to remain sustainable, it must meet both user and manufacturer needs (Fig. 17.1). Contact lenses must have stable, consistent tear films, resist the adhesion of tear proteins and lipids, maintain proper water content, and allow sufficient oxygen permeability to support corneal metabolism. Therefore, contact lenses must possess surface properties ensuring excellent biocompatibility, being neither fully hydrophilic nor hydrophobic [4].

Attempts to correct vision date back to Leonardo da Vinci's 1508 experiment, where a patient submerged their head in a water-filled container. Though impractical, this paved the way for advancements. Glass lenses were introduced in the 1800s, but significant material progress began in 1934 with the production of poly(methyl methacrylate) (PMMA) [5]. PMMA was later followed by the development of poly(hydroxyethyl methacrylate) (PHEMA) hydrogels, which offered lightweight, durable, and highly oxygen-permeable lenses for extended wear [4].

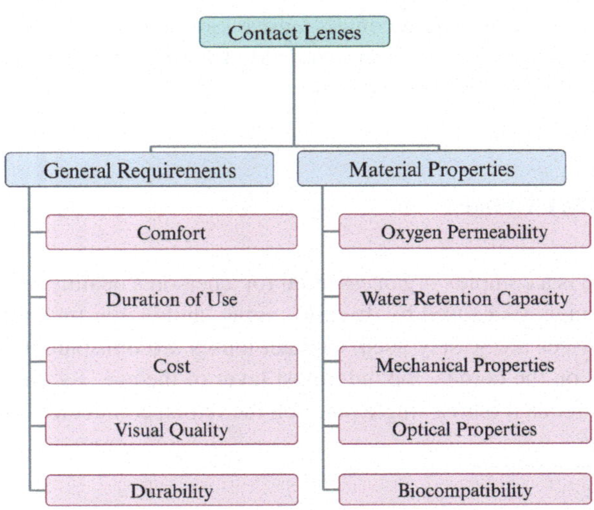

Fig. 17.1 Essential features of a contact lens. The features of contact lenses are divided into two main categories: general requirements for user comfort and the optical, biophysical, and chemical properties required in a contact lens material

Contact lenses can be categorized in various ways but are primarily divided into rigid, soft, and hybrid categories based on the materials used. Soft lenses are further subdivided into hydrophilic and hydrophobic types.

17.1.1 Rigid Contact Lenses

Rigid contact lenses were first produced in 1948 by Kevin Touhey using PMMA and were commercially launched in 1979. PMMA is obtained through the polymerization of methyl methacrylate monomer (Fig. 17.2a). While PMMA offers excellent transparency, UV resistance, weather resilience, acceptable surface wettability, and lightweight properties, its primary disadvantage is poor oxygen permeability (Dk = 1.35 Barrer). This limitation can lead to hypoxia, causing serious ocular problems [6]. The oxygen permeability is generally measured by Dk, a universally accepted standard. Here, D represents diffusion, and k represents the solubility of oxygen. As the Dk value increases, the material's oxygen permeability also increases. For example, pure water's oxygen permeability is 80 barrers, and for a lens to be worn safely, its oxygen permeability must be at least 87 barrers. However, it is reported that to completely avoid problems caused by low oxygen permeability, oxygen transmission should be greater than 125 barrers [4, 7]. For this reason, the low oxygen permeability of PMMA has prevented the long-term use of hard lenses [8]. To overcome this issue, PMMA lenses are usually produced with a small diameter, which allows the cornea to be oxygenated with tears formed during blinking and lens movement [1].

To address this, PMMA lenses are usually made with smaller diameters to allow oxygenation of the cornea via tears during blinking and lens movement. In the late 1970s, rigid gas-permeable (RGP) lenses with higher oxygen permeability (Dk = 10–30) were developed by copolymerizing methyl methacrylate (MMA) with siloxane-functionalized methacrylates, such as methacryloxypropyl tris(trimethylsiloxy)silane (TRIS). The ratio of TRIS and MMA determines the

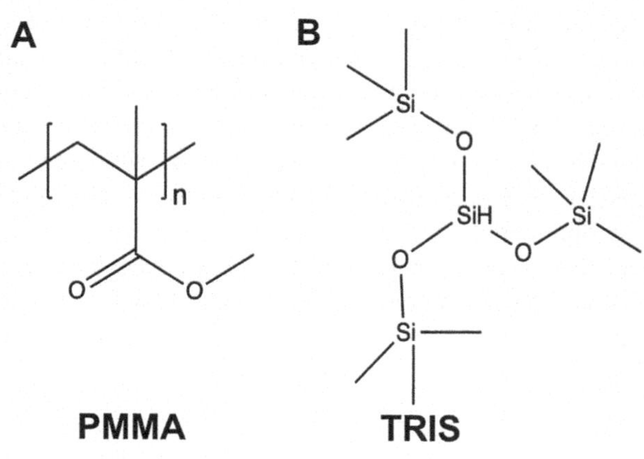

Fig. 17.2 Materials used in the production of rigid contact lenses. Poly(methyl methacrylate) (PMMA) (**a**) and the cross-linker TRIS (**b**)

oxygen permeability, elastic modulus, hardness, and water retention capacity of the lenses. Although new types of rigid contact lenses were superior to PMMA in terms of oxygen permeability, this characteristic was not sufficient for these lenses to be used for long durations. Subsequent developments, such as fluoromethacrylate copolymerization with TRIS cross-linkers, improved mechanical properties and oxygen permeability (Dk = 30–160 barrers) [1, 9]. Some products with high Dk values are FDA-approved for extended use of up to 7 days [1].

Although soft lenses dominate prescriptions today, rigid lenses remain in use for conditions like keratoconus, trauma, or corneal grafts, where they mask irregular corneal structures for clear vision [10]. Some practitioners also prefer rigid lenses in uncomplicated cases due to their lower risk of microbial keratitis [11].

17.1.2 Soft Contact Lenses

Soft lenses are distinguished by their high water retention capacity, which gives them flexibility and allows them to conform to the eye's shape. These lenses are primarily made from silicone-based elastomers and hydrogels.

17.1.2.1 Silicone-Based Elastomers

Silicone-based lenses, made from polydimethylsiloxane (PDMS), were designed in the 1950s for extended use. However, their low water content caused inadequate tear wettability, protein adhesion, and lens adhesion to the cornea, rendering them unsuitable for long-term wear [12]. Despite their excellent optical properties and high oxygen permeability, these lenses failed in extended wear applications. Research into improving water retention through plasma treatment yielded unstable results [1]. Innovative approaches combined silicone elastomers with hydrogels to merge high oxygen permeability with water retention. Studies revealed different behaviors between conventional hydrogels and silicone-based hydrogels: while oxygen permeability in conventional hydrogels increased with water content, it decreased in silicone-based hydrogels. Additionally, the oxygen permeability of silicone-based hydrogels is much higher than that of conventional hydrogels up to at least 80% water content. This is because the main material transmitting oxygen is flexible silicone rather than water [13, 14]. However, silicone's low surface energy caused wetting problems, leading to protein and lipid adhesion. Interpenetrating hydrophilic polymer networks were introduced to address this, resulting in second- and third-generation silicone hydrogel lenses launched by companies like CIBA Vision, Johnson & Johnson, and Cooper Vision [15].

17.1.2.2 Hydrogels

Since contact lenses are placed directly on the cornea, they inhibit the natural physiological oxygen delivery to the eye's surface. Therefore, preventing hypoxic stress is critical in lens design, which requires using materials that allow maximum oxygen permeability [16].

While silicone lenses offer superior oxygen permeability, hydrogel lenses dominate the global market due to other advantages, such as high water content, minimal tissue irritation, and low mechanical stress on the cornea. Mechanical stress can also lead to the same problems as hypoxic stress. Some of these issues include increased mitosis of epithelial cells, increased protease and glycosidase activities, and changes in the cornea's hydration and transparency [15, 17, 18]. To minimize the impact of contact lenses on corneal metabolism, it is important and necessary to reduce the hypoxic and mechanical stresses that may originate from the lens [19].

Hydrogel lenses meet most requirements for comfortable wear, aside from oxygen permeability. While oxygen dissolves well in water, producing lenses with extremely high water content for maximum permeability poses manufacturing challenges and compromises mechanical strength. For comfortable use, hydrogel contact lenses need to have a well-balanced water content, mechanical properties, oxygen permeability, surface wettability, and optical properties (refractive index) and hydrolytic stability. Additionally, the material must be biocompatible, nontoxic, and resistant to sterilization and disinfectants [4].

17.1.2.3 Materials Used in Hydrogel Contact Lens Production

The synthetic polymers most commonly used in hydrogel production and their properties are summarized below.

17.1.2.3.1 HEMA

The first soft contact lens material, poly(hydroxyethyl methacrylate) (PHEMA) (Fig. 17.3a), was developed by Wichterle and Lim in 1961 and published in *Nature* in 1960 under the article titled *"Hydrophilic Gels for Biological Use."* [20]. The initial biocompatibility studies on PHEMA were published in 1959 and 1960. Subsequently, Wichterle collaborated with local companies to manufacture soft contact lenses. Using a homemade device built from his grandson's toy, Prof. Wichterle produced the first contact lens. Clinical tests conducted by Dr. Dreifus at Prague's Second Ophthalmology Clinic confirmed that PHEMA lenses improved vision without causing side effects. This innovation marked the beginning of the soft lens era, and in 1972, Bausch & Lomb introduced it to the global market [21].

Like all hydrophilic contact lenses, PHEMA polymer networks are created with slight cross-linking. A common cross-linker used is ethylene glycol dimethacrylate (EGDMA) (Fig. 17.3b), added at less than 1 mol% relative to the monomer.

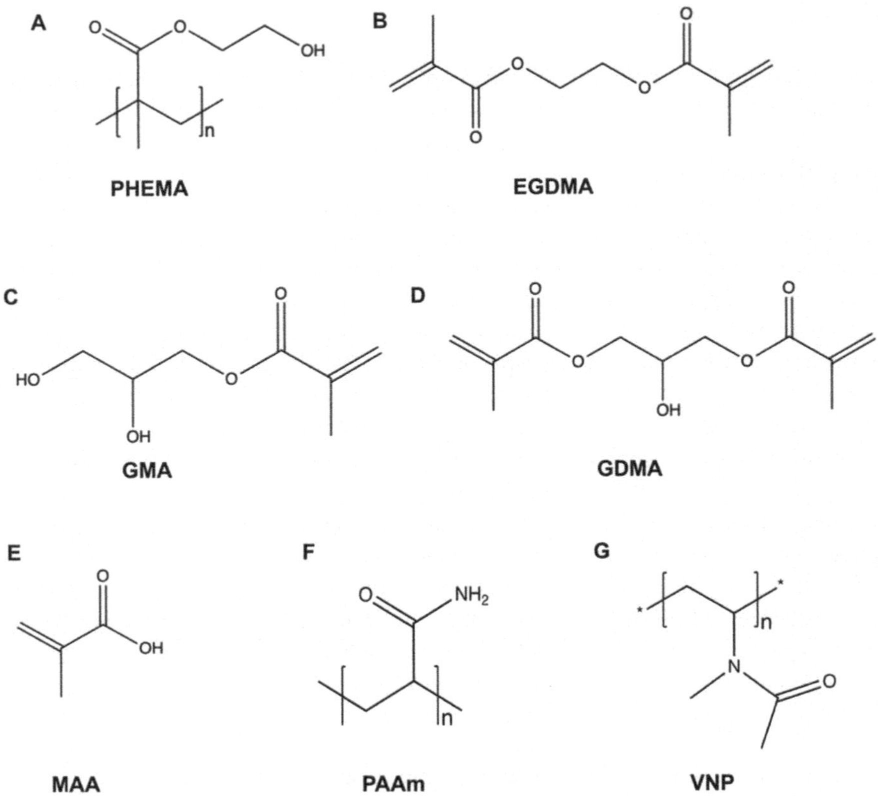

Fig. 17.3 Commonly used soft contact lens materials and cross-linkers: (**a**) poly(hydroxyethyl methacrylate) (PHEMA), (**b**) cross-linker ethylene dimethacrylate (EGDMA), (**c**) glycerol methacrylate (GMA), (**d**) glycerol dimethacrylate (GDMA), (**e**) methacrylic acid (MAA), (**f**) polyacrylamide, and (**g**) N-vinylpyrrolidone (NVP)

However, due to EGDMA's hydrophobic nature, hydrophobic regions can form during polymerization, leading to a heterogeneous structure. In some cases, more polar and hydrophilic cross-linkers like triethylene glycol dimethacrylate (TEGDMA) are used instead [15, 22, 23].

Another method involves creating copolymers using HEMA and another monomer with higher hydrophilicity, yielding polymer networks with greater water retention capacity. For instance, copolymers formed by combining HEMA with diethylene glycol methacrylate (DEGMA) or triethylene glycol methacrylate (TEGMA) are examples of such hydrogels [24, 25]. These polymers were tested by Professor Wichterle and colleagues in Prague during the 1980s [26]. Although HEMA-DEGMA lenses were once considered optimal by some users, their production lacked consistency, and their mechanical properties were subpar, leading to their discontinuation [15].

17.1.2.3.2 Dihydroxy Methacrylates

Incorporating dihydroxy methacrylate comonomers, such as glycerol methacrylate (GMA) (Fig. 17.3c), significantly increases the water retention capacity of lenses [15, 27]. Glycerol dimethacrylate (GDMA) (Fig. 17.3d) is commonly used in producing such lenses. These lenses demonstrate substantially higher water retention (57–60% compared to PHEMA's 40%) and offer oxygen permeability with Dk = 20 barrers [15].

17.1.2.3.3 Methacrylic Acid

Methacrylic acid (MAA) (Fig. 17.3e) is the most commonly used comonomer to enhance water retention in lenses, present in many commercial contact lenses. Even at low concentrations (1–3% by weight), MAA allows lenses to achieve high water content (55–85% by weight) [15]. This is due to the hydrogen bonds formed between MAA's carboxylic acid groups and water molecules. For example, the water retention capacity of the HEMA–MAA copolymer is greater than that of the HEMA polymer alone, and this capacity increases as the concentration of the comonomer MAA increases. However, increasing MAA concentration lowers the refractive index due to its low molar refraction [28].

Carboxylic acid groups in MAA are also used to activate polymer surfaces, facilitating ionic transformations to improve swelling properties. For example, the carboxylic acid groups of MAA are converted to ionic sodium salts to increase the swelling of the gel. Yet, MAA ionization introduces negative charges to the polymer, attracting positively charged proteins like lysozyme, leading to lens surface adhesion [12, 29].

17.1.2.3.4 Acrylamides

Acrylamide-based hydrogels are valued for their exceptional swelling properties. Polyacrylamides (PAAm) (Fig. 17.3f) are typically used as thickeners in wastewater treatment and paper manufacturing due to their water solubility. PAAm lenses are created by cross-linking N,N-dimethylacrylamide monomer with N,N′-methylenebisacrylamide. Highly swollen, soft gels derived from cross-linked PAAm serve various biomedical applications. Unlike methacrylic acid salts, PAAm swelling is pH-independent. Acrylic acid is also known as propenoic acid and has carboxylic acid groups. These groups are attached to vinyl groups. Because carboxylic acid can ionize, it increases the ionic strength and pH sensitivity of the hydrogel [30]. Combining acrylic acid with other polymers enables the development of diverse hydrogels. For example, Ejeromedoghene and colleagues reported that the poly(hydroxyethyl acrylate) (pHEA), PAAm, and tungsten oxide (WO3) nanocomposites they created showed high light transmittance and exhibited very good antibacterial effect against keratitis bacteria, which can be seen in contact lens solutions [31].

17.1.2.3.5 N-Vinyl-2-Pyrrolidone

N-vinylpyrrolidone (NVP) (Fig. 17.3g) is a typical monomer used for nonionic hydrogels, with high water retention (>50%) and oxygen permeability [32]. While NVP finds applications in pharmaceuticals and cosmetics, its weak mechanical properties, tendency to absorb tear proteins, and potential to cause corneal dryness limit its use as a homopolymer. To improve its performance, NVP is copolymerized with other monomers like HEMA or alkyl methacrylates [15, 33]. For instance, Xu et al. developed a soft contact lens using HEMA, NVP, and methyl methacrylate (MMA), which demonstrated high water retention, good mechanical properties, and efficacy against glaucoma. While NVP increases the water retention capacity of this structure, MAA increases its mechanical strength [34]. The copolymer of NVP with methyl methacrylate is known on the market as lidofilcon.

17.1.2.4 Hybrid Contact Lenses

Hybrid contact lenses combine a rigid gas-permeable central optical zone with a silicone hydrogel peripheral region. These lenses aim to overcome the disadvantages of soft and rigid lenses, offering better optical properties and comfort [35]. Their larger diameter provides superior focus and vision compared to other lens types [8]. The advantages and disadvantages of hard, soft, and hybrid lenses are summarized in Table 17.1.

17.2 Hydrogels in Intraocular Lens Production

Cataracts, the leading cause of blindness worldwide, affect the lens of the eye, impairing vision. Cataract surgeries involve replacing the opaque lens with an artificial intraocular lens (IOL). Initially, rigid poly(methyl methacrylate) materials were used for IOLs. Today, hydrophobic and hydrophilic acrylate polymers, siloxanes, and collamer (a copolymer of collagen and HEMA) are employed [36]. By utilizing elastic materials, it has become feasible to implant intraocular lenses (IOLs) that can flex through a small incision in the cornea (Fig. 17.4a, b). Consequently, it is essential for hydrogels to possess adequate mechanical properties, flexibility, and the ability to slowly unfold without wrinkling after insertion [1, 37]. Research indicates that acrylic-based hydrogels unfold in a more controlled and gradual manner than silicone-based hydrogels. Moreover, the high refractive index of acrylic materials allows for the production of thinner IOLs [36, 38].

Table 17.1 Advantages and disadvantages of hard, soft, and hybrid contact lenses

Lens	Materials	Advantages	Limitations	References
Rigid contact lenses	PMMA	Low cost with good mechanical and optical properties	Minimal oxygen permeability and water retention capacity; lacks flexibility	[8]
	RGP	High oxygen permeability and durability	High cost and contains abrasive components	[1]
Soft contact lenses	Silicon hydrogels	High oxygen permeability and water retention capacity, offering comfort and durability	Potentially abrasive; requires the use of hydrophilic monomers	[13, 14]
	HEMA hydrogels	Comfortable and flexible with low cost	Limited oxygen permeability and water retention capacity; prone to protein accumulation	[15, 21]
Hybrid contact lenses		Comfortable usage, clear vision, and ease of focus	Expensive and handling is challenging	[8]

During cataract surgery, the most common complication arises from the abnormal proliferation of residual lens epithelial cells, which leads to posterior capsule opacification (PCO). This condition, known as "secondary cataract," can be treated by clearing the tissue using a YAG laser. However, the laser's imprecise focusing can cause dislocation, fragmentation, or indentation of PMMA IOLs (polymethyl methacrylate intraocular lenses) [39]. Compared to PMMA lenses, acrylic, silicone, and HEMA-based hydrogels are more resistant to such damage.

Additionally, studies have shown that acrylic implants result in less capsule formation compared to other materials. To prevent capsule formation, hydrogels can be coated with bioactive polymers such as polyethylene glycol (PEG) or those containing sulfonate and carboxylate groups, which inhibit cell adhesion [40, 41].

Another concern is the potential damage caused by UV radiation to the retina after the natural crystalline lens is removed. This has prompted manufacturers to incorporate UV-absorbing chromophores into IOL materials. Benzonitrile and benzophenone-based chromophores, which absorb UV radiation below 400 nm, are commonly used for this purpose [38].

17.3 The Future of Lenses

Since their first patent in 1960, contact lenses have been considered a potential vehicle for drug delivery to the eye. Hui and Willcox reviewed materials and methods for delivering pharmaceutical agents via lenses in vivo and in clinical trials up to 2016 [42]. While current lenses provide some benefits, modified materials for

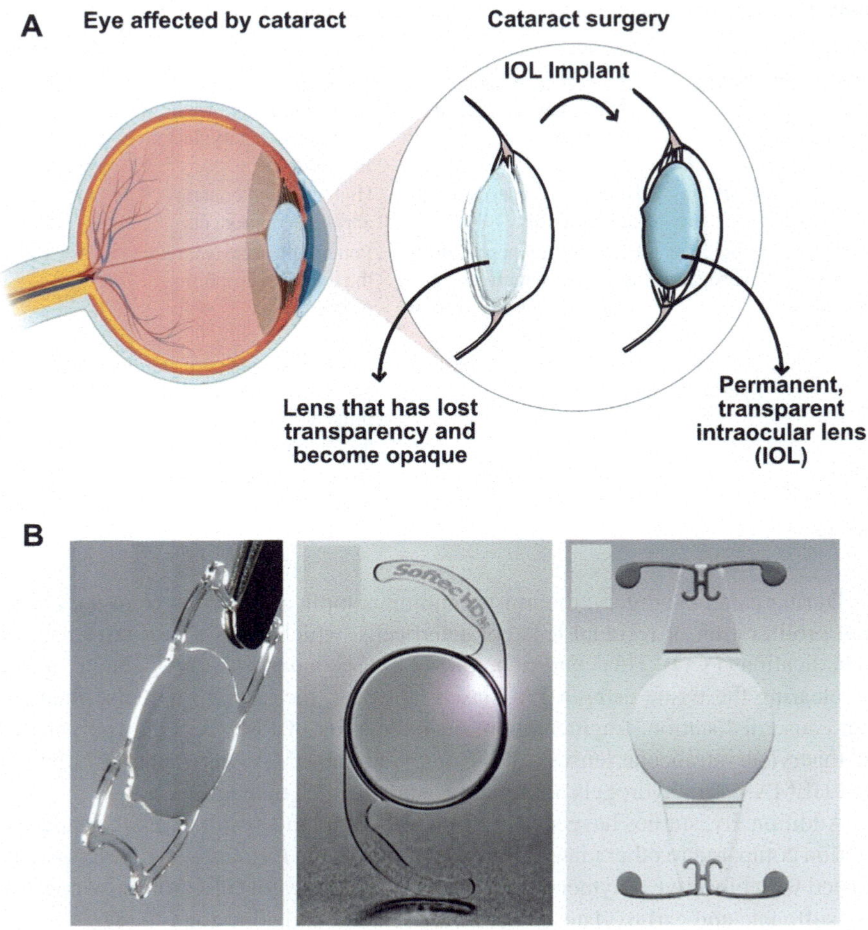

Fig. 17.4 Schematic representation of cataract surgery and intraocular lenses (IOL). (**a**) The eye's vision is corrected by replacing the lens, which loses its transparency due to aging, UV exposure, trauma, or systemic chemical effects, with a transparent, permanent artificial lens. The figure was created using BioRender.com. (**b**) IOLs released by different companies on the market. (Figure **b** is used with permission from source 37; copyright belongs to ACS Publications)

drug delivery show promising dose kinetics. However, more intraocular studies are necessary to confirm these advantages.

Among the various methods for modifying materials, molecular imprinting stands out as a particularly promising approach for controlled drug delivery. This technique involves creating template-shaped cavities within the polymer matrix with predefined selectivity and affinity. These cavities act as macromolecular memory sites for the drug, enabling controlled and sustained release [23, 43, 44]. For instance, Byrne et al. employed molecular imprinting to develop lenses capable of releasing a cocktail of drugs over extended periods, improving wear comfort. Their study demonstrated that careful design could enable the controlled release of multiple molecules over days or weeks [45]. However, a significant limitation of such drug delivery tests is their inability to fully replicate the in vivo environment. These tests are typically conducted under static conditions, which fail to account for natural eye movements, blinking, and the continuous renewal of tears and air. These factors explain the differences observed between in vitro and in vivo results. To address this, researchers have developed systems that better mimic the eye's natural environment and found that standard static systems release drugs more rapidly than the simulated physiological conditions [46]. This finding highlights the importance of using tear-mimicking systems for reliable testing.

Beyond drug delivery, contact lenses also provide a platform for delivering stem cells to the ocular surface in cases of disease [47]. For example, Girolamo and colleagues used an FDA-approved contact lens as a substrate to transfer epithelial cells to the ocular surface for treating Limbal Stem Cell Deficiency (LSCD). Patients monitored over 1 year showed a 100% success rate [48].

Another emerging field is the use of contact lenses as biosensors to detect ocular and systemic diseases (Fig. 17.5). This innovation could significantly aid doctors in diagnosing and monitoring diseases. Tears, being easily accessible and less complex, are frequently used to detect ocular conditions, diabetes, and even cancer. Changes in the concentration of molecules within tears can provide insights into disease progression and pave the way for personalized treatments [49]. Contact lenses integrated with electronics capable of detecting such changes have been extensively researched, with some even being commercialized [50, 51].

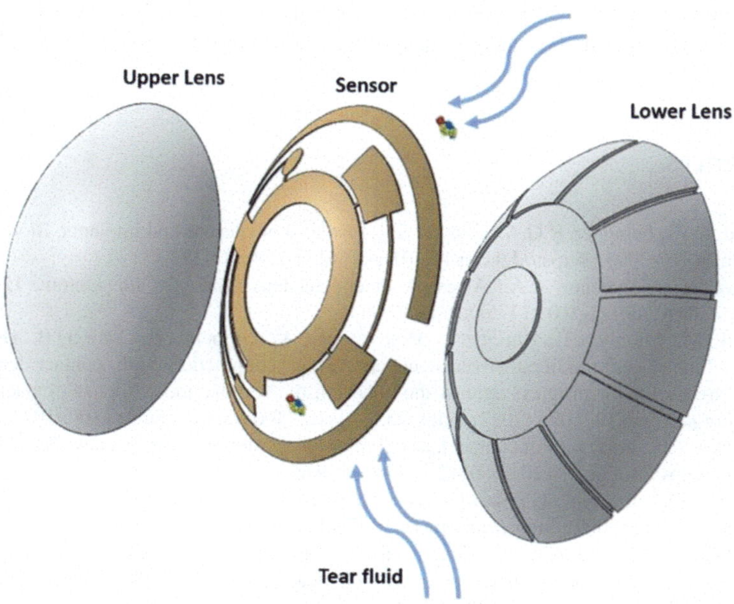

Fig. 17.5 Structure of a contact lens biosensor. (Used with permission from source [49])

Research on intraocular lenses (IOLs) has focused on achieving perfect vision for patients after cataract surgery, eliminating the need for glasses. Another exciting area of research involves high-tech microelectronic biosensors capable of measuring intraocular pressure (IOP) at any time [52, 53]. A recently introduced device, the *Eyemate-IO*, is designed for glaucoma patients undergoing cataract surgery, allowing regular monitoring of IOP. However, this biosensor is not yet integrated into the IOL itself; instead, it is placed in the sulcus (the space between the peripheral cornea and the iris) during cataract surgery [54, 55]. The next step for this technology is to integrate IOP-measuring biosensors directly into the IOL.

17.4 Conclusion

Contact lenses and intraocular lenses have evolved and improved as our understanding of ocular physiology deepens and feedback from users is incorporated. While contact lenses are often used for cosmetic purposes, they are sometimes the only option for correcting vision. From the ideas of Leonardo da Vinci in the 1500s to advancements enabling 30-day wear today, this field has come a long way since the limited-use lenses of the 1800s.

Collaborations among biologists, chemists, engineers, and medical professionals have led to innovations such as drug delivery systems and biosensors integrated into lenses, making the field even more intriguing. Ultimately, contact lenses used to correct vision and IOLs used to treat cataracts are not only products that significantly enhance quality of life but also pave the way for the development of advanced materials and implants featuring cutting-edge technologies.

References

1. Lloyd, A.W., Faragher, R.G.A., Denyer, S.P.: Ocular biomaterials and implants. Biomaterials. **22**, 769–785 (2001). https://doi.org/10.1016/s0142-9612(00)00237-4
2. Pucker, A.D., Tichenor, A.A.: A review of contact lens dropout. Clin. Optom. **12**, 85–94 (2020). https://doi.org/10.2147/opto.s198637
3. Padzik, M., Chomicz, L., Bluszcz, J., Maleszewska, K., Grobelny, J., Conn, D.B., Hendiger, E.B.: Tannic acid-modified silver nanoparticles in conjunction with contact lens solutions are useful for progress against the adhesion of Acanthamoeba spp. to contact lenses. Microorganisms. **10**, 1076 (2022). https://doi.org/10.3390/microorganisms10061076
4. Nicolson, P.C., Vogt, J.: Soft contact lens polymers: an evolution. Biomaterials. **22**, 3273–3283 (2001). https://doi.org/10.1016/s0142-9612(01)00165-x
5. Key, J.E.: Development of contact lenses and their worldwide use. Eye Contact Lens Sci. Clin. Pr. **33**, 343–345 (2007). https://doi.org/10.1097/icl.0b013e318157c230
6. McMahon, T.T., Zadnik, K.: Twenty-five years of contact lenses: the impact on the cornea and ophthalmic practice. Cornea. **19**, 730–740 (2000). https://doi.org/10.1097/00003226-200009000-00018

7. Holden, B.A., Mertz, G.W.: Critical oxygen levels to avoid corneal edema for daily and extended wear contact lenses. Invest. Ophthalmol. Vis. Sci. **25**, 1161–1167 (1984)
8. Alam, F., Elsherif, M., AlQattan, B., Ali, M., Ahmed, I.M.G., Salih, A., Antonysamy, D.S., Yetisen, A.K., Park, S., Butt, H.: Prospects for additive manufacturing in contact lens devices. Adv. Eng. Mater. **23**, 2000941 (2021). https://doi.org/10.1002/adem.202000941
9. Singh, J., Agrawal, K.K.: Polymeric materials for contact lenses. J. Macromol. Sci., Part C. **32**, 521–534 (1992). https://doi.org/10.1080/15321799208021431
10. Efron, N., Morgan, P.B., Helland, M., Itoi, M., Jones, D., Nichols, J.J., van der Worp, E., Woods, C.A.: International rigid contact lens prescribing contact lens anterior eye. J. Br. Contact Lens Assoc. **33**, 141–143 (2009). https://doi.org/10.1016/j.clae.2009.11.005
11. Morgan, P.B., Efron, N., Hill, E.A., Raynor, M.K., Whiting, M.A., Tullo, A.B.: Incidence of keratitis of varying severity among contact lens wearers. Br. J. Ophthalmol. **89**, 430 (2005). https://doi.org/10.1136/bjo.2004.052688
12. Luensmann, D., Jones, L.: Protein deposition on contact lenses: the past, the present, and the future. Cont. Lens Anterior Eye. **35**, 53–64 (2012). https://doi.org/10.1016/j.clae.2011.12.005
13. Efron, N., Morgan, P.B., Cameron, I.D., Brennan, N.A., Goodwin, M.: Oxygen permeability and water content of silicone hydrogel contact lens materials. Optom. Vis. Sci. Off. Publ. Am. Acad. Optom. **84**, 328–337 (2007). https://doi.org/10.1097/opx.0b013e31804375ed
14. Morgan, P.B., Efron, N.: The oxygen performance of contemporary hydrogel contact lenses. Cont. Lens Anterior Eye. **21**, 2–6 (1998). https://doi.org/10.1016/s1367-0484(98)80016-7
15. Michalek, J., Hobzova, R., Pradny, M., Duskova, M.: Biomedical Applications of Hydrogels Handbook, pp. 303–315 (2010). https://doi.org/10.1007/978-1-4419-5919-5_16
16. Bruce, A.S., Brennan, N.A.: Corneal pathophysiology with contact lens wear. Surv. Ophthalmol. **35**, 25–58 (1990). https://doi.org/10.1016/0039-6257(90)90046-x
17. Chalmers, R.: Overview of factors that affect comfort with modern soft contact lenses. Cont. Lens Anterior Eye. **37**, 65–76 (2014). https://doi.org/10.1016/j.clae.2013.08.154
18. Bahram, M., Mohseni, N., Moghtader, M.: Emerging Concepts in Analysis and Applications of Hydrogels. IntechOpen, London (2016). https://doi.org/10.5772/64301
19. Lin, M.C., Soliman, G.N., Song, M.J., Smith, J.P., Lin, C.T., Chen, Y.Q., Polse, K.A.: Soft contact lens extended wear affects corneal epithelial permeability: hypoxic or mechanical etiology? Cont. Lens Anterior Eye. **26**, 11–16 (2003). https://doi.org/10.1016/s1367-0484(02)00088-7
20. Wichterle, O., Lím, D.: Hydrophilic gels for biological use. Nature. **185**, 117–118 (1960). https://doi.org/10.1038/185117a0
21. Kopecek, J.: Hydrogels: from soft contact lenses and implants to self-assembled nanomaterials. J. Polym. Sci. Part A: Polym. Chem. **47**, 5929–5946 (2009). https://doi.org/10.1002/pola.23607
22. Pedley, D.G., Skelly, P.J., Tighe, B.J.: Hydrogels in biomedical applications. Br. Polym. J. **12**, 99–110 (1980). https://doi.org/10.1002/pi.4980120306
23. Hajirasouliha, E., Zandi, M., Tabatabaei, M.H., Zarrinbakhsh, P.: Ocular contact lenses: smart materials for biomedical applications. Polym. Bull. **81**, 7791–7832 (2024). https://doi.org/10.1007/s00289-023-05113-9
24. Cejková, J., Lojda, Z., Brůnová, B., Vacík, J., Michálek, J.: Comparison of tolerance to hydrophilic contact lenses made of Hema (37% H2O) and Hema-Degma (55%, 65% H2O) in the rabbit eye. I. Changes in corneal transparency due to disturbed hydration. Cesk. Oftalmol. **45**, 401–407 (1989)
25. Cejková, J., Lojda, Z., Brůnová, B., Vacík, J., Michálek, J.: Comparison of tolerance to hydrophilic contact lenses made of Hema (37% H2O) and Hema-Degma (55%, 65% H2O) in the rabbit eye. II. Changes in corneal transparency due profound degenerative processes. Cesk. Oftalmol. **45**, 408–411 (1989)
26. Wichterle, O.: The Task: "Contact lenses," Recollection, pp. 128–141. Impreso, Prague (1992)
27. Khan, A., Rana, T., Naeem, Z., Javaid, S., Sajid, N., Habib, A., Fatima, M., Khan, M.: Hydrogels: a novel drug delivery system. J. Biomed. Res. Environ. Sci. **1**, 439–451 (2020). https://doi.org/10.37871/jbres1176

28. Chhabra, P., Gupta, R., Suri, G., Tyagi, M., Seshadri, G., Sabharwal, S., Niyogi, U.K., Khandal, R.K.: Studies on development of polymeric materials using gamma irradiation for contact and intraocular lenses. Int. J. Polym. Sci. **2009**, 1–9 (2009). https://doi.org/10.1155/2009/906904
29. Garrett, Q., Milthorpe, B.K.: Human serum albumin adsorption on hydrogel contact lenses in vitro. Invest. Ophthalmol. Vis. Sci. **37**, 2594–2602 (1996)
30. Sennakesavan, G., Mostakhdemin, M., Dkhar, L.K., Seyfoddin, A., Fatihhi, S.J.: Acrylic acid/acrylamide based hydrogels and its properties—a review. Polym. Degrad. Stab. **180**, 109308 (2020). https://doi.org/10.1016/j.polymdegradstab.2020.109308
31. Ejeromedoghene, O., Hu, Y.P., Oderinde, O., Yao, F., Akinremi, C., Akinyeye, R., Adewuyi, S., Fu, G.: Transparent and photochromic poly(hydroxyethyl acrylate–acrylamide)/WO3 hydrogel with antibacterial properties against bacterial keratitis in contact lens. J. Appl. Polym. Sci. **139**, 51815 (2022). https://doi.org/10.1002/app.51815
32. Sun, Z., Zhang, R., Sun, Y., Wang, X., Duan, L., Gao, G.: Methyl methacrylate rigid contact lenses doped with N-Vinylpyrrolidone and 3-[Tris(trimethylsiloxy)silyl] propyl methacrylate for improved oxygen permeability. Polymer. **309**, 127439 (2024). https://doi.org/10.1016/j.polymer.2024.127439
33. Ma, X., Zhang, Y., Cai, S., He, X.: Multi-responsiveness N-vinylpyrrolidone and methyl acrylate copolymer with wide tunable range of response temperature. J. Mol. Liq. **360**, 119437 (2022). https://doi.org/10.1016/j.molliq.2022.119437
34. Xu, J., Li, X., Sun, F.: Preparation and evaluation of a contact lens vehicle for Puerarin delivery. J. Biomater. Sci. Polym. Ed. **21**, 271–288 (2010). https://doi.org/10.1163/156856209x415774
35. Liu, X., Ye, Y., Ge, Y., Qu, J., Liedberg, B., Zhang, Q., Wang, Y.: Smart contact lenses for healthcare monitoring and therapy. ACS Nano. **18**, 6817–6844 (2024). https://doi.org/10.1021/acsnano.3c12072
36. Kirchhof, S., Goepferich, A.M., Brandl, F.P.: Hydrogels in ophthalmic applications. Eur. J. Pharm. Biopharm. **95**, 227–238 (2015). https://doi.org/10.1016/j.ejpb.2015.05.016
37. Karayilan, M., Clamen, L., Becker, M.L.: Polymeric materials for eye surface and intraocular applications. Biomacromolecules. **22**, 223–261 (2021). https://doi.org/10.1021/acs.biomac.0c01525
38. Werner, L., Namalis, N.: Cataract and Refractive Surgery Essentials in Ophthalmology, pp. 63–84 (2005). https://doi.org/10.1007/3-540-26678-x_4
39. Tetz, M., Jorgensen, M.R.: New hydrophobic IOL materials and understanding the Science of Glistenings. Curr. Eye Res. **40**, 969–981 (2015). https://doi.org/10.3109/02713683.2014.978476
40. He, R., Hu, X., Tan, H.C., Feng, J., Steffi, C., Wang, K., Wang, W.: Surface modification of titanium with curcumin: a promising strategy to combat fibrous encapsulation. J. Mater. Chem. B. **3**, 2137–2146 (2015). https://doi.org/10.1039/c4tb01616e
41. Bozukova, D., Pagnoulle, C., Pauw-Gillet, M.-C.D., Desbief, S., Lazzaroni, R., Ruth, N., Jérôme, R., Jérôme, C.: Improved performances of intraocular lenses by poly(ethylene glycol) chemical coatings. Biomacromolecules. **8**, 2379–2387 (2007). https://doi.org/10.1021/bm0701649
42. Hui, A., Willcox, M.: In vivo studies evaluating the use of contact lenses for drug delivery. Optom. Vis. Sci. **93**, 367–376 (2016). https://doi.org/10.1097/opx.0000000000000809
43. White, C.J., Byrne, M.E.: Molecularly imprinted therapeutic contact lenses. Expert Opin. Drug Deliv. **7**, 765–780 (2010). https://doi.org/10.1517/17425241003770098
44. Jones, L.W., Byrne, M., Ciolino, J.B., Legerton, J., Markoulli, M., Papas, E., Subbaraman, L.: Revolutionary future uses of contact lenses. Optom. Vis. Sci. **93**, 325–327 (2016). https://doi.org/10.1097/opx.0000000000000847
45. DiPasquale, S.A., Uricoli, B., DiCerbo, M.C., Brown, T.L., Byrne, M.E.: Controlled release of multiple therapeutics from silicone hydrogel contact lenses for post-cataract/post-refractive surgery and uveitis treatment. Transl. Vis. Sci. Technol. **10**, 5 (2021). https://doi.org/10.1167/tvst.10.14.5

46. Phan, C.-M., Bajgrowicz, M., Gao, H., Subbaraman, L.N., Jones, L.W.: Release of fluconazole from contact lenses using a novel in vitro eye model. Optom. Vis. Sci. **93**, 387–394 (2016). https://doi.org/10.1097/opx.0000000000000760
47. Bobba, S., Girolamo, N.D.: Contact lenses: a delivery device for stem cells to treat corneal blindness. Optom. Vis. Sci. Off. Publ. Am. Acad. Optom. **93**, 412–418 (2015). https://doi.org/10.1097/opx.0000000000000699
48. Girolamo, N.D., Bosch, M., Zamora, K., Coroneo, M.T., Wakefield, D., Watson, S.L.: A contact lens-based technique for expansion and transplantation of autologous epithelial progenitors for ocular surface reconstruction. Transplantation. **87**, 1571–1578 (2009). https://doi.org/10.1097/tp.0b013e3181a4bbf2
49. Tseng, R.C., Chen, C.-C., Hsu, S.-M., Chuang, H.-S.: Contact-lens biosensors. Sensors (Basel). **18**, 2651 (2018). https://doi.org/10.3390/s18082651
50. Mansouri, K., Shaarawy, T.: Continuous intraocular pressure monitoring with a wireless ocular telemetry sensor: initial clinical experience in patients with open angle glaucoma. Br. J. Ophthalmol. **95**, 627 (2011). https://doi.org/10.1136/bjo.2010.192922
51. Mansouri, K., Medeiros, F.A., Tafreshi, A., Weinreb, R.N.: Continuous 24-hour monitoring of intraocular pressure patterns with a contact lens sensor: safety, tolerability, and reproducibility in patients with glaucoma. Arch. Ophthalmol. **130**, 1534–1539 (2012). https://doi.org/10.1001/jamaophthalmol.2013.1350
52. Dick, H.B., Gerste, R.D.: Future intraocular lens technologies. Ophthalmology. **128**, e206–e213 (2021). https://doi.org/10.1016/j.ophtha.2020.12.025
53. Sarac, B., Yücer, S., Sahin, H., Unal, M., Ciftci, F.: Wearable and implantable bioelectronic: biosensing contact lens and applications. Chem. Eng. J. **491**, 152016 (2024). https://doi.org/10.1016/j.cej.2024.152016
54. Koutsonas, A., Walter, P., Roessler, G., Plange, N.: Implantation of a novel telemetric intraocular pressure sensor in patients with glaucoma (ARGOS study): 1-year results. Invest. Ophthalmol. Vis. Sci. **56**, 1063–1069 (2015). https://doi.org/10.1167/iovs.14-14925
55. van den Bosch, J.J.O.N., Pennisi, V., Rao, H.L., Mansouri, K., Weinreb, R., Thieme, H., Hoffmann, M.B., Choritz, L.: Reproducibility of consecutive automated telemetric noctodiurnal IOP profiles as determined by an intraocular implant. Br. J. Ophthalmol. **108**, e323080 (2024). https://doi.org/10.1136/bjo-2022-323080

Chapter 18
Regulations for Hydrogels and Bioinks in Clinical Trials

Shahla Bagheri Fam

Abstract Regulatory guidelines are essential for the clinical translation of hydrogels and bioinks. Such regulations have been defined to ensure that these innovative materials fulfill the required standards. Any product intended for clinical application must be designed based on manufacturing and regulatory requirements from the start. The FDA (Food and Drug Administration) and MDR (Medical Device Regulation) are two major regulatory systems for clinical approval in the United States and the European Union, respectively. While both systems aim to ensure the safety, efficacy, and quality of products, MDR imposes stricter requirements. Most of the medical devices drop out due to issues of safety, lack of efficacy, or not meeting the standards defined by the regulatory authorities. According to statistics, the average timeframe for bringing a new device to market is about 7–10 years. This is longer for a new drug. This highlights the difference in approval pathways for medical devices versus drugs. Guidelines written by authorities for new drugs or devices are to ensure they are safe and effective and exhibit high quality before going to the clinic.

Keywords Regulatory guidelines · Food and Drug Administration · FDA · Medical device · Medical device regulation · MDR · Regulatory authorities

18.1 Classification of Hydrogels and Bioinks

Hydrogels are cross-linked materials consisting of water-swollen polymers. Their tunable and stimuli-responsive properties enable them to mimic different extracellular environments inside the body, and they are highly utilized in tissue engineering, cosmetic, and regenerative medicine due to their biocompatibility, utility, and responsive behavior [1, 2]. Despite extensive R&D efforts conducted every year, by

S. B. Fam (✉)
Department of CMC Radiopharm, Bayer AS, Oslo, Norway

N. Hasirci, V. Hasirci (eds.), *Hydrogels and Bioinks in Tissue Engineering*, Biological and Medical Physics, Biomedical Engineering,
https://doi.org/10.1007/978-3-031-90966-5_18

2024, just around 100 hydrogels have been approved for clinical applications [3]. This means that not every product developed in R&D labs can successfully navigate the regulatory pathways. According to statistics, the average timeframe for bringing a new device to market is 7 years and the approval of a new drug can take up to 12 years [4, 5].

Products used in clinical applications are primarily classified as either "drugs" or "devices." Both FDA and MDR provide quite similar definitions for medical devices mentioned in "Federal Food, Drug, and Cosmetic (FD&C) Act in section 201(h)" and "Regulation (EU) 2017/745 (MDR)," respectively [6, 7].

The FD&C Act (21 U.S.C. 321) defines a "device" as given below:

> *an instrument, apparatus, implement, machine, contrivance, implant, in vitro reagent, or similar article, including any component or accessory, which: (A) Recognized in the official National Formulary or United States Pharmacopeia, or any supplement to them; (B) Intended for use in the diagnosis, cure, mitigation, treatment, or prevention of disease in man or other animals; or (C) Affecting the structure or any function of the body but does not achieve its primary intended purposes through chemical action or metabolism.*

From a regulatory perspective, "hydrogels and bioinks" are considered medical devices if their therapeutic mode of action results from the inherent physical, chemical, or mechanical properties without involving any medicinal component or chemical mechanisms of action. The MDR has a broader scope when it comes to hydrogels. Hydrogels used for nontherapeutic purposes like cosmetic implants must adhere to all regulatory pathways applicable to medical devices.

For many products classified as devices, the exclusionary clause—"does not achieve its primary intended purposes through chemical action within or on the body"—is often confusing. A product that has chemical action could also be a device if it does not achieve its primary intended purposes through that chemical action [8]. Chemical action can be shown as part of a device if it's not the general mechanism of action.

As per Article 1, Paragraph 5 (EU Regulation 2017/745), a device intended to stay within the human body is defined as an implantable device. This is the case for many hydrogels and hydrogel-based bioinks used within tissue engineering [9].

The FDA's Center for Devices and Radiological Health (CDRH) regulates devices according to their intended uses, regardless of their composition. Likewise, the MDR classifies medical devices based on their intended use, attributes, and risks.

As established by the Medical Device Amendments (MDA) of 1976 (P.L. 94-295), FDA has divided medical devices into three categories based on their purpose, characteristics, and risks from lowest to highest risk [10].

- **Class I:** Class I devices carry the lowest risk to consumers. These lower-risk devices, including things like oxygen masks or surgical instruments, are regulated by "general controls" to make sure they are safe and effective. General controls include:

  - Complying with good manufacturing practices (GMPs).
  - Healthcare Economics: Standards for adverse event reporting to the FDA.
  - Requirements for device registration and general recordkeeping.

- **Class 2:** Class 2 devices provide greater risk, and there are some "special controls" specific to them. These controls include:
  - Detailed labeling requirements that describe the information required on the product.
  - Performance standards specific to devices.
  - Device class-specific testing protocols.
- **Class 3:** Class 3 devices usually are life-sustaining or life-supporting, are implanted in the body, or present a potential for illness or injury. Pacemakers, breast implants, and HIV diagnostic tests are some examples. These devices need to go through "general control and premarket approval (PMA)," which is a series of extensive clinical trials to show safety and effectiveness.

The MDR takes a more risk-oriented approach in its classification of devices. In EU Regulation 2017/745, Class II devices are separated into two categories: Class IIa, exhibiting low risk (such as injectors, surgical gloves, and tooth implants), and Class IIb, having high risk (such as blood bags, tracheal tubes, and catheters).

18.2 Device Development Process

Designing a new medical device production is a team effort that involves a cohort of researchers working in a close collaboration with physicians. They follow the science and engineering design process, known as design-build-test-redesign. This involves extensive tests in the laboratory and experiments on animals. This preclinical stage can take approximately 2–3 years, and it can cost up to 10–20 million U.S. dollars, depending on the device's complexity. The initial costs are high, and not every startup can afford this phase. Instead, they approach venture capitalists for funding and to cover the costs. Most new device ideas are derived from academic research centers and are funded mainly through federal or other grants. However, these centers do not have sufficient resources to advance a project beyond the early prototype stage. To commercialize these ideas, the intellectual property (IP) right is typically licensed to a company or startup that will develop it. In order to motivate inventions from federally sponsored laboratories, mostly academia, Congress enacted the Bayh-Dole Act, which provides university researchers the patent ownership of intellectual property developed through government funding [11]. Drugs and medical devices are developed through established procedures that ensure their safety and effectiveness prior to their availability to the public. The development flows for drugs and devices are similar from inception to approval, consisting of five basic steps. However, devices have a simplified approval process and can take less time to approve than drugs and other products [12].

18.2.1 Step 1: The Concept and Device Discovery

Device discovery processes generally begin when researchers realize that an unmet medical need exists. Then, they develop a concept for a new device. They prepare a "proof of concept," a detailed document that describes the steps required to test if the idea will succeed. Most of these ideas prove wildly impractical, but the good ones make it into the later stages of development.

18.2.2 Step 2: Preclinical—Ready the Prototype

The proof of concept step involves a prototype of the device, which is an "early version" of the medical device. Preclinical studies take place after the proof of concept has been sufficiently successful. This phase provides validated data regarding the safety and efficiency of biomaterial-based devices before clinical testing [13]. In the product development processes defined by the FDA, there are two types of preclinical research studies: in vitro and in vivo [14]. In vitro is defined as a process that occurs in a laboratory vessel or other controlled experimental environment whereas in vivo is an assessment of product safety and efficacy within a living organism or natural setting. FDA requires device developers to follow good laboratory practices (GLP), which is a set of rules and requirements to establish quality systems defined in medical product development regulations, for preclinical laboratory studies. The GLP regulations are found in 21 CFR Part 58.1.

18.2.3 Step 3: Pathways to Clinical Approval

In the United States, clinical approval is granted directly by the FDA, whereas in the European Union, based on the commission implementing regulation EU-2017/2185, conformity assessments are conducted by "Notified Bodies" under the framework of the Medical Device Regulation (MDR). A notified body is an organization authorized by an EU Member State (or other countries under specific agreements) to assess the conformity of a product before entering the market. An updated list of notified bodies can be found on the official website of the European Union [15].

18.2.3.1 Pre-market Approval

According to the FDA, the Pre-Market Approval (PMA) process is mandatory for new medical devices that have no existing equivalent. Manufacturers must notify the FDA of their intent to market a device at least 90 days in advance. The PMA process is the strictest and requires sufficient scientific evidence proving the device's safety and effectiveness. Early collaboration with the FDA, through "Pre-submission" meetings, is recommended.

In some cases, a modified PMA, such as a modular or streamlined PMA, may be allowed. Clinical evidence, typically in the form of Level I or II evidence, is necessary to obtain approval. Clinical trials, which may involve thousands of subjects and span multiple years, must be conducted under an Investigational Device Exemption (IDE). A major factor influencing approval timelines is the time needed for regulatory and institutional review board (IRB) approvals, which often extend the approval process by several months.

The 21st Century Cures Act allows for device approvals based on observational studies, which could speed up approval but also raises concerns about safety and efficacy. Early engagement with the FDA is crucial for optimizing the approval process. The PMA review process includes an initial 45-day determination of application completeness, followed by a 180-day review period. If a PMA is insufficient, the FDA will provide feedback and reset the review timeline upon resubmission [5]. The equivalent process to Pre-Market Approval under the MDR is the "conformity assessment," carried out by a Notified Body for high-risk devices (Class III or Class IIb). This process has been designed based on EU Medical Device Regulation (EU MDR) 2017/745, Annex IX, and includes an audit to evaluate the Quality Management System (QMS), as outlined in ISO 13485. Since hydrogels and bioinks are classified as Class IIb or III devices in the EU, they will follow the conformity assessment pathway under the MDR regulations to obtain final clinical approval.

18.2.3.2 510(K) Application

The 510(k) application is a fast-track FDA process for medical devices that demonstrate substantial equivalence to an existing, approved device. If a device can be shown to be equivalent to a predicate device, it does not need to go through the more rigorous Pre-Market Approval (PMA) process. However, if the FDA determines the device is not equivalent, a full PMA will be required (https://www.fda.gov/medical-devices/premarket-approval-pma/pma-application-methods#streamlined).

The 510(k) process has faced criticism due to concerns that assuming equivalence between devices may not always be valid. One example is the problematic Pinnacle metal-on-metal hip implant, which was fast-tracked but later had to be discontinued due to patient reactions. Another issue is the "serial predicate" problem, where newer devices are approved based on older devices, sometimes leading to approvals without current, sufficient clinical evidence.

In 2011, the Institute of Medicine recommended eliminating the 510(k) process due to these concerns but faced opposition from manufacturers who argued it would hinder innovation. Additionally, critics argue that the 510(k) process encourages manufacturers to make only minor improvements to existing devices rather than creating true innovations, leading to higher costs without added benefits for patients. Finally, the fast-track process may reduce manufacturers' incentive to complete long-term clinical studies, as they can market devices without full evidence of efficacy [5].

18.2.3.3 The Humanitarian Device Exemption (HDE)

A Humanitarian Use Device (HUD) is designed to treat or diagnose conditions affecting fewer than 4000 individuals annually in the United States. HUDs are regulated through the FDA's Office of Orphan Products Development. To use an HUD, approval from both the FDA and a local Institutional Review Board (IRB) is required. The application for an HDE is similar to a PMA, but unlike a PMA, scientific evidence of efficacy is not necessary. The sponsor only needs to demonstrate that the device offers a probable health benefit that outweighs the risks, focusing on safety rather than efficacy [16].

18.2.4 Step 4: FDA Device Review

After the Advisory Committee meeting, the FDA determines whether the device is approved, needs additional information, or is not approved. By law, the decision and supporting evidence must be published in the Federal Register.

18.2.5 Step 5: FDA Post-market Device Safety Monitoring

The clinical trials conducted before a medical device is approved (premarket trials) provide important information about the safety and effectiveness of the device. However, not all potential issues can be predicted. Unexpected problems and risks might be observed during the clinical application of the device. To address this, the FDA keeps track of devices even after they are approved. This ongoing monitoring helps maintain public safety and confidence in medical devices [17].

FDA officials routinely inspect medical device manufacturing sites. Inspections may be scheduled in advance or occur unannounced. Inspections can be routine or prompted due to a specific issue. The purpose of these inspections is to ensure that manufacturers adhere to good manufacturing practices and rules. If standards are not met, the FDA has the authority to shut down the facility. Reporting defects can be one of the issues that prompt an inspection. The FDA has several programs for reporting problems related to approved medical devices. MedWatch, the FDA's adverse event reporting program, acts as a channel for reporting issues with medical products and receiving updates on new safety information. This channel's subscribers can receive regular MedWatch safety alerts. The Medical Product Safety Network (MedSun) is an adverse event reporting program that monitors the safety and effectiveness of medical devices. The FDA partners with 350 healthcare providers across the United States to report serious injuries or deaths linked to medical devices. Each month, the FDA publishes the MedSun newsletter, which provides important safety updates for consumers.

The MDR requires continuous post-market surveillance, with mandatory reporting into the EUDAMED database, making post-market compliance more intensive than the FDA system.

18.3 Clinical Trial Phases and Requirements

Clinical trials are research studies conducted on humans to evaluate the safety, effectiveness, and potential benefits of new medical devices or drugs. These trials are organized into four distinct phases, each with specific objectives and requirements to ensure the research is reliable and ethically sound [18].

- **Phase I**: This initial phase involves a small group (typically 20–100 healthy volunteers or patients) and focuses on assessing safety, determining optimal dosage, and understanding how the treatment is processed in the body. Potential side effects are also identified. Regulatory approval from agencies like the FDA or EMA is required before starting Phase I.
- **Phase II**: This phase includes a larger group of participants, often ranging from dozens to several hundred, who have the condition the treatment is designed to address. The goal is to evaluate the treatment's effectiveness, further monitor safety, and refine dosing strategies while establishing initial proof of concept.
- **Phase III**: Thousands of participants are enrolled, often across multiple sites and countries. This phase is critical for confirming efficacy, monitoring side effects, and comparing the new treatment to existing ones. Regulatory agencies require robust statistical analyses, predefined endpoints, and strict adherence to good clinical practices (GCP). Positive results from Phase III are essential for obtaining marketing approval.
- **Phase IV**: Conducted after the treatment is approved for public use, these post-marketing studies assess long-term safety, effectiveness, and rare side effects in real-world settings. Regulatory authorities may require Phase IV trials, particularly for high-risk or innovative treatments.

18.4 Global Regulatory Perspectives

Regulatory affairs professionals are essential in clinical research, ensuring compliance with the rules defined by agencies like the FDA (USA), EMA (Europe), MHRA (UK), and PMDA (Japan). They guide clinical trials to meet regulatory standards and navigate the approval process.

The regulatory systems in different regions differ in ways that affect the time and cost of introducing new medical devices. In the EU, independent organizations called Notified Bodies (NBs) handle device certification and issue the CE mark. This process is often quicker and less bureaucratic than the FDA's, but it can lead to inconsistencies. For example, some devices, like abdominal aorta stent grafts, were approved in Europe but did not pass US requirements.

The criteria for approval are also different. In the United States, high-risk devices must prove safety and effectiveness through large clinical trials. In the EU, manufacturers only need to show the device is safe and works as intended, allowing for smaller, faster studies. For example, the Guard Wire device had a small study in the EU but required a much larger trial in the United States to prove its effectiveness. While the EU process is faster and cheaper, the US system is more thorough, reducing risks but taking longer time to approve devices [5].

18.5 Conclusion

The successful clinical translation of hydrogels and bioinks relies on strict regulatory frameworks. Both US-FDA and EU-MDR systems have developed robust pathways for the classification, testing, and approval of hydrogels and bioinks-based medical devices. While both organizations aim to ensure the safety, effectiveness, and quality of medical devices, there are significant differences in their requirements and processes. FDA approval is often faster due to direct engagement with the regulatory body, whereas MDR can take longer due to involvement with Notified Bodies and stringent documentation requirements. The FDA provides a clear pathway (510(k) or PMA), while the MDR heavily relies on Notified Bodies for conformity assessment. For global manufacturers, understanding these differences is critical for navigating regulatory pathways effectively.

References

1. Advances in engineering hydrogels. https://doi.org/10.1126/science.aaf3627
2. Biomedical applications of functional hydrogels: innovative developments, relevant clinical trials and advanced products. https://doi.org/10.1016/j.biomaterials.2024.122718
3. Hydrogels in the clinic: an update. https://doi.org/10.1002/btm2.10680
4. Congressional budget office. https://www.cbo.gov/publication/57126. Accessed 5 Nov 2024
5. Drugs, devices, and the FDA: part 2: an overview of approval processes: FDA approval of medical devices. https://doi.org/10.1016/j.jacbts.2016.03.009
6. Classification of products as drugs and devices and additional product classification issues guidance for industry and FDA staff. https://www.fda.gov/regulatory-information/search-fda-guidance-documents/classification-products-drugs-and-devices-and-additional-product-classification-issues. Accessed 5 Nov 2024
7. Design and clinical application of injectable hydrogels for musculoskeletal therapy. https://doi.org/10.1002/btm2.10295

8. Classification of products as drugs and devices & additional product classification issues: guidance for industry and FDA staff "final guidance". https://www.fda.gov/media/80384/download
9. Regulation (EU) 2017/745 of the European Parliament and of the Council of 5 April 2017 on medical devices. https://eur-lex.europa.eu/legal-content/EN/TXT/?uri=CELEX:32017R0745
10. Classify Your Medical Device. https://www.fda.gov/medical-devices/overview-device-regulation/classify-your-medical-device. Accessed 15 Jan 2025
11. Medical device development from prototype to regulatory approval. https://doi.org/10.1161/01.CIR.0000134695.65733.64
12. Drug and device development processes. https://www.fda.gov/patients/learn-about-drug-and-device-approvals. Accessed 27 Dec 2024
13. An overview of 3D bioprinting and its regulatory framework. Int. J. Res. Publ. Rev. **5**(6), 6674–6682 (2024)
14. Preclinical research. https://www.fda.gov/patients/drug-development-process/step-2-preclinical-research. Accessed 27 Dec 2024
15. List of notified bodies under Regulation (EU) 2017/745 on medical devices. https://webgate.ec.europa.eu/single-market-compliance-space/notified-bodies/notified-body-list?filter=legislationId:34,notificationStatusId:1. Accessed 10 Feb 2025
16. Humanitarian device exemption. https://www.fda.gov/medical-devices/premarket-submissions-selecting-and-preparing-correct-submission/humanitarian-device-exemption. Accessed 15 Jan 2025
17. The device development process. https://www.fda.gov/patients/learn-about-drug-and-device-approvals/device-development-process. Accessed 15 Jan 2025
18. Clinical research. https://www.fda.gov/patients/drug-development-process/step-3-clinical-research. Accessed 15 Jan 2025

Index

N. Hasirci, V. Hasirci (eds.), *Hydrogels and Bioinks in Tissue Engineering*, Biological and Medical Physics, Biomedical Engineering,
https://doi.org/10.1007/978-3-031-90966-5

The manufacturer's authorised representative in the EU is Springer Nature Customer Service Centre GmbH, Europaplatz 3, 69115 Heidelberg, Germany. If you have any concerns regarding our products, please contact ProductSafety@springernature.com

Printed and bound by CPI Group (UK) Ltd, Croydon, CR0 4YY
30/09/2026
02225588-0014